# Digital Mapping and Indigenous America

Employing anthropology, field research, and humanities methodologies as well as digital cartography, and foregrounding the voices of Indigenous scholars, this text examines digital projects currently underway, and includes alternative modes of "mapping" Native American, Alaskan Native, Indigenous Hawaiian, and First Nations land. The work of both established and emerging scholars addressing a range of geographic regions and cultural issues is also represented. Issues addressed include the history of maps made by Native Americans; healing and reconciliation projects related to boarding schools; language and land reclamation; Western cartographic maps created in collaboration with Indigenous nations; and digital resources that combine maps with narrative, art, and film, along with chapters on archaeology, place naming, and the digital presence of elders.

This text is of interest to scholars working in history, cultural studies, anthropology, Native American studies, and digital cartography.

**Janet Berry Hess**, JD, PhD, is Professor of Art History and Project Director of the National Endowment for the Humanities project "Mapping Indigenous Cultures and Living Resources" at Sonoma State University.

## Routledge Research in Art and Race

Routledge Research in Art and Race is a new series focusing on race as examined by scholars working in the fields of art history and visual studies. Proposals for monographs and edited collections on this topic are welcomed.

**The Black Arts Movement and the Black Panther Party in American Visual Culture**
*Jo-Ann Morgan*

**Henry Ossawa Tanner**
Art, Faith, Race, and Legacy
*Naurice Frank Woods, Jr.*

**Race, Anthropology, and Politics in the Work of Wifredo Lam**
*Claude Cernuschi*

**Theodore Gericault, Painting Black Bodies**
Confrontations and Contradictions
*Albert Alhadeff*

**Travel, Art and Collecting in South Asia**
Vertiginous Exchange
*Natasha Eaton*

**Digital Mapping and Indigenous America**
*Edited by Janet Berry Hess*

For more information about this series, please visit: https://www.routledge.com/Routledge-Research-in-Art-and-Race/book-series/RRAR

# Digital Mapping and Indigenous America

Edited by Janet Berry Hess

Routledge
Taylor & Francis Group
NEW YORK AND LONDON

First published 2021
by Routledge
52 Vanderbilt Avenue, New York, NY 10017

and by Routledge
2 Park Square, Milton Park, Abingdon, Oxon, OX14 4RN

*Routledge is an imprint of the Taylor & Francis Group, an informa business*



*Library of Congress Cataloging-in-Publication Data*
A catalog record for this title has been requested

ISBN: 978-0-367-27217-3 (hbk)
ISBN: 978-0-367-74737-4 (pbk)
ISBN: 978-0-429-29554-6 (ebk)

Typeset in Sabon
by MPS Limited, Dehradun

For Everett Berry, Jr., and the ancestors
honoring the relationship between the Berrys and the Echo-Hawks
and for Kai.

# Contents

# Figures

# Contributors

**David G. Anderson** is a professor in the Department of Anthropology at the University of Tennessee, Knoxville. He has done fieldwork in the Southeastern, Midwestern, and Southwestern United States, and in the Caribbean, primarily in the US Virgin Islands. Professional interests include exploring the development of cultural complexity in Eastern North America, climate change and its impact on human societies past and present, improving the nation's cultural resource management program, and writing syntheses of archaeological research. He founded the Paleoindian Database of the Americas (PIDBA) in 1990, and has long been an advocate for large scale, openly accessible data analysis and reporting in archaeology, as exemplified by efforts like the Digital Index of North American Archaeology (DINAA).

**Daryl Baldwin** (Kinwalaniihsia), a citizen of the Miami Tribe of Oklahoma, was born and raised in the Great Lakes area and currently resides with his wife Karen in Liberty, Indiana. Baldwin's forefathers were active in the political affairs of the Miami Nation dating back to the 18th century, and he continues this dedication toward tribal self-determination through his efforts in language and cultural revitalization. Baldwin was born during the mid-20th century, at a time when the last speakers of his heritage language were passing. This loss motivated him to begin seeking documented language resources and linguistic support, which ultimately led him to pursue an MA in English linguistics at the University of Montana. With the support of his wife Karen, together they embarked in 1991 on the difficult work of raising their four children with the language in a homeschool environment. Growing community interest for language and cultural education prompted Miami Tribal leaders, in 2001, to approach their allies at Miami University in Oxford, Ohio, to create the Myaamia Center. Baldwin was asked to be the founding director, and since its inception, he and his staff have continued capacity building for this effort. The work of the Miami Tribe of Oklahoma, with its partner Miami University, has become nationally and internationally recognized for its research, planning, and implementation of community language and cultural revitalization efforts. In the fall of 2016, Baldwin's lifelong efforts, and arguably those of his family, tribal community, and university, were recognized with an award from the MacArthur Foundation (https://www.macfound.org/fellows/955/).

**Holly Elwell Bostrom** is the Director of Operations and Planning at The Trust for Public Land. Previously, she served as the Director of the Climate-Smart Cities Program.

Having joined The Trust for Public Land in 2012, she has fostered the growth of the Climate-Smart Cities team and the development of a program that is nationally recognized as a leader in urban greening for climate solutions and climate justice. The Climate-Smart Cities program develops deep partnerships with local governments and other community leaders by integrating applied science, geographic information systems (GIS) decision support, conservation finance, local state and federal policy, and on-the-ground parks, open space, and green infrastructure design and development. Bostrom has an MA in Urban and Environmental Policy and Planning from Tufts University, Massachusetts, and a BA in Environmental Studies: Community and International Development from the University of Vermont.

**Daniel G. Cole** is the GIS Coordinator of the Smithsonian Institution (SI). He has worked in this position since 1990, and since 1986 has served as the research cartographer at SI. Prior to working at the Smithsonian, he worked as a geographer/cartographer at Woods Hole Oceanographic Institution, NOAA (National Oceanic and Atmospheric Administration), University of Maryland, and Oregon State University. He co-edited (with Imre Sutton) a multi-authored book on cartography in Native American affairs, supervised field mapping and GIS analyses of archaeological sites in Mongolia, worked in investigations of biodiversity and species range prediction through the GIS/Remote Sensing interface, and is involved with multiple exhibits at SI. From June 2009 to June 2010, Cole was president of the Canadian Cartographic Association; and from April 2018 to April 2019, he was the president of the Cartography and Geographic Information Society (CaGIS). On the side, he serves as a judge in the CaGIS Map Design competition.

**David J. Costa** is the Program Director for the Language Research Office at the Myaamia Center at Miami University, Oxford, Ohio. A third-generation northern Californian, he grew up in San Jose, California, completing his BA in linguistics at UCLA in 1985, and his PhD in linguistics at UC Berkeley in 1994, with his dissertation on the Miami-Illinois language. A revised version of his dissertation was published through the University of Nebraska Press in 2003. He has been studying the Miami-Illinois language since 1988, and has worked extensively with the Miami Tribe of Oklahoma on language revitalization since 1995. In his capacity as Program Director for the Language Research Office, Costa conducts continuing research on the Miami-Illinois language, as well as helping to design language curricula for the Myaamia Center, and answering language inquiries from members of the Center and the Miami Tribe. Costa is also now involved in a long-term project to analyze and annotate the data from the Miami-Illinois language manuscripts that have been uploaded into MIDA (the Miami-Illinois Digital Archive), a process that will continue for many years to come. As part of his continuing linguistic research into the Miami-Illinois language, Costa is also working on a fully annotated collection of interlinearized Miami-Illinois texts, as well as completing papers documenting his research on the syntax and word order of Miami-Illinois. His goal is to publish the results of both of these projects within the next few years. In addition to his work on Miami-Illinois, Costa has also done extensive research on the Shawnee language, the Algonquian languages of southern New England, and comparative Algonquian.

**James W. Gerencser** is the College Archivist at Dickinson College and co-director of the Carlisle Indian School Digital Resource Center. He holds an MLS from the University of Pittsburgh and an MA in History from Shippensburg University, Pennsylvania. With a strong interest in making unique archival content more easily discoverable and accessible to wide audiences, he has spearheaded digital humanities efforts at Dickinson for more than 20 years.

**Steven W. Hackel** was born and raised in California, and earned his BA at Stanford University, California, and his PhD in American History from Cornell University, Ithaca, New York, with specializations in early America and the American West. From 1994 to 1996 he was a post-doctoral fellow at the Omohundro Institute of Early American History and Culture in Williamsburg, Virginia. He taught at Oregon State University from 1996 to 2007 and joined the faculty at the University of California, Riverside (UCR) in the fall of 2007. Within the larger field of American history, Hackel's research specializes on the Spanish Borderlands, the American West, colonial California, and California Natives. He is especially interested in Native responses to colonialism, the effects of disease on colonial encounters, and new ways of visualizing these processes through digital history. His publications include a biography on Fray Junípero Serra, a monograph on Native life in the California missions, numerous essays, a textbook, and two edited volumes. He is the General Editor of the Early California Population Project and the Project Director for the Early California Cultural Atlas. He co-curated the Huntington Library's international exhibition, "Junípero Serra and the Legacy of the California Missions." He currently is co-chair of the Early Modern Studies Institute's Seminar on the Spanish Borderlands. His current work involves a study of immigration and community formation in California before 1850. He is also completing The Pobladores Database, a database on some 18,000 immigrants who settled in California before 1850.

**Janet Berry Hess**, a European-American settler on Pomo and Miwok lands, is Project Director for the National Endowment for the Humanities Digital Advancement Grant, "Mapping Indigenous American Cultures and Living Histories" (MIAC-LH). She received her JD at the University of Iowa College of Law, her MA at Columbia University, New York, and her PhD in Art History at Harvard University. She currently teaches at Sonoma State University, California, and has authored numerous publications addressing post-liberation art and architecture in Ghana, Tanzania, and South Africa, as well a text addressing the intersection of settler/invader and Indigenous history (*Osage and Settler: Reconstructing Shared History Through an Oklahoma Family Archive*). Hess' family has lived in Oklahoma since 1850. She was born in Oklahoma and annually visits her ancestral hometowns of Pawnee and Hominy, Oklahoma.

**Lea Hong** has been the Hawai'i State Director of The Trust for Public Land, a national non-profit organization dedicated to conserving land for people, since 2006. She previously chaired the environmental and cultural resource law practice group at the Honolulu law firm of Alston Hunt Floyd & Ing (now Dentons), and worked for the Earthjustice Legal Defense Fund. Hong is a graduate of Rice University, Houston, Texas, and the William S. Richardson School of Law at the University of Hawai'i, where she serves on the Friends of the Law School Board and has taught classes on environmental litigation, environmental regulatory compliance, and conservation transactions. She has received a number of recognitions that she credits to the communities and partners with whom she works, including *Honolulu Magazine*'s Islander of the Year (Environment), the Hawai'i Conservation Alliance Outstanding Leadership Award, and the Hawai'i Women Lawyers Outstanding Woman Lawyer of the Year. She is from Wahiawā, on the island of O'ahu.

**Rebekah Ingram** received her PhD from the School of Linguistics and Language Studies at Carleton University, Ottawa, Ontario, Canada. Her dissertation, entitled *Naming Place in Kayen'kéha*, outlined a new model and methodology for the study of place names and the application of this model and methodology to Kanyen'kéha place names in cooperation with Kanyen'kéha communities. Presently, Ingram works in collaboration with these same communities using the Nunaliit mapping framework, developed by Carleton's Geomatics and Cartographic Research Centre, to create an atlas of Kanyen'kehà:ka place names and conceptions of space as well as an atlas hub for Indigenous knowledge and interdisciplinary, land-based research.

**Eric Kansa** (PhD, Harvard University) leads development of Open Context (http://opencontext.org). His research explores Web architecture, service design, and how these issues relate to the social and professional context of the digital humanities. Kansa also researches policy issues relating to intellectual property, including text-mining and cultural property concerns. He actively participates in a number of open science, open government, cyberinfrastructure, text-mining, and scholarly user needs initiatives. He has taught and practiced project management and information service design in the UC Berkeley School of Information's Clinic program. He has been a principal investigator and co-investigator on projects funded by the William and Flora Hewlett Foundation, National Endowment for the Humanities (NEH), International Museum of Library Services (IMLS), Hewlett-Packard, Sunlight Foundation, Google, National Science Foundation (NSF), and Alfred P. Sloan Foundation.

**Sarah Whitcher Kansa, RPA, PhD,** is Executive Director of the Alexandria Archive Institute, the non-profit organization that develops *Open Context*. She has a PhD in Archaeology and has spent twenty-five years conducting zooarchaeological research at sites in the Near East and Europe. She served as Vice President of the International Council for Archaeozoology (ICAZ) from 2014-2018, and was elected President for the 2018-2022 term. She is co-chair of the Digital Technology Committee of the Archaeological Institute of America (AIA) and served on the Publications Committee for the Society for American Archaeology (SAA) and the American Schools of Overseas Research (ASOR). She is also the Series Co-Editor for *Archaeobiology* (Lockwood Press) and Executive Editor of *Open Context*.

**Christine Diindiisi McCleave, MA,** enrolled citizen of Turtle Mountain Ojibwe Nation, is Executive Director for the National Native American Boarding School Healing Coalition. McCleave was the Coalition's first employee and has grown the organization's programs, staff, and budget since 2015. McCleave's grandfather attended Marty Catholic Indian Boarding School in South Dakota (SD) and Haskell Indian Boarding School in Kansas, and her great grandfather attended

Carlisle Indian School, Pennsylvania, where he played football with Jim Thorpe. Boarding school's inter-generational impacts on her personal life and children's lives led McCleave to complete her MA in Leadership thesis on the spectrum of spiritual practices between traditional Native American spirituality and Christianity and the legacy of the boarding schools on spiritual activities and Indian activism today. McCleave is an American Indian Graduate Center fellow, a Tiwahe Foundation awardee, and a fellow of the Nexus Boards and Commissions Leadership Institute. McCleave also has a Mini-MBA in Non-Profit Management from St. Thomas University, Miami, Florida, and a BS in Communication Studies from Northwestern College, Orange City, Iowa. She previously worked in Communications and Marketing at Indian Land Tenure Foundation and Human Resources Management at Aon Hewitt. She lives in Minneapolis, Minnesota, with her family.

**Rose Miron** is the Director of the D'Arcy McNickle Center for American Indian and Indigenous Studies at the Newberry Library in Chicago, Illinois. She holds a BA in History with a minor in Spanish and a PhD in American Studies from the University of Minnesota. Prior to joining the Newberry in July 2019, Miron served as the Program Manager for the National Native American Boarding School Healing Coalition, and she continues to serve on their Research Advisory Council. Her current manuscript project, titled "Indigenous Archival Activism: Narrating Nationalism in the Mohican Tribal Archive and Beyond," examines how Indigenous peoples use tribal archives to claim authority over the creation, assembly, and retrieval of their historical materials and frames this work as a distinct type of Indigenous activism that reshapes narratives of Native history. Her project is based in nine years of close collaboration with the Stockbridge-Munsee Mohican Nation.

**Sarah Montoya** is a Mexican-American settler on Tonvga and Payaya lands and a PhD candidate in UCLA's Gender Studies Department. Her dissertation, *Electronic Empires and Digital Domains*, examines representations and material histories of technological and computational development in the US through the lens of settler colonial studies. Her work traces the relationship between settler colonial ontologies and spatial regimes, the creation and maintenance of information and communication infrastructures by the US settler state, and the development of geographic information systems. She serves as a content management specialist, designer, and programmer for several digital humanities projects, including The Racial Violence Hub (racialviolencehub.com) and Race and Deaths in Custody (raceanddeathsincustody.com).

**Kelsey Noack Myers, RPA, PhD**, is an Archaeologist and the District Tribal Liaison for the US Army Corps of Engineers' Rock Island District of the Mississippi River Division. She previously served as the Senior Archaeologist for the Chippewa Cree Tribe of Rocky Boy's Indian Reservation, Montana, and her doctoral dissertation focused on cultural perseverance and increasing the visibility and inclusion of historic Native groups in the study and practice of colonial archaeology. She has participated in archaeological projects across the Mid-Atlantic, Southeastern, Midwestern, and Northern Plains regions of the United States in the private, academic, and government sectors. As a co-convener for the Digital Data Interest

Group of the Society for American Archaeology, she facilitates technological capacity building and open-source ethics in professional development and practice. Her interests include culture contact and colonialism, combining archaeological methods with traditional indigenous knowledge, public archaeology education, legacy collections, and 3D data applications in archaeology.

**Anna Naruta-Moya** is Project Director of the Indigenous Digital Archive, a project of the Museum of Indian Arts and Culture (MIAC) in collaboration with the New Mexico State Library Tribal Libraries Program and the Indian Pueblo Cultural Center. She is a MIAC Research Associate, an Associate Research Professor at the University of New Mexico, Knight Foundation Prototyping Grant awardee, and twice winner of IMLS (Institute of Museum and Library Services) National Leadership Grants (2016–2019 and 2019–2022). She has been an archivist with the Hoover Institution Archives, Stanford University, and the US National Archives. Of Slavic heritage, with her husband Daniel Moya (Tewa, P'o Suwae Ge Owingeh) she was a 2017–2018 Digital Knowledge Sharing Fellow of the American Philosophical Society. They also conducted research for National Native American Boarding School Healing Coalition(NABS) on deaths in, and the unmarked cemeteries of, boarding schools in New Mexico. Her writings include Polish cochineal's historic role as a cultural connector of Europe and Central Asia. Anna earned her PhD from University of California, Berkeley.

**Breece Robertson** is Vice President and Chief Research and Innovation Officer for The Trust for Public Land. She joined The Trust for Public Land in 2001 to create a comprehensive, coordinated planning and geographic information system (GIS) program for the national land conservation organization. Today, she provides leadership for the Trust's Enterprise GIS, a technology and research program that is the leading provider of "Land for People" science in the country. Robertson regularly presents at conferences around the country and is featured in and has authored many publications. Prior to working at The Trust for Public Land she was an Instructor and Course Developer for AllpointsGIS, Inc.

**Susan D. Rose** is Charles A. Dana Professor of Sociology at Dickinson College and Co-director of the Carlisle Indian School Digital Resource Center, both in Pennsylvania. She is co-editor with Jacqueline Fear-Segal of *Carlisle Indian Industrial School: Indigenous Histories, Memories, and Reclamations* (University of Nebraska Press, 2016), named by the *Philadelphia Inquirer* as one of the best books of 2016.

**Samantha Steindel-Cymer** has worked as an archaeological site supervisor with the Maya Research Program, delegate to the United Nations representing the World Federation for Mental Health (WFMH), and held the position of Assistant Curator at the Museum of the San Fernando Valley. With a background in historical research, archaeological fieldwork, and the intersections of volunteer outreach and non-profit volunteer coordination, she has further served as museum educator, community organizer, archaeological educator at Yosemite National Park's facelift event, and with the Los Angeles Music Center Education Guild. Ms. Steindel-Cymer has a BA in Anthropology from Smith College, and is currently a Candidate for the MA in Cultural Resources Management at Sonoma State University.

**Fraser Taylor** is the Chancellor's Distinguished Research Professor and Director of the Geomatics and Cartographic Research Centre (GCRC), Geography and Environmental Studies, at Carleton University, Ottawa, Canada. A cartographer and award winner of international renown, his current research is focused on developing the field of cybercartography, which he first defined in 1997. Established in 2003, GCRC undertakes extensive research and publication, technological development of the Nunaliit digital atlas mapping framework, and national and international inter-disciplinary project development with academic institutions, Indigenous communities, and others. A field of current focus includes language mapping and the reclamation and preservation of Indigenous languages and place names. Taylor's most recent books are entitled *Developments in the Theory and Practice of Cybercartography: Applications and Indigenous Mapping*. Elsevier (2014). *Further Developments in the Theory and Practice of Cybercartography: International Dimensions and Language Mapping*. Elsevier (2019).

**Victor Temprano** is a programmer and web developer from British Columbia, Canada. He has received degrees in History and Religious Studies from Mount Allison University and McGill University, both in Canada. His project Native Land has grown to international recognition in the 2010s, and he continues to maintain the website within its new non-profit structure. He runs a small technology startup, Mapster Technology, in Vancouver, British Columbia, based around mapping in the digital sphere. He also spends time working on numerous initiatives related to Indigenous language revitalization and maintains many other hobby projects.

**Romola V. Thumbadoo** has worked formally in criminal, social, and Indigenous justice in Canada (20 years), and at the grassroots level with the *Circle of All Nations* global eco-peace community founded by Late Indigenous Elder William Commanda (20 years). In January 2018, she completed her thesis entitled "Ginawaydaganuc *and the* Circle of All Nations*: The Remarkable Environmental Legacy of Elder William Commanda*" under the supervision of Dr. D. R. Fraser Taylor, Cartographer and Director of the Geomatics and Cartographic Research Centre (GCRC), at Carleton University, Ontario, Canada. Her research examined the ongoing relevance of the *Circle of All Nations* work of late Algonquin Elder Commanda and his contested vision for an eco-peace center at the Asinabka Sacred Chaudiere site, in Ottawa, Canada. Thumbadoo serves as Coordinator of the *Circle of All Nations*, is Research Associate at the GCRC, and is undertaking postdoctoral research to create a cybercartographic atlas on William Commanda's discourse and legacy.

**Douglas Troy**, Myaamia Center Technical Specialist, oversees computer science graduate assistants assigned to the Center and provides technical direction for the Miami-Illinois Digital Archive (MIDA), the Myaamia online dictionary and mobile app, and several other apps created for the Center by undergraduate students since 2010. He holds a PhD in Computer Science from the University of Waikato, New Zealand. Troy is a Professor Emeritus at Miami University, Florida, and currently serves as Director of the Miami University College of Engineering's graduate programs. Prior to his current roles, he served as Associate Dean in the College of Engineering and Computing and, before that, as Department Chair of the Department of Computer Science and Software Engineering, and was a tenured professor in the department. Troy has over forty years of experience in the research and development of software.

**Frank Vitale IV** is a project researcher for the Carlisle Indian School Digital Resource Center at Dickinson College. He received his MSc in the History of Science, Medicine, and Technology from the University of Oxford, England, where he explored colonial histories of medicine and mortality. His current research focuses on the intersections between the histories of medicine, mortality, and colonialism at the Carlisle Indian Industrial School.

**Joshua J. Wells** is an Associate Professor at Indiana University South Bend, with joint appointments in the Anthropology and Informatics programs and serves as director for the campus Center for Excellence in Research and Scholarship. Through the Digital Index of North American Archaeology, he researches interoperability issues between large-scale heritage management databases, promoting data reusability and reproducible research. He sits on the editorial boards of *Open Context*. Wells has led and been a member of numerous interest groups and task forces on digital data issues for the Society for American Archaeology. He has consulted on big data, and open science issues in Australia, North America, and the European Union.

**Stephen J. Yerka, MA, RPA**, collaborates on all technical aspects of the DINAA project. He is the Historic Preservation Specialist for the Eastern Band of the Cherokee Indian Tribal Historic Preservation Office. He is also Co-Director of the Paleoindian Database of the Americas (PIDBA) and a member of the editorial board for *Open Context*. A former GIS/IT manager for the Tennessee Archaeology Laboratory, Yerka worked on the first phase of DINAA and has published in American archaeology and informatics.

**Natale Zappia** is Associate Professor of History and the Director of the Institute for Sustainability at California State University, Northridge. His work explores the intersection of food systems, Indigenous political economies, and ecological transformations across early North America. His recent book, *Traders and Raiders: The Indigenous World of the Colorado Basin* (UNC Press, 2014; paperback 2016), tells the early history of the Indigenous Lower Colorado River, a watershed that looms large over the modern urban landscapes of Los Angeles. Zappia is now at work on several new projects including "Food Frontiers: Native Space and Power in Early North America" which explores the Indigenous *longue durée* of America's food history.

**Jeanette Zerneke** brings to the Early California Cultural Atlas (ECCA) experience in developing innovative programs and methodologies that facilitate humanities scholarship through spatial analysis. As Director of Technology for Electronic Cultural Atlas Initiative (ECAI), she works with a diverse group of technology experts to develop tools and methodologies to support the ECAI mission. This work involves developing infrastructure, programs, methodologies, working groups, and workshops to support ECAI affiliates in project development and integration. Zerneke has also worked directly with many projects to develop websites and ePublications highlighting the range of possibilities of new technologies to organize and present cultural information in innovative ways.

# Acknowledgments

I would like to thank Pbonchai Tallman and Victor Temprano as the guiding lights of this project. The National Endowment for the Humanities (NEH) awarded me a Digital Advancement grant which helped inspire this text, and I am grateful to the NEH, Mary Downs, and Sheila Brennan. Walter Echo-Hawk, and the friendship of the Berrys and Echo-Hawks, have been my central inspiration, and I am grateful to have met Chairman Echo-Hawk, the great leader, tireless attorney, and spiritual force and inspiration. Paula Farid welcomed me into her home in Pawhuska, Oklahoma. We sat out the tornado together; I know you are proud.

Early supporters of both "Mapping Indigenous American Cultures and Living Histories" and this text include Hollis Robbins, Stephanie Dyer, Benjamin Madley, and Erica Tom, and I am grateful for the contributions of Samantha Steindel-Cymer and Elise Blindauer. The constructive advice of Mishuana Goeman and Desiree Martinez was essential; thank you. Nicole Ream was a ray of sunshine throughout my grant process; thank you, Nicole. Thank you to Lisa Boyd Brashear, Michelle Everson, and Tammie Richards for years of support. I appreciate the welcoming spirit of Bob Heath, Bent Claw, Badger, and Micah, and the early input of Shawn Dumont. Yellow Crow, we miss you so; thank you for your respect, and for calling me Miss Janet. Thank you, Isabella Vitti and Katie Armstrong for your encouragement and kindness, and thank you, Daryl Baldwin, for your important moment of encouragement at NEH, which I will never forget.

Uncle Everett Berry, Mama (Catherine Berry Hess) and Daddy (Ross Hess)—I know you are proud. Kai will perhaps discover our history someday. I have left a trail for him.

Chapter 5 has been reprinted courtesy of Nick Thieberger and the *Language Documentation & Conservation Journal.* Originally printed in 2016; Baldwin, Daryl, David J. Costa and Douglas Troy. "Myaamiaataweenki eekincikoonihkiinki eeyoonki aapisaataweenki: A Miami Language Digital Tool for Language Reclamation." *Language Documentation & Conservation* 10. 394–410, https://scholarspace.manoa.hawaii.edu/handle/10125/24713.

# Introduction

## Digital Mapping—Ethics, the Law, and the Sacred

*Janet Berry Hess*

Examining Indigenous histories, locations, languages, and other information through digital mapping, and creating maps related to Indigenous peoples, is shaped by cultural understandings and ways of being in the world.[1] The ways in which Indigenous lands and cultures are understood, described, and represented by Indigenous people may differ from Western representations of Indigenous locations and cultures.[2] Digital mapping related to Indigenous nations does not constitute an ethics-neutral transfer of data, but rather a decision made on the part of those responsible for creating digital maps to respect tribal cultural values, purposes and worldviews.[3] This introduction will address the legal and ethical context of Native American, Alaskan Native, and Native Hawaiian-related digital mapping, provide a backdrop to the rise of digital projects related to Native Nations in America, and acknowledge digital protocols for safeguarding community control in research, mapping, and repatriation/return.[4]

Current legal challenges mounted on behalf of tribal Nations frequently spring from challenges to the violation or loss of tribal sovereignty.[5] Many challenges are based on the consequences of the loss or exploitation of land.[6] These losses, and the ongoing poverty and exploitation of Indigenous people worldwide, are addressed in the United Nations Declaration on the Rights of Indigenous Peoples (UNDRIP), which expresses "minimum standards for the survival, dignity and well-being of the indigenous peoples of the world," including prohibitions against forcible removal and the protection of "land-rights standards" and associated "environmental and subsistence standards," as well as political autonomy and the right to cultural survival.[7] The historical and ongoing violations of the spirit of UNDRIP in legal decisions and actions within the United States, as well as the failure to extend or enforce in a timely fashion the rights guaranteed under the Native American Graves Protection and Repatriation Act (NAGPRA), Section 106 of the National Historic Preservation Act, and the American Indian Religious Freedom Act, among other statutory protections,[8] illustrate what legal scholar Walter Echo-Hawk has termed "the dark side of federal Indian law."[9]

At the same time that Native American nations are subjected to the vagaries of an "unjust legal regime,"[10] non-Native scholars are ethically obligated to come to terms with a Western history of research. This history has had negative consequences for Indigenous people ranging from misrepresentation and uncompensated appropriation, to violence, trauma and cultural destruction. In *Decolonizing Methodologies: Research and Indigenous Peoples*, Linda Tuhiwai Smith describes "research as an institution of knowledge that is embedded in a global system of imperialism and power."[11] The negative consequences of Western research are historically evident in the realm of mapping, and ongoing: Mishuana Goeman, in *Mark My Words: Native*

*Women Mapping Our Nations*, summarizes the Western history of "mappings of Native lands and bodies" as "essential to colonial and imperial projects."[12] Goeman details how "the 'real' of settler colonial society is built on the violent erasures of alternative modes of mapping and geographic understanding,"[13] and presents the stories of Native women as one such alternative mode, discussing "narratives that mediate and refute ... gendered colonial structures that continue to dominate and enact violence[.]"[14]

For the reasons Smith and Goeman, among many others, have indicated, ethical digital mapping of Indigenous spaces conducted by a non-Indigenous individual or group must take account of the destructive history of Western intervention. Ethical mapping takes account of the manner in which, as Kathryn Shanley argues, "concepts of cultural identity primarily or inextricably [are] based on situated knowledge"[15]—knowledge contained in place-based language, land-based knowledge, spiritual practices and art identified with location, embodied and gendered spaces,[16] and beliefs and experiences associated with ancestors—as opposed to the Western conception of knowledge as everywhere universal, transportable,[17] and appropriate (or even possible) to share with all.[18] The place-based nature of knowledge and identity, a cultural view that characterizes much Native American scholarship and literature,[19] is also noted in the work of cultural outsiders, as in Keith Basso's *Wisdom Sits in Places: Landscape and Language Among the Western Apache*,[20] and Hugh Brody's *Maps & Dreams*.[21] A "decolonizing methodology," to use Smith's term, acknowledges this situated nature of knowledge and the "unmappability" of aspects of Indigenous cultures and experiences.[22] To avoid disrespect, appropriation, and even tragedy, digital resources "should be designed by those understanding the culture to ensure that Indigenous peoples' knowledge is the foundation" and "disseminated back to people in a language they can understand[.]"[23]

In addition to these legal and methodological considerations, scholars addressing Indigenous knowledge are ethically bound to follow the Protocols for Native American Archival Materials. As Jennifer O'Neal states,

> tribal communities have continually fought for their land, legal rights, self-determination, and the restoration of their tribal status after termination—the United States policy from the mid-1940s to the mid-1960s that ended the government's recognition of sovereignty of tribes, trusteeship of Indian reservations, and exclusion of Indians from state laws. After that time, activism expanded to focus on Native American religious and cultural practice rights.[24]

Following this effort, "engagement increased surrounding the care and preservation of Native American cultural heritage, most notably collections housed at non-tribal museums, culminating in" the passage of NAGPRA.[25] However, not only did NAGPRA not address "human remains, funerary and sacred objects, and objects of cultural patrimony" held in private collections, but it also "did not provide guidance or regulations regarding the care and preservation of Native American archival collections at non-native repositories[.]"[26] For this reason, a group of "archivists, librarians, museum curators, historians, and anthropologists" met in 2006 to identify best professional practices related to archival materials, and to incorporate these practices into "ten policy, legal and human rights topics including: consultation, understanding of Native American values and perspectives, accessibility and use, cultural sensitivity, culturally responsive context, intellectual and cultural property, repatriation, research protocols, reciprocal education and training, [and] archival educational awareness."[27] The Protocols have not been universally endorsed or implemented. As O'Neal argues, a paradigm shift will be required for institutions to respect and collaborate closely to foreground Indigenous ways of understanding the world.[28]

Digital mapping of Indigenous lands, histories, cultural practices, and resources should be considered within this methodological, legal and ethical context. Native Americans historically have made, and in the present continue to make, maps which differ in format from Western cartography, embedding information in language, narratives, music, visual art, and other practices.[29] The widespread use of digital mapping by tribal Nations for management purposes is exemplified in the many editions of *Tribal GIS: Supporting Native American Decision-Making*, in which numerous authors discuss the use of geographic information system (GIS) technology to address "natural resources and the environment, transportation, cultural and historical preservation, economic development, health, education, public safety, agriculture, and enterprise[.]"[30] In addition, as Joshua Bell, Kimberly Christen, and Mark Turin state, "Indigenous communities, museums, archives, and libraries, as well as individuals and family groups, are increasingly using digital materials in sophisticated and intersecting ways... this use productively builds on and is in dialogue with the complex ways communities have been and continue to use visual media to assert their sovereignty, challenge the terms and nature of representation, and create new intercultural dynamics."[31] Native American individuals, organizations, and academic departments have created hundreds of digital resources that describe their cultural centers and collections, and address such issues as cultural geography and practices; oral history and digital storytelling; repatriation; counter- or guerrilla mapping as a means for pushing back against imperialism and resource acquisition;[32] and memory, healing, and reconciliation.

Many Indigenous-led digital projects are assisted by Mukurtu, an open source platform at the Center for Digital Scholarship and Curation at Washington State University designed to "meet the needs of diverse communities who want to manage and share their digital cultural heritage in their own way[.]"[33] A central feature of Mukurtu is its use of "cultural protocols," which enable communities "to determine fine-grained levels of access to your digital heritage materials based on your community needs and values." As the website for Mukurtu states, "Protocols make it possible to define a range of access levels for digital heritage objects and collections from completely open to strictly controlled[.]"[34] That not all knowledge is available to all scholars everywhere is a point of controversy for some non-tribal scholars, particularly the restriction of specified archival materials, sound and film recordings, and images of art by tribally specified layers of access. Moments of friction may arise over suggestions that non-tribal scholars should not be involved in specific kinds of research (mapping sacred information or locations, for example); over requests for repatriation/return of sacred objects or information held by non-Native individuals and institutions; and over requests that research "give back to the community in the way of teaching, tribal scholarships, infrastructure building, and sustainability."[35]

At the same time that some digital resources—which may involve the representation of materials held in archives, libraries, and/or museums—overlook tribal protocols, collaboration has grown in the wake of NAGPRA. Museums, for example, are increasingly turning to digital resources as a mechanism for repatriation and more inclusive engagement with Native American nations (sending digital copies of documents, sound recordings, films, and other objects and information), although this process is characterized by "ever more complex dilemmas and institutional barriers with which they [digital or analog objects] were associated."[36] Thus, for example, a digital resource may contain a public link to an image or document that tribal communities do not wish to share, or expose the location of artifacts, architecture, or sacred sites. Tribal communities and/or councils may oppose sharing such information with non-tribal members, advocate restricting access to select tribal members (according to status, age, gender, and other categories), or request that art or information remain in Western institutions or archives (or, more commonly, request sole ownership and return).[37] Many, although not all, non-Native scholars and institutions are undertaking the challenging, individualized, and processual work of collaboration, repatriation/return, and the creation of resources that acknowledge the primacy of Indigenous voices, worldviews and sovereign rights in a context of colonial possession and collection.[38]

This volume provides a glimpse at the world of digital resources related to Native American history and culture created by Native and non-Native scholars and professionals, along with chapters addressing First Nations.[39] Although inclusion here does not signify endorsement, digital mapping projects created by non-Native scholars nonetheless increasingly involve what O'Neal has described as "collaborative and respectful stewardship."[40] In providing a glimpse at the vast range of digital mapping resources, the work of both established and emerging/new scholars, in academic and non-academic language, addressing a range of geographic regions and cultural issues, is represented. This selection foregrounds a range of modes of intellectual and self-representation, and includes the history of maps made by Native Americans; healing and reconciliation projects related to boarding schools;[41] language and land reclamation; Western cartographic maps created in collaboration with Indigenous nations; and resources that combine maps with narrative, art, and film, along with chapters on archaeology, place naming, and the digital voice of Elders.

Collaboration is "something to be striven for … amid a legacy of colonial relations,"[42] although it is sometimes dismissed or, more frequently, underfunded and adopted as an institutional gesture after projects are complete. As reviewers of the groundbreaking "After the Return" series note, "[c]olonial legacies embedded in collections; the expense of hardware and software maintenance and migration; geographical remoteness of many communities from urban centers; Indigenous restrictions of access to traditional knowledge; and the limited digital literacy of many in these communities represent just a few of the many challenges in implementing digital 'return.'"[43] It is my hope that this volume promotes the centrality of Indigenous voices and encourages repatriation/return by individuals and institutions, while simultaneously encouraging the re-centering of Indigenous voices in ongoing projects and providing an overview of digital resources related to Indigenous lands, languages, and histories.

## Notes

1 The phrase "Being-in-the-World" originated with Martin Heidegger, who discussed one's embodied experience of the world, and offered a challenge to the split between mind and body, nature and culture (an understanding already understood in Indigenous communities). The phrase is widely used in academic communities: see, for example, Belcourt,

Annjeanette E., Cyda Swaney and Allyson Kelley. "Indigenous Methodologies in Research: Social Justice and Sovereignty as the Foundations of Community-Based Research." In Kathryn Shanley and Bjorg Evjen (eds.). *Mapping Indigenous Presence: North Scandanavian and North American Perspectives*. Tucson: University of Arizona Press, 2015, 63; Goeman, Mishuana. *Mark My Words: Native Women Mapping Our Nations*. Minneapolis: University of Minnesota Press, 2013, 9, 214. For an overview of government, academic, and Indigenous mapping in the U.S., see Cole, Daniel and Imre Sutton (eds.). *Mapping Native American Cartographic Interactions Between Indigenous Peoples, Government, and Academia: Cartography and the Government* (Vols. 1–3). Createspace Independent Publishing Platform, 2014. For an overview of cartography related to Indigenous land in Canada, see Taylor, D.R. Fraser (ed.). *Developments in the Theory and Practice of Cybercartography: Applications and Indigenous Mapping*, 2nd edition (Modern Cartography Series, Vol. 5). Oxford, U.K., 2014.

2 For an overview of cultural mapping, spatial representation, and the ontological nature of mapping, as well as the use of Indigenous maps to challenge Western cartographic conventions, see Duxbury, Nancy, W.F. Garrett-Petts and David MacLennan, "Cultural Mapping as Cultural Inquiry: Introduction to an Emerging Field of Practice." In *Cultural Mapping as Cultural Inquiry*. New York: Routledge, 2015.

3 For extended analysis of digital mapping as a form of knowledge production interwoven with issues of power, politics, and agency, see Barques-Pedreny, Pol, David Chandler and Elena Simon (eds.). *Mapping and Politics in the Digital Age*. New York: Routledge, 2018.

4 For a discussion of digital repatriation and return, see Bell, Joshua A., Kimberly Christen and Mark Turin. "After the Return: Digital Repatriation and the Circulation of Indigenous Knowledge." *Museum Anthropology* 7, no. 1–2 (2013). *Digital Mapping and Indigenous America* does not have space to address, but ideally would address, every tribal Nation, including the 36 federally recognized tribal Nations that exist across the colonial border of the U.S. and Mexico (including the Apache, Cocopah, Kickapoo, Kumeyaay, Pai, O'Odham, and Yaqui, among others). Tribal Nations south of the U.S. also deserve extended representation.

5 For a discussion of the recent loss of sovereignty and suggestions for reform, see Echo-Hawk, Walter R. *In the Courts of the Conqueror: The 10 Worst Indian Law Cases Ever Decided*. Golden, CO: Fulcrum Publishing, 2010, 423–460.

6 Ibid., 457–458.

7 Ibid., 27–29; see also Anaya, S. James. *Indigenous Peoples in International Law*, 2nd edition. New York: Oxford University Press, 2004.

8 Native American Graves Protection and Repatriation Act, 25 U.S.C., Sections 3001-3003 (2006). See also Trope, Jack F. and Walter R. Echo-Hawk, "The Native American Graves Protection and Repatriation Act: Background and Legislation." In Mihesuah Devon A. (ed.). *Repatriation Reader: Who Owns Native American Remains*. Lincoln: University of Nebraska Press, 2000, 123–168. For an accessible guide to understanding NAGPRA and obstacles to realizing its objectives, see Atalay, Sonya, Jen Shannon and John Swogger. *Journeys to Complete the Work... and Changing the Way We Bring Native American Ancestors Home*. PRC, 2017.

9 Echo-Hawk, Walter R. *The Sea of Grass: A Family Tale from the American Heartland*. Golden, CO: Fulcrum Publishing, 2018, 386.

10 Ibid.

11 Smith, Linda Tuhiwai. *Decolonizing Methodologies: Research and Indigenous Peoples*, 2nd edition. London: Zed Books, 2012 (originally published in 1999), ix.

12 Goeman, *Mark My Words*, 16.

13 Ibid., 2.

14 Ibid., 3, 19.

15 Shanley, Kathryn. "'Mapping' Indigenous Presence: The Declaration on the Rights of Indigenous Peoples at Rhetorical Turns and Tipping Points." In *Mapping Indigenous Presence*, 10.

16 For an account of the bias of disembodied discourse in Western epistemology, see Stoller, Paul. *Sensuous Scholarship*. Philadelphia: University of Pennsylvania Press, 1997.

17 Ibid.

18 Aside from the experiential, place-based, and spiritual aspects of knowledge and the challenges with translation, the conversion of knowledge from practice or perception to language may be reductive. See Kroskrity, Paul V. and Margaret Field. *Native American Language Ideologies: Beliefs, Practices, and Struggles in Indian Country*. Tucson: University of Arizona Press, 2009, 18 ("In the transition to literacy ... language is often flattened, secularized ... reduced to a symbolic code largely lacking in constitutive agency.").
19 For an example of the connection between land and epistemology as it applies to women and gender studies, see Arvin, Maile, Eve Tuck and Angie Morrill, "Decolonizing Feminism: Challenging Connections Between Settler Colonialism and Heteropatriarchy." *Feminist Formations* 25, no. 1 (December 2012): 8–34.
20 Basso, Keith H. *Wisdom Sits in Places: Landscape and Language Among the Western Apache*. Albuquerque: University of New Mexico Press, 1996.
21 Brody, Hugh. *Maps & Dreams*. Long Grove, IL.: Waveland Press, 1981. There is an enormous body of work revealing and discussing situated knowledge by Indigenous scholars. See, for example, Cajete, Gregory. "Philosophy of Native Science." In Anne Waters (ed.). *Native American Thought*. Malden, MA: Blackwell Publishing, 2004, 45–57.
22 For a discussion of the "unmappable," see Becker, Amy. "Mapping the Unmappable in Indigenous Digital Cartographies." MA Thesis, Department of Anthropology, University of Victoria, 2014.
23 Evjen, Bjorg and David R.M. Beck. "Growing Indigenous Influence on Research, Extended Perspectives, and a New Methodology: A Historical Approach." In *Mapping Indigenous Presence*, 48.
24 O'Neal, Jennifer. "Respect, Recognition, and Reciprocity: *The Protocols for Native American Archival Materials*." In Dominique Daniel and Amalia Levi (eds.), *Identity Palimpsests: Archiving Ethnicity in the U.S. and Canada*. Sacramento: Litwin Books, 2013, 137, 126–127.
25 Ibid., 127.
26 Ibid.
27 Ibid., 131–132.
28 Ibid., 136.
29 See Castor, Laura. "Crossroads on the Path to Mental Decolonization: Research, Traditional Knowledge, and Joy Harjo's Music." In Shanley and Evjen, 229–249.
30 Taylor, Anne, David Gadsden, Joseph Kerski and Heather Guglielmo. *Tribal GIS: Supporting Native American Decision-Making*, 2nd edition. New York: Esri Press, 2017.
31 Bell, Christen and Turn, "After the Return," 1.
32 See Hunt, Dallas and Shaun A. Stevenson, "Decolonizing Geographies of Power: Indigenous Digital Counter-Mapping Practices on Turtle Island." *Settler Colonial Studies* 7, no. 3 (June 17, 2016): 372–392.
33 Mukurtu CMS. https://mukurtu.org.
34 Ibid.
35 Belcourt, Swaney and Kelley, "Indigenous Methodologies in Research," 72. For an examination of the complexities and contradictions of defending cultural claims, and a critique of selected cultural sovereignty claims and requests for repatriation, see Brown, Michael F. *Who Owns Native Culture?* Cambridge: Harvard University Press, 2003.
36 Bell, Christen and Turin, "After the Return," 1; see also Golding, Viv and Wayne Modest. *Museums and Communities: Curators, Collections and Collaboration*. New York: Bloomsbury, 2013.
37 For rich discussions of the contoverial nature of sharing spiritual aspects of Native American culture with non-Native individuals, see Irwin, Lee (ed.). *Native American Spirituality: A Critical Reader*. Lincoln: University of Nebraska Press, 2000, 11–36.
38 See Bell, Christen and Turin, "After the Return."
39 A unique set of legal and methodological protocols apply to First Nations. See Bell, Catherine and Robert K. Paterson (eds.). *Protection of First Nations Cultural Heritage: Laws, Policy, and Reform* (Law and Society). Vancouver: UBC Press, 2009; Aird, Karen and Gretchen Fox. *Indigenous Living Heritage in Canada*. Ottawa, ON: Canadian Commission for UNESCO's IdeaLab, April 2020.

40 O'Neal, "Respect, Recognition, and Reciprocity," 136; see Bell, Christen and Turin, "After the Return."
41 For information on residential boarding schools and cartography, see Pyne, Stephanie and D.R. Fraser Taylor (ed.). *Cybercartography in a Reconciliation Community: Engaging Intersecting Perspectives* (Modern Cartography Series, Vol. 8). Cambridge, MA., 2019.
42 Material World, "Review Essay: After the Return: Digital Repatriation and the Circulation of Indigenous Knowledge" (October 10, 2014). https://materialworldblog.com/2014/10/review-essay-after-the-return-digital-repatriation-and-the-circulation-of-indigenous-knowledge/
43 Ibid.

## References

Aird, Karen and Gretchen Fox. *Indigenous Living Heritage in Canada*. Ottawa, ON: Canadian Commission for UNESCO's IdeaLab, April 2020.

Anaya, S. James. *Indigenous Peoples in International Law*, 2nd edition New York: Oxford University Press, 2004.

Arvin, Maile, Eve Tuck, and Angie Morrill, "Decolonizing Feminism: Challenging Connections Between Settler Colonialism and Heteropatriarchy." *Feminist Formations 25*, 1 (December 2012): 8–34.

Atalay, Sonya, Jen Shannon, and John Swogger. *Journeys to Complete the Work... and Changing the Way We Bring Native American Ancestors Home*. United States: PRC, 2017.

Barques-Pedreny, Pol, David Chandler and Elena Simon, ed., *Mapping and Politics in the Digital Age*. New York: Routledge, 2018.

Basso, Keith H. *Wisdom Sits in Places: Landscape and Language Among the Western Apache*. Albuquerque: University of New Mexico Press, 1996.

Becker, Amy. "*Mapping the Unmappable in Indigenous Digital Cartographies*." MA Thesis, Department of Anthropology, University of Victoria, 2014.

Belcourt, Annjeanette E., Cyda Swaney, and Allyson Kelley, "Indigenous Methodologies in Research: Social Justice and Sovereignty as the Foundations of Community-Based Research," in Kathryn Shanley and Bjorg Evjen, *Mapping Indigenous Presence: North Scandanavian and North American Perspectives*. Tucson: University of Arizona Press, 2015, 57–77.

Bell, Joshua A., Kimberly Christen and Mark Turin, "*After the Return: Digital Repatriation and the Circulation of Indigenous Knowledge*." *Museum Anthropology* 7, no. 1–2 (2013).

Bell, Catherine and Robert K. Paterson, eds., *Protection of First Nations Cultural Heritage: Laws, Policy, and Reform* (Law and Society Series). Vancouver: UBC Press, 2009.

Bjorg Evjen and David R.M. Beck, "Growing Indigenous Influence on Research, Extended Perspectives, and a New Methodology: A Historical Approach," in Kathryn Shanley and Bjorg Evjen, *Mapping Indigenous Presence: North Scandanavian and North American Perspectives*. Tucson: University of Arizona Press, 2015: 27–56.

Brody, Hugh. *Maps & Dreams* Long Grove, IL: Waveland Press, 1981.

Brown, Michael F. *Who Owns Native Cultures?* Cambridge: Harvard University Press, 2003.

Cajete, Gregory, "Philosophy of Native Science," in Anne Water, ed., *Native American Thought*. Malden, MA: Blackwell Publishing, 2004, 45–57.

Duxbury, Nancy, W.F. Garrett-Petts, and David MacLennan, "Cultural Mapping as Cultural Inquiry: Introduction to an Emerging Field of Practice, in *Cultural Mapping as Cultural Inquiry*. New York: Routledge, 2015.

Echo-Hawk, Walter R., *In the Courts of the Conqueror: The 10 Worst Indian Law Cases Ever Decided*. Golden, CO: Fulcrum Publishing, 2010.

Echo-Hawk, Walter R., *The Sea of Grass: A Family Tale from the American Heartland*. Golden, CO: Fulcrum Publishing, 2018.

Goeman, Mishuana. *Mark My Words: Native Women Mapping Our Nations*. Minneapolis: University of Minnesota Press, 2013.

Golding, Viv and Wayne Modest, *Museums and Communities: Curators, Collections and Collaboration*. New York: Bloomsbury, 2013.

Dallas Hunt and Shaun A. Stevenson, "Decolonizing Geographies of Power: Indigenous Digital Counter-Mapping Practices on Turtle Island," *Settler Colonial Studies* 7, 3 (June 17, 2016): 372–392.

Irwin, Lee, ed., *Native American Spirituality: A Critical Reader*. Lincoln: University of Nebraska Press, 2000: 11–36.

Kroskrity, Paul V. and Margaret Field In *Native American Language Ideologies: Beliefs, Practices, and Struggles in Indian Country*. Tucson: University of Arizona Press, 2009:

Material World, "Review Essay: After the Return: Digital Repatriation and the Circulation of Indigenous Knowledge" (October 10, 2014), https://materialworldblog.com/2014/10/review-essay-after-the-return-digital-repatriation-and-the-circulation-of-indigenous-knowledge/.

Mukurtu CMS. https://mukurtu.org.

Native American Graves Protection and Repatriation Act, 25 U.S.C., Sections 3001-3003 (2006).

O'Neal, Jennifer. "Respect, Recognition, and Reciprocity: *The Protocols for Native American Archival Materials*," in Dominique Daniel and Amalia Levi, *Identity Palimpsests: Archiving Ethnicity in the U.S. and Canada*. Sacramento: Litwin Books, 2013.

Pyne, Stephanie and D.R. Fraser Taylor, ed., *Cybercartography in a Reconciliation Community: Engaging Intersecting Perspectives* (Elsevier: Modern Cartography Series, Vol. 8). Cambridge, MA: Elsevier, 2019.

Shanley, Kathryn, "'Mapping' Indigenous Presence: The Declaration on the Rights of Indigenous Peoples at Rhetorical Turns and Tipping Points," in Kathryn Shanley and Bjorg Evjen, eds., *Mapping Indigenous Presence: North Scandanavian and North American Perspectives*. Tucson: University of Arizona Press, 2015, 5–26.

Smith, Linda Tuhiwai. *Decolonizing Methodologies: Research and Indigenous Peoples*, 2nd edition (London: Zed Books, 2012 (originally published in 1999)).

Stoller, Paul. *Sensuous Scholarship*. Philadelphia: University of Pennsylvania Press, 1997.

Taylor, Anne, David Gadsden, Joseph Kerski, and Heather Guglielmo *Tribal GIS: Supporting Native American Decision-Making*. New York: Esri Press, 2017, 2nd edition.

Taylor, D.R. Fraser, ed., *Developments in the Theory and Practice of Cybercartography: Applications and Indigenous Mapping* (Elsevier Science: Modern Cartography Series, Vol. 5) (Oxford, UK: 2014), 2nd edition.

Trope, Jack F and Walter R. Echo-Hawk, "The Native American Graves Protection and Repatriation Act: Background and Legislation," in Devon A. Mihesuah, ed., *Repatriation Reader: Who Owns Native American Remains*. Lincoln: University of Nebraska Press, 2000, 123–168.

# 1 Alive with Story

## Mapping Indigenous Los Angeles and Carrying Our Ancestors Home

*Sarah Montoya*

### Introduction

As part of a continuum of resistance, digital projects and repositories offer platforms to reckon with settler representations of space and peoples. Two projects, Mapping Indigenous Los Angeles (MILA) and Carrying Our Ancestors Home (COAH), redress the relationship between Native and Indigenous bodies, communities, and land as they provide pedagogical tools for tribal communities, institutions, and settlers to dismantle settler colonial narratives. Mapping Indigenous LA, accessible at mila.ss.ucla.edu, unsettles colonial cartographic and geographic knowledge as it weaves together the stories of the Native and Indigenous inhabitants who originally occupied and have come to occupy Los Angeles as the result of complex relocations and diasporic processes. Indigenous digital counter-mapping projects confront and re-orient the colonial gaze through a nuanced rendering of relationships with space, place, and memory. Carrying Our Ancestors Home, accessible at coah-repat.com, addresses the history and intricacies of repatriation and offers insight into how tribal communities and institutions address the Native American Graves Protection and Repatriation Act (NAGPRA). Digital repositories and archives facilitate the preservation of cultural heritage and lifeways while simultaneously evidencing the legacy of political mobilization and current activism by Native communities. Though largely based on Gabrielino/Tongva lands (Los Angeles), these projects seek to build solidarity amongst and serve as repositories for global Native, Indigenous, and Aboriginal communities.

### Land is Life: Mapping Indigenous LA

The colonial construction of Native and Indigenous peoples is deeply entangled with techno-scientific development and the establishment of settler colonial property regimes. Within the bounds of a colonial fantasy and settler colonial imaginary, land is configured as an exploitable resource and Native and Indigenous peoples are flattened into "flora and fauna" as objects to be managed by the settler state.[1] The rhetoric of scientific objectivity and cartographic practices emphasizing "accuracy" and utility were weaponized by imperial and colonial powers to guise settler violence as space was transformed from territory into property.[2] The creation of a settler state legal mechanism for dispossession solidified the notion of "lawfully" owned property as a cornerstone of settler societies. Aboriginal scholar Aileen Moreton-Robinson (Koenpal, Quandamooka Nation) offers the term "possessive logics" to understand the link between property to a white, patriarchal settler identity which, in turn, is intimately connected with the political project of establishing and maintaining settler conceptions of ownership.[3] Denied personhood, humanity, and history through strategic colonial cartographic erasures, Native and Indigenous communities were and are un-placed and displaced.

Mapping Indigenous LA disrupts colonial erasure and displacement via digital counter-mapping in the form of story mapping. Mapping Indigenous Los Angeles utilizes Esri's Story Maps via ArcGIS to re-map complex relationships amongst Native and Indigenous communities and serve as a repository for community histories. It features Spanish and English content and offers a series of pedagogical resources for educators and community members alike. The project is envisioned as a collaborative research project, guided by co-principle investigators who represent a variety of departmental affiliations at UCLA including Dr. Maylei Blackwell, Professor in the César E. Chávez Department of Chicana/o and Central American Studies, Dr. Mishuana Goeman (Tonawanda Band of Seneca), Professor in Gender Studies, and Dr. Wendy G. Teeter, Curator of Archaeology for the Fowler Museum and UCLA NAGPRA Coordinator. UCLA, as a land grant institution in one of the largest cities in the settler state of the U.S., functions an interdisciplinary hub for scholars working with and in Native and Indigenous communities. The project is supported by the Institute of American Cultures, California Humanities, University of California Humanities Research Institute, University of California Center for New Racial Studies, The UCLA Center for Digital Humanities, UCLA American Indian Studies Research Center, Social Science Computing, and the UCLA College of Social Sciences. Mapping Indigenous Los Angeles represents alliances and solidarities within the university and amongst Native and Indigenous communities within LA. The scope of the project includes Gabrielino/Tongva and Fernandeño Tataviam Band of Mission Indians, Pacific Islanders, peoples of Oceania, American Indian narratives of relocation to LA, peoples of the Latin American Indigenous Diaspora, and maps tracing Indigenous relationships with land and waterways. As part of its community collaboration, the site actively seeks to build relationships with Native and Indigenous peoples in LA and offers detailed guides on how to create community-authored maps to be featured on the site.

## Source Materials

Story maps of MILA are the result of community-driven and community-generated content from Native and Indigenous peoples with relationships to LA. The project actively petitions communities for story maps and provides a lengthy, step-by-step screenshot tutorial and guide for communities to outline the entire process—from securing permissions and curating visual content to narrative development. In order to have a community story map featured on MILA's site, communities are asked to reach out the MILA team with a story map concept and key points, their team/community, and a research timeline. After the initial contact and acceptance, MILA offers technical aid to those involved with the creation and maintenance of a story map. Esri's Story Maps platform offers a readily available software, accessed via browser, that does not require high-level technical coding skill and can host narrative and visual content, in addition to map coordinates. Participants are encouraged to

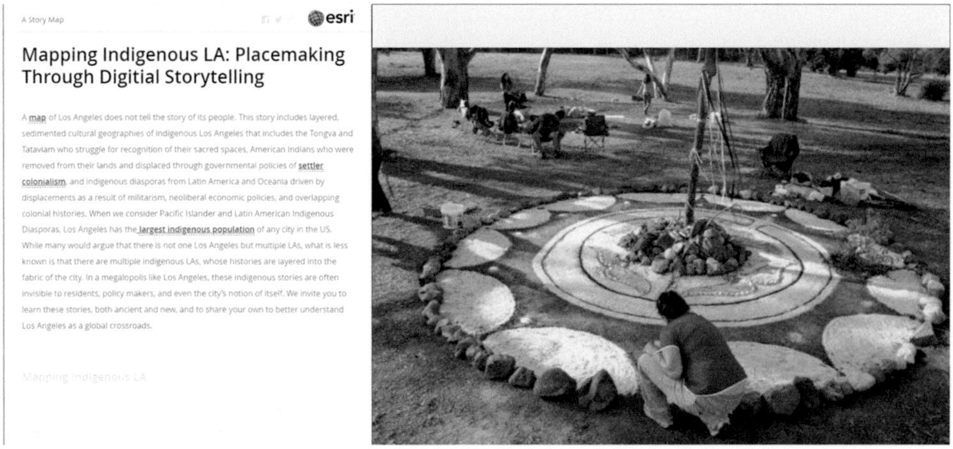


*Figure 1.1* From "Mapping Indigenous LA: Placemaking Through Digital Storytelling."

gather community histories in the form of photographs, interviews/oral histories, and videos to create dynamic and interactive experiences for viewers.

Esri's Story Maps features two-paneled content to provide historical and contemporary visual representations of Native and Indigenous placemaking. The community-authored narrative panel provides opportunities to link to definitions, concepts, and related content.

The story maps currently featured on the site represent the significant effort of Native and Indigenous peoples to create community-authored representations and disorient settler configurations of space. Many of the maps overtly address settler colonial violence and erasures and draw out complex historical racialization processes. "Mapping Indigenous LA: Placemaking Through Digital Storytelling" provides an in-depth overview of the project, highlighting local original inhabitants and the promise for cross-cultural exchanges in the densely networked cityscape of Los Angeles. "Latin American Indigenous Diaspora" documents crucial events, community organizations, and places of community gathering for both a variety of Mayan and Indigenous Oaxacan peoples. "Indigenous Urbanity in Los Angeles: 1910s–1930s" deftly brings together migration and labor histories, cinema and media studies, Native and Indigenous scholarship and activism, and offers an oral history of a Chumash/Tohono O'odham Elder, whose family has lived in the Wilmington/Carson for generations. Maps like the "Fernandeño Tataviam Map" and "Perspectives on A Selection of Gabrieleño/Tongva Places" offer intricate histories of dispossession and survival guided by Elders and community members, preserving Indigenous place-naming and the reclamation of historical place and presence. "American Indian Education Timeline & Resources" provides an overview of Indian education from its assimilationist, settler colonial roots to the establishment of tribal colleges and the current state of American Indian education. This map, like "American Indian Health Resources," offers a series of currently accessible resources for American Indian community members. "Los Angeles Waterways" outlines the varied presence of waterways, both artificial and naturally occurring, along the LA landscape; the map regards water not simply as a resource but as an Indigenous lifeway and indicates current revitalization projects.

## Resources and Pedagogical Materials

Mapping Indigenous Los Angeles hosts pedagogical materials and additional reading materials approved by Native and Indigenous community members and faculty. Again, the project encourages communities to offer their own community-created or community-approved resources. Teaching materials are divided by topic area and sub-divided into age ranges from K–12 resources to college-level reading materials. Topics encompass a variety of museum field trip guides, Indigenous and Native approaches to teaching, and resources for teaching Native American history ethically and responsibly in the state of California including critical approaches to the California Mission System. The site also provides resources for assessing American Indian materials and resources for Native and Indigenous educators. As such, MILA offers several avenues to dismantle settler state-sponsored narratives of conquest.

## Challenges

As with many digital projects, data sovereignty[4] and information privacy pose concerns. For Native and Indigenous peoples who have historically had data pilfered and turned over to the state,[5] the right and ability to govern and keep secure knowledge and information is something each community must consider when deciding what information to share in a publicly available map through a university-affiliated project. The project currently requires that participants arrange information and data hosting through a variety of sites (for instance, Esri's ArcGIS and image-hosting or video-hosting sites) and thus it is critical for participating Indigenous communities to carefully orchestrate community access and establish security protocol. MILA's team offers workshops and technical support for Indigenous communities navigating the project.

## Remapping Return: Carrying Our Ancestors Home

Until the passage of the Native American Grave Protection and Repatriation Act (NAGPRA) in the 1990s, the legal language surrounding Native human remains recapitulated settler violence. Archaeology and anthropology in the U.S. has a fraught relationship with scientific racism and routine grave desecration. Perhaps most infamous are Thomas Jefferson's excavation of Indigenous burial mounds by enslaved peoples and Dr. Samuel Morton's crania studies which required the regular decapitation of disinterred bodies.[6] Morton's phrenology study data codified scientific racism as it weaponized the trope of the "Vanishing Indian."[7] Scientific racism worked in tandem with the establishment of property regimes, and property regimes extended, too, to the remains of Native and Indigenous peoples. The relegation of Native peoples into landscape and association with primitivism enabled the display of sacred items and the remains of ancestors within the venue of Natural History Museums and established settler "ownership" over ancestral remains "discovered" on private property. Consider, for instance, the 1906 Federal Antiquities Act which utilized a legal language whereby archaeological resources become federal property.

While it did attempt to limit looting, it communicated a worldview in which Native remains were the property of the United States government.[8]

Carrying Our Ancestors Home addresses both the violence of desecration and the victories and complexities of repatriation as it pertains to NAGPRA. This legislation was the hard-earned result of decades of organizing by Native and Indigenous peoples to reclaim sacred objects and human remains which had been violently and unethically taken, housed, and displayed. The law requires federal agencies and institutions having received federal funds to repatriate, or return, human remains, funerary objects, sacred objects, and objects of cultural patrimony.[9] Much of the available materials on NAGPRA are authored by state-sanctioned entities or academic institutions, but COAH offers Indigenous and Native authored perspectives on repatriation for the benefit of both tribal community members and institutional representatives. The project again sees a UCLA-based team serving Native, Indigenous, and Aboriginal communities, directed by Dr. Wendy G. Teeter and Dr. Mishuana Goeman and managed by Sedonna Goeman-Shulsky (Tonawanda Band of Seneca), the Archaeology Collections Manager at the Fowler Museum. The project provides primary sources authored and approved by Indigenous and Native community members working with intuitions to trace the difficulties and varied protocols of repatriation. The site is built on Mukurtu CMS, an open-source content management system designed to house community digital heritage projects, and hosts a series of videos and relevant, curated literature as the project places institutions and tribal communities into conversation with one another.

## Source Materials

In order to successfully repatriate ancestral remains and cultural items, institutions must develop respectful relationships with tribal communities and understand the context from which tribal and intuitional relationships emerged. To support this, COAH provides a socio-historical context for understanding NAGPRA and repatriation. The site hosts "Fighting for Our Ancestors," a documentary tracing the American Indian Student Movement for Repatriation at UCLA in 1990s. The piece details on-campus and community American Indian political mobilization and the antagonistic response toward repatriation from the departments of Anthropology and Archaeology. Despite a pronounced history of violence in the fields, the passage of NAGRA resulted in a series of settler colonial critiques of repatriation which cast Native, Indigenous, and Aboriginal communities as adversaries to scientific study and advancement. These critiques re-entrenched both scientific racism and recirculated an imaginary in which Native and Indigenous peoples are relegated to a moment in the past or represented only in the present as a demonstration of backwardness. These combative responses are also documented on the site's featured timeline, built on Northwestern University Knight Lab's TimelineJS platform, which offers a chronological history of repatriation on a national and local level from the late 1960s to present.

Carrying Our Ancestors Home's media grapples with a catalog of institutional violence while demonstrating a willingness to work with institutions and alongside scientific communities to accomplish repatriation. Carrying Our Ancestors Home provides "Repatriation Stories" as video content which features interviews with tribal community members and elders, Native and Indigenous scholars and activists, and

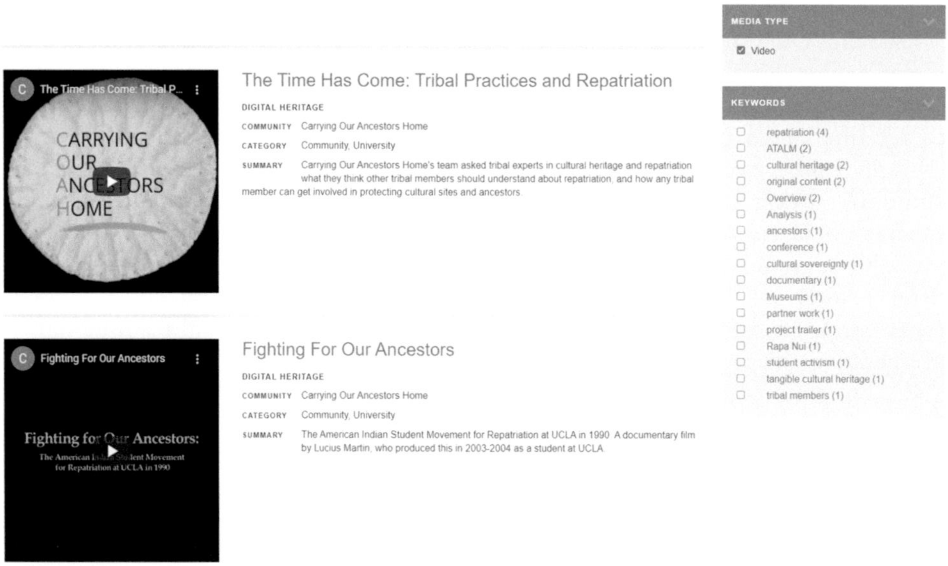


*Figure 1.2* From "Repatriation Stories." Videos and documents are organized and accessible through Mukurtu's meta-data tags which include content type, community, category, and keywords (pictured here).

repatriation coordinators. "What is NAGPRA?" outlines the scope of the law while noting the difficulties of implementation, the variance amongst institutions and tribes, and the short-comings of the legal language utilized in the legislation—including the politics of federal recognition. When NAGPRA was put into effect, it created the legal necessity for NAGPRA Coordinators, working for respective institutions that were under the jurisdiction of NAGPRA, to consult with tribal members or representatives of Native communities. In "Why is There Variation in Implementation of Repatriation?," interviewees explain that Native cultures and spiritual practices are not monolithic; thus NAGPRA, while providing the legal impetus, represents only a cursory step in a much longer process. The site archives and makes widely available a set of repatriation-related documents, including Indigenous and Native-authored repatriation literature. The content emphasizes that institutions must remain flexible in understanding that repatriation protocol and procedure can vary greatly.

Carrying Our Ancestors Home archives concerted Native and Indigenous repatriation efforts, offering hope and guidance to other Native and Indigenous communities. In "Rapa Nui Repatriation 2018," authored by Ma'u Henua News, community members return two ancestors to their homelands. Additional video content comes in the form of conference presentations and round tables where tribal community members discuss their experiences in museum/institutional environments and their experiences with navigating repatriation efforts. Supporting literature includes guides for both museums and communities authored by Native and Indigenous communities. In many ways, COAH serves as a hub for resource-distribution as Native communities continue to navigate the legal and social implications of NAGPRA.

## Conclusion

Mapping Indigenous Los Angeles and Carrying Our Ancestors Home provide Native and Indigenous communities with vital resources to connect with one another and preserve community histories. Future development of both projects depends largely on continued funding and sustained community involvement. Mapping Indigenous Los Angeles' story maps serve as a form of community archive, preserving the voices and work of community elders and documenting the lived experiences and Native and Indigenous peoples in the area. Carrying Our Ancestors Home highlights and addresses the need to center Native and Indigenous voices and contributions to repatriation efforts. Each site draws clear links between the historical treatment of Native and Indigenous communities and our present moment and asks settlers to rethink our relationships with space, as well as our accountability and responsibility to Indigenous communities. Both MILA and COAH confront and correct colonial and settler colonial representations of Native and Indigenous peoples committed via the map and the museum, recasting and re-storying the relationship between Native and Indigenous life and land.

## Notes

1 Goeman, Mishuana. *Mark My Words: Native Women Mapping Our Nations*. Minneapolis: University of Minnesota Press, 2013, 18.
2 Blomley, Nicholas, "Law, Property, and the Geography of Violence: The Frontier, The Survey, and The Grid." *Annals of the Association of American Geographers* 93, no. 1 (2003): 125, 129.
3 Moreton-Robinson, Aileen. *The White Possessive: Property, Power, and Indigenous Sovereignty*. Minneapolis: University of Minnesota Press, 2015, 11.
4 Sunia, Michele and Maggie Walter, "Indigenous Data, Indigenous Methodologies and Indigenous Data Sovereignty." *International Journal of Social Research Methodology* 22, no. 3 2019: 236–237.
5 See Bryan, Joe and Denis Wood, *Weaponizing Maps: Indigenous Peoples and Counterinsurgency in the Americas*. New York: Guilford Press, 2015, pp. 157–161, who trace the disastrous impact of the Bowman Expedition's *México Indígena* project (2005–2008), led by Peter Herlihy (University of Kansas) and funded by institutions such as the Foreign Military Studies Office of the U.S. Department of Defense, the American Geographical Society (AGS), the University of Kansas' Center of Latin American Studies, and Mexico's *Secretaría de Medio Ambiente y Recursos Naturales*. Eventually, the project was accused of geopiracy by Indigenous communities who were not informed that the data supplied would be made available to the government.
6 Thomas, David Hurst, *Skull Wars: Kennewick Man, Archaeology, and the Battle for Native American Identity*. New York: Basic Books, 2000, pp. 33–34; 38–40.
7 Trope, Jack F. and Walter R. Echo-Hawk. "The Native American Graves and Repatriation Act: Background and Legislative History." In Tamara L. Bray (ed.). *The Future of the Past: Archaeologists, Native Americans, and Repatriation*. Garland Publishing, 2001, p. 11.
8 Trope, Jack F. and Walter R. Echo-Hawk, "The Native American Graves and Repatriation Act: Background and Legislative History." In Tamara L. Bray (ed.). *The Future of the Past: Archaeologists, Native Americans, and Repatriation*. Garland Publishing, 2001, p. 12.
9 "Native American Graves Protection and Repatriation Act: Law and Policy." *National Park Service*, U.S. Department of the Interior. Accessed August 2, 2020. www.nps.gov/subjects/nagpra/law-and-policy.htm.

## References

Blomley, Nicholas. "Law, Property, and the Geography of Violence: The Frontier, The Survey, and The Grid." *Annals of the Association of American Geographers* 93, no. 1 (2003): 121–141.

Bryan, Joe, and Denis Wood. *Weaponizing Maps: Indigenous Peoples and Counterinsurgency in the Americas*. New York: Guilford Press, 2015.

Goeman, Mishuana. *Mark My Words: Native Women Mapping Our Nations*. Minneapolis: University of Minnesota Press, 2013.

Moreton-Robinson, Aileen. *The White Possessive: Property, Power, and Indigenous Sovereignty*. Minneapolis: University of Minnesota Press, 2015.

"Native American Graves Protection and Repatriation Act: Law and Policy." *National Park Service*. U.S. Department of the Interior, Accessed August 2, 2020. www.nps.gov/subjects/nagpra/law-and-policy.htm.

Suina, Michele and Maggie Walter. "Indigenous Data, Indigenous Methodologies and Indigenous Data Sovereignty." *International Journal of Social Research Methodology*, 22, no. 3 (2019): 233–243.

Thomas, David Hurst. *Skull Wars: Kennewick Man, Archaeology, and the Battle for Native American Identity*. New York: Basic Books, 2000.

Trope, Jack F. and Walter R. Echo-Hawk. "The Native American Graves and Repatriation Act: Background and Legislative History." In *The Future of the Past: Archaeologists, Native Americans, and Repatriation*. Tamara L. Bray (ed.). New York: Garland Publishing, 2001.

# 2 Digitally Re-presenting the Colonial Archive

## Resources for Researching and Teaching the Carlisle Indian Industrial School and the Native American Boarding School Movement

*Frank Vitale IV, Susan Rose, and James Gerencser*

The official story of the Carlisle Indian School as a progressive, benevolent experiment was originally told through documents and images used for public relations and fundraising. Pratt's time serving in the U.S. Cavalry was reflected in the military manner in which he planned, documented, and operated the school. Casting these activities as supporting discipline and industry, Pratt purposefully molded a positive image of the school in widely-disseminated photographs and texts that belied ongoing challenges. This image often contrasts sharply with the thoughts and experiences of the students themselves, as well as the real challenges of operating the school. The Carlisle Indian School Digital Resource Center (CISDRC) has made both publicity materials and internal administrative documents freely and easily accessible so that the carefully crafted narrative of the school's proponents can be reconsidered by those interested in learning more about Carlisle, its students, and the broader Indian Boarding School Movement.

The CISDRC is an effort to aid research by bringing together, in digital format, a variety of resources related to the Carlisle Indian School that are physically preserved in various locations.[1] Through making these resources easily accessible, we seek to increase knowledge and understanding of the school and its complex legacy, while also facilitating efforts to uncover the stories of the many thousands of students who were enrolled there. With the CISDRC, the intention is not merely to share archival material, but to further build and develop the archival record by offering a space where descendants of Carlisle students may "talk back" to the official record by offering corrections and adding both their voices and their personal documentary collections to the conversation. Doing so enriches the preserved historical record while supporting increased opportunities for learning, understanding, and healing.

Research of any kind, including historical research, is limited by source availability. When studying boarding schools like Carlisle, scholars' ability to access and compare information shapes the kinds of questions that can be asked and answered. Genevieve Bell, whose 1998 doctoral thesis provided the first detailed study of the Carlisle Indian School, spent months at the National Archives in Washington, D.C., reviewing student records and building a basic database to organize the information she gleaned from these sources.[2] These same records and many more are now accessible from anywhere in the world, at any time, to anyone with Internet access.

The ease of access to this wealth of material often masks the effort required to collect, organize, and carefully describe this content, so it is worth highlighting how digitization helps to reshape the colonial archives of boarding schools. Aside from reducing research obstacles and access costs for descendants and scholars, digitization presents historical records in new and powerful ways. Enabled by the hundreds of thousands of pages of material that can now be easily read, evaluated, and compared, researchers have begun rewriting the history of the Carlisle Indian School. By re-presenting these once-hidden analog records in digital formats, the CISDRC has helped to redefine how primary sources and colonial archives can be approached, understood, and utilized. Descendants have used this information to better understand the experiences of their ancestors who attended Carlisle. Teachers have begun using the school's unique primary sources in classroom instruction ranging from the primary to post-secondary levels. Historical information has been used to guide efforts to return the remains of students from the Carlisle Indian School cemetery to their home communities. And scholars from across the globe are regularly accessing these resources to inform their research.

## The Primary Sources Themselves

Initial planning for the CISDRC focused on the known extant school records, which have been frequently utilized and cited by scholars through the years. These core materials are publicly accessible in just a handful of repositories—the U.S. National Archives, the Cumberland County Historical Society, the Beinecke Library at Yale University, and the U.S. Army Heritage and Education Center. The types of records held by each archive, and thus the information that can be gleaned from them, is highly variable, depending on the original purposes of each document. By understanding the difference provenances and uses of these materials, their information becomes more powerful, as hidden factors including authorial purpose, inherent bias, and even historical chronology come to light.

When the Carlisle Indian School closed in 1918, the administrative files maintained by the school were shipped to the offices of the Bureau of Indian Affairs (BIA) in Washington, D.C. Documents within the Bureau's archives show that these files were occasionally consulted to respond to research inquiries from the 1920s through the 1940s. Probably sometime in the 1950s, these records were then deposited at the National Archives. Once there, the files were inventoried, organized, and cataloged. These materials formed the foundation of the CISDRC's digitization efforts, which began in 2013.[3]

The body of records at the National Archives includes thousands of individual student files and notecards, which school administrators had used to keep track of students and alumni during the school's years of operation. Bound ledgers reveal additional student-centered information such as admission data, attendance, outing placements, financial records of earnings and spending, discharge from the school, and even death. Other extant school records include financial ledgers for the school's operations, copies of outgoing correspondence, a handful of personnel files, a few magazines and programs printed by the school, and even meeting minutes for one of the student debating societies.

Interestingly, these records from the school reveal little about its day-to-day operations. Such information was found elsewhere through further research at the National Archives. One grouping of files from the BIA relates to Carlisle between 1907 and 1918, and these materials have been separately housed and organized by subject.[4] These files include correspondence between staff at Carlisle and BIA officials dealing with individual student concerns, various operational matters, and even broader policy issues.

Information about daily operations dating between 1881 and 1907 was found in yet another location. These files are part of the incoming correspondence of the BIA, all organized chronologically by date received.[5] A rough subject index that was created at the time the letters were filed more than 100 years ago is still needed to identify individual items related to Carlisle from among the tens of thousands of documents received in any given year. The process to identify and then digitize relevant files is extremely laborious, but the records found in this manner reveal previously unknown information about the school, opening new avenues for research. Similar information is also being identified among BIA correspondence files dating from before 1881.[6]

Besides these Carlisle-related records found at the National Archives (and others yet to be identified there), a wealth of important materials can be found at the Cumberland County Historical Society (CCHS), located in Carlisle, Pennsylvania. With a mission to preserve and provide access to material of local interest, CCHS has received numerous donations over the years that document the school and its students. Several thousand photographs and negatives provide a visual record of the school, and a large collection of newspapers and magazines printed at the school shed light on its activities and how they were shared with other audiences. Personal scrapbooks, oral histories, artifacts, and countless ephemeral pieces round out holdings that often reflect the more public aspects of the school, rather than the internal administrative files located at the National Archives.[7]

The Beinecke Rare Book and Manuscript Library at Yale University holds the personal papers of Richard Henry Pratt, donated by his children in 1959.[8] This large collection includes letters, speeches, photographs, diaries, scrapbooks, writings, student drawings, and other materials. The extensive correspondence with current and former Carlisle students, with students' parents, with supporters of Carlisle, and with numerous government and military figures provides a detailed picture of Pratt's ideas and activities. In addition, this material reflects the kinds of relationships he maintained with his students and his peers and provides some contrast to the more business-like nature of the administrative files at the National Archives.

Finally, a collection of loose photographs and photo albums is maintained at the U.S. Army Heritage and Education Center (AHEC), along with other materials on the history of the Carlisle Barracks.[9] The U.S. Army Heritage and Education Center also holds papers that have been gathered together to document the activities of Richard Henry Pratt as well as papers that document the barracks cemetery. The collection even includes student attendance books for the paint and harness shops at the school.[10] While somewhat less coherent than the archival collections housed at other repositories, some unique photos and documents provide valuable research potential.

## Primary Sources and Their Contexts

Primary sources kept in archives and other repositories were for the most part never intended for use by today's researchers. Rather, most surviving records had specific purposes when they were created, which rarely align with modern usages. The colonial archives, and specifically collections of boarding school records, are no exception. Alistair Tough notes that records produced by imperial administrators were intended in the first instance to aid colonial governance. As a result, Tough asserts that modern use of colonial records must be grounded in hermeneutics, or rather, that primary sources must be read and interpreted based on an understanding of their context in the past.[11] Administrative records for boarding schools, created and maintained by school and federal officials, are just that: information gathered by mostly white, largely male colonial administrators for the purposes of managing and running boarding schools. Understanding this provenance, and its impact on the types of data preserved in various records, is thus necessary when researching and representing boarding schools using biased administrative documents.

Surviving files from and about the Carlisle Indian School are typical of extant records from other contemporaneous boarding schools. Student files detail academic, medical, personal, and disciplinary information. These files are supplemented by student financial and enrollment ledgers as well as official correspondence. Ephemeral records, especially school newspapers and event programs, also frequently survive, along with the occasional cache of student class work, autobiographical memoirs, and personal scrapbooks. All told, these documents form a complex network of information about each individual in which distinctions in provenance, and thus information reliability, can become obscured. This archival web thus poses a significant challenge for scholars researching any Indian boarding school and its students.

When faced with an analog collection of these records, scholars are left with little recourse other than long hours spent *in situ* wading through hundreds of boxes and folders, piecing information together bit by bit.[12] Digitization changes the very process of this search, re-presenting analog records within a guided framework that enhances scholars' and students' ability to search and understand them. Aside from eliminating the physical obstacles to accessing remote archives, projects like the CISDRC allow users to concurrently consult as many documents as they can open on their computers. In a way only made possible through digitization, users can simultaneously access records from numerous different collections and repositories. By re-presenting records side by side, users can more easily piece together stories and chronologies from information in disparate documents. Tasks that once took months, such as consulting all existing student files or examining which students came from a particular Nation or on a particular date, can now be done anytime, anywhere, with the click of a button.

Ease of access does not, however, make the archival web easier to untangle, or provenance easier to establish. This is where a digitization project's metadata framework comes into play. Common elements to all digital humanities projects, including search functions, item descriptions, and various tags, all help to frame the myriad records available to users. Drawing upon data inter-relatability, this frame gives users the information they need to establish document provenance and context. Search functions, encompassing both staff-created descriptions and computer-generated transcriptions, replace mere collection inventories, allowing users to find even passing mentions of their keyword instantaneously. Document summaries, written by staff with a knowledge of the school and its history, help users to browse records and establish authorial purpose without relying on paleography and prior knowledge. Tags then pull together related documents based on people, places, organizations, topics, and archival repositories. Links between records piece together previously separated chains of correspondence and recordkeeping, automatically providing users with more aspects of a story or event. Topic tags bring together

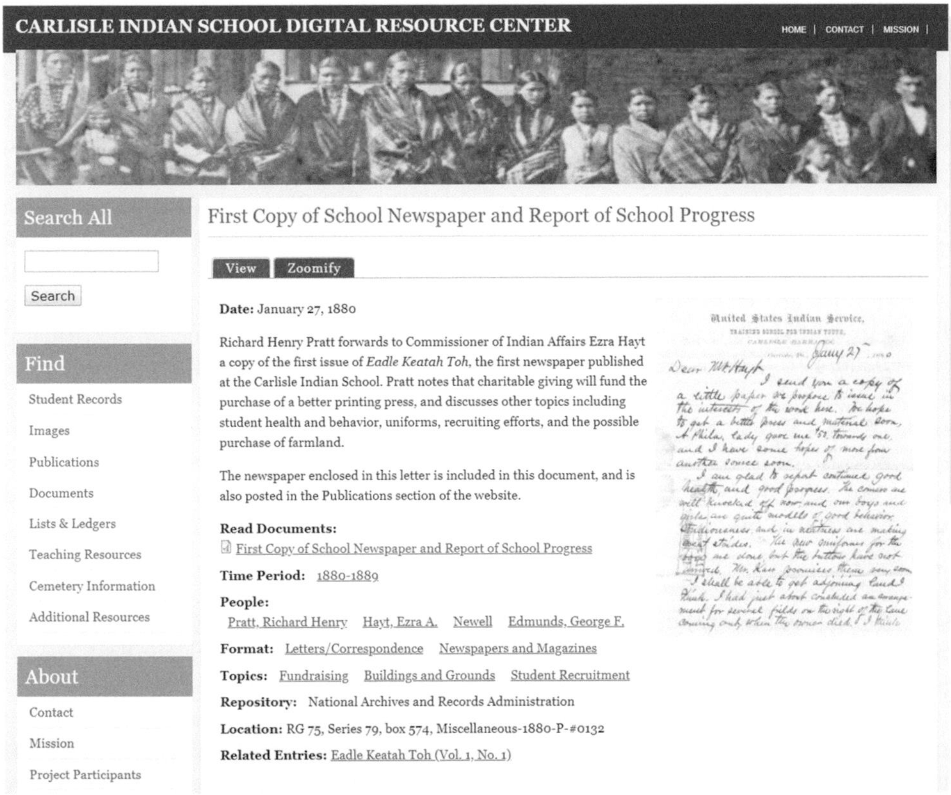


*Figure 2.1* Example of a document post, including a summary and tags.

distinct stories within a broader historical trend, identifying contextual chronologies that influenced both record creation and people's actions. Digitization is thus a form of re-presentation, in which individual analog documents are made to communicate with one another toward an end of increased information accessibility and contextualization. This increased juxtaposition between related documents is valuable for scholars, students, and descendants, as it lowers the threshold of entry for historical research.

Furthermore, document inter-relatability sheds new light on old questions about the Carlisle Indian School, while also giving rise to new questions and hypotheses about its operations. Consider, for example, long-held arguments about student censorship. Scholars have noted that school publications, even when claiming to faithfully present student voices, were heavily edited and censored by school administrators. Jacqueline Fear-Segal's investigation of the shadow editor of Carlisle's school newspapers, known as the Man-on-the-Bandstand, plays a central role in her larger discussions of surveillance at the school.[13] Using the CISDRC, scholars can trace the development of this symbol of censorship over time, identifying instances of moralization, surveillance, and discipline as presented in print. With the aid of summaries and tags, it also becomes clear that student voices were frequently co-opted to support the moralizing, Christianizing, assimilationist mission of the school.[14] This trend of censorship in school newspapers is thus easily recognizable thanks to the large number of examples brought together through digitization and metadata comparison.

The actual implementation of this censorship is also now identifiable thanks to the digitization of administrative correspondence that, until recently, have never had a proper item-level inventory. Scholars now have access to letters between school and federal officials noting how the newspaper will be an effective means of spreading "the interests of the work" at Carlisle.[15] Similar correspondence also shows that administrators sought to quash all negative news about the school, even when it appeared in the local and national press, as a means of "doing everything possible to build up, rather than pull down, to help to preserve the good name of the [Indian] Service, and to 'lend a hand.'"[16] These brief examples illustrate that censorship by school and federal officials in school newspapers was part of larger, purposefully orchestrated efforts designed to shape and protect Carlisle's public image while simultaneously furthering its colonialist activities. This finding might not be surprising for researchers, but it has only been adequately evidenced through digitization, and the links created by a digital re-presentation of the school's administrative archives.[17]

In addition to uncovering new evidence of the colonialist operations of the Carlisle Indian School, digitization through the CISDRC has helped amplify the voices of some of the school's students, preserved through a variety of official and ephemeral records. Doing so has shown how digitization provides a framework for both the re-presentation of the colonial archive and the perspectives of the colonized. One significant contribution of digitization in this vein is the uncovering of a heretofore understudied form of student expression: official complaints. Recently digitized official correspondence, again originally meant solely for internal use, preserves many instances of students complaining about censorship, surveillance, and discipline. Mamie Vilcan (Chitimacha) complained about mail surveillance in 1913, claiming that the school matron had personally targeted her letters for examination and confiscation despite Vilcan's right to privacy both under the law and as an adult.[18]

Other students complained of harsh discipline and poor treatment at the hands of school administrators, poor quality food and healthcare, and non-consensual enrollment.[19] Taken individually, these sources provide rare student perspectives on conditions at the Carlisle Indian School; when brought together through the metadata framework of the CISDRC, they reveal otherwise obscure trends in the school's history, including methodical attempts to surveil mail, frequent complaints against school officials, and patterns of favoritism for specific high-profile student groups including athletes.

Furthermore, digitally re-presenting student perspectives within a larger framework helps recontextualize information found elsewhere in the colonial archives, reshaping what we know about events and actions in the history of the school. An example of this is the case of Wesley Two Moons (Cheyenne), who died at the school in 1911 of pneumonia. Two Moons's student file is sparse, like that of most students who died at the school.[20] In these records, administrators attribute the death of Two Moons to a medical history of illness. However, an anonymous student complaint about discipline and the school's guardhouse adds a new layer to that story. Metadata for dates, topics, and people link Two Moons's student file to this document, in which an individual identified by school officials as Two Moons complains about harsh

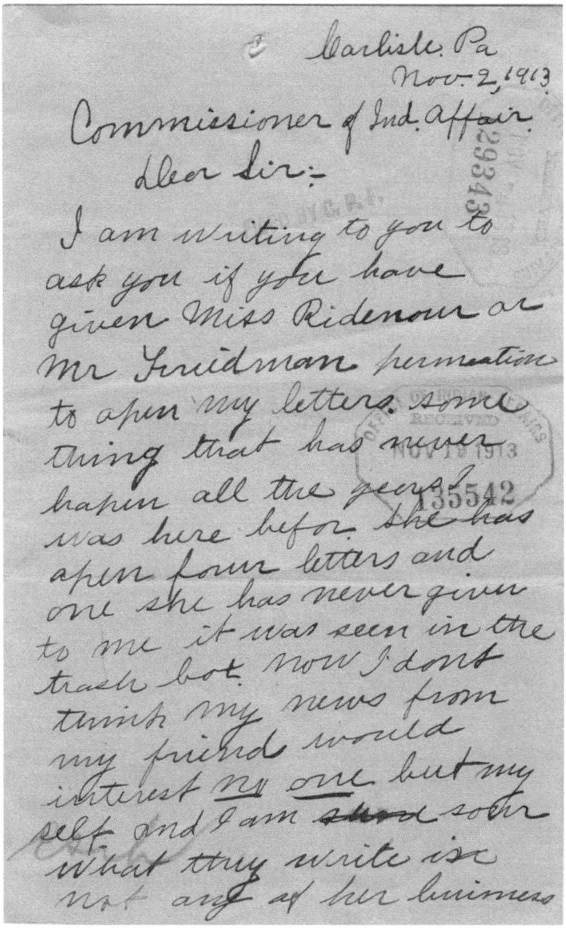
Carlisle. Pa
Nov. 2, 1913
Commissioner of Ind. Affair.
Dear Sir:-
I am writing to you to
ask you if you have
given Miss Ridenour or
Mr Friedman permation
to apen my letters. some
thing that has never
hapen all the years I
was here befor. she has
apen four letters and
one she has never given
to me it was seen in the
trash bot. now I dont
think my news from
my friend would
interest no one but my
self and I am ~~shurd~~ sure
What they write is
not any of her business

*Figure 2.2* Letter from Mamie Vilcan to the Commissioner of Indian Affairs, November 2, 1913.

treatment at the hands of multiple school officials. Two Moons states that he was thrown in the "cool and damp" guardhouse for a minor infraction, where he "contracted pneumonia and was neglected till almost dead."[21] Written by Two Moons days before his death, this complaint is the only surviving testimony of his treatment at Carlisle. Presenting either Two Moons's student file or his "anonymous" complaint individually tells us very different things about events in Two Moon's personal history and the institutional history of the Carlisle Indian School. When re-presented together in an inter-linked framework, these documents speak to one another, extending our understanding of Two Moons's personal experience at the school while simultaneously challenging prevailing notions about discipline and health as presented in the colonial archive.

Finally, it is worth noting that a digitized re-presentation of archival records can help highlight and counteract gaps in source and information survival. Carlisle students and alumni who died while the school was still in operation serve as good examples of these gaps. The student files that survive today seem to have been maintained by school administrators primarily to keep track of living alumni. As a result, the files of students and alumni who died prior to Carlisle's closing in 1918 were likely discarded by school officials as part of their record-keeping practices.[22] This historic action significantly impacts modern-day efforts to identify and study many of Carlisle's former students. When consulting analog records, a sparse or non-existent student file significantly hinders research efforts, since one must then search within dozens of minimally indexed ledger books. Digitization and the power of search functions changes that research process, as all documents mentioning a person can easily be brought together. Tagging people's names is especially powerful in this regard, as the records for individuals who were known by more than one name or name spelling can be easily correlated and compared. Digital re-presentation thus serves as an important means of recovering information about all of Carlisle's former students, leading to significant shifts in how the school is understood. Numerous individuals now interred under headstones reading "Unknown" have been identified thanks to previously unsearchable enrollment and financial ledgers.

Whether in analog or digital format, the colonial archive poses significant limitations to our understanding of students' experiences and Indian boarding schools like Carlisle. However, digitization and the re-presentation of these records helps counteract internal bias and information absence. Through metadata tags and the resulting ease with which records can be cross-referenced, additional information about student experiences becomes available. Many descendants of Carlisle's students now are able to find more information about their ancestors that was previously hidden within bound ledgers and publications.

## Descendants, Oral Histories, and Citizen Archivists

In some families and communities, stories of the Carlisle Indian School were shared, but in others there was a great silence. The traumatic past that many experienced has affected future generations who want to know more about their history and reclaim it. Some do so through academic research, others like Shan Goshorn and Emily Arthur through art and curation, and still others through storytelling as a means to both express themselves and educate others.[23]

The Carlisle Indian School digital archive tell us only so much. While it provides valuable documentation, it represents primarily the "official" story of school, recorded by government officials who chose what to document and how. The voices of the students themselves are largely absent, excepting the few published memoirs of Luther Standing Bear (Sioux), Jason Betzinez (Apache), and Michael Burns (Apache).[24] More stories and more voices need to be heard. W. E. B. DuBois argued that the state of historical memory may be defined as the state of cultural struggle; of contested truths, of moments, events, and texts in history that thresh out rival versions of the past which can in turn be put to the service of the present.[25] How do we thresh out those rival versions, the counter narratives?

Funded by a grant from the National Historical Publications and Records Commission (NHPRC), three members of the CISDRC project team had the opportunity to visit with numerous Native communities, giving presentations about the online resource for community members, teachers and students, and librarians and archivists. The learning was reciprocal, with descendants both exploring the website for more information about their loved ones and also offering stories, documents, and photographs that had been passed down through several generations. At two symposia held in Carlisle in 2012 and again in 2018, organized by Dickinson College and the Cumberland County Historical Society, many people spoke of their ancestors' experiences at Carlisle. Native perspectives are essential for piecing together the story of the young people sent to Carlisle and the impact it had on them, their families, and their communities.

One family story, revealed in the documentary *The Lost Ones: Long Journey Home*, was able to be produced only through the sharing of oral histories by the Lipan Apache Band of Texas and archival research.[26] Through the work and teachings of many Native and allied scholars, artists, and filmmakers, multiple stories are emerging to reveal the often tragic history of Carlisle and the Indian Boarding School Movement more generally. Brenda Child argues that the Indian boarding school is often used as a metaphor that stands in for so many attempts at annihilation and devastation.[27] It is true that the schools were only one of the wars waged against Indians as settler colonists attempted to dispossess them of their land and culture. It is also true that some students at Carlisle reported having positive experiences, actively choosing to enroll at Carlisle and completing their own admission applications. Other families report mixed experiences of their ancestors being taken from their families, but then returning home to build a three-story house having learned masonry at Carlisle. Still others know little about what their ancestors may have experienced since there was often a great silence about what happened at Carlisle. As Michel Rolph-Trouillot reminds us, "Any historical narrative is a particular bundle of silences, the result of a unique process, and the operation required to deconstruct these silences will vary accordingly."[28] This is a complex and contested history that involves all Americans. How do we teach our children about our intersecting pasts and the ways they affect our lives, relationships, and policies in the present? In re-searching and re-representing the past, how do we work toward shared futures?

## Teaching with a Digitized Archive

The CISDRC supports the positioning of Native American boarding school history within the wider history of the United States and American education. In exploring the ways in which this troubled and veiled history can be used to understand and illuminate current educational concerns, practices, and community struggles, the CISDRC provides ways for this history to be taught within the context of Native and non-Native schools and colleges, and within community organizations such as libraries, cultural centers, and museums. The Teaching Resources section of the website offers options for teachers, students, and others interested in exploring this history.[29] Teaching guides and student worksheets are posted in Microsoft Word format as well as Adobe PDF so that educators can easily and freely access and adapt them. While the site provides over 200,000 pages of primary source documents, which can be a daunting volume of material for busy teachers to explore and distill into assignments for their students, the guides assemble a reasonable amount of relevant material and specific assignments to address particular questions or skill sets. Lesson plans provide selected materials to examine questions such as the goals of the school's founder, the impact of the school on indigenous communities, and the interconnection between the school and national ethnic and race relations. These teaching resources thus offer secondary and post-secondary teachers some ideas for how to bring this important part of American history into their classrooms.

While members of the CISDRC project team developed the majority of the modules, participants in the 2016 Carlisle Indian School Teachers' Institute, funded by the NHPRC, also contributed materials based on their expertise. Their lesson plans focus both on themes (e.g. athletics, gender, land allotment and tenure, comparing Carlisle with contemporary Indian boarding schools, etc.) and on skills with secondary school standards in mind. For example, teachers wanting students to understand the differences between primary and secondary sources can find a module that guides students in analyzing both types of documents while learning more about Pratt's vision for the school and critiques of his educational experiment. Students can also explore the site by name, Nation, or date to find more information about a particular person or group of students, and then search for secondary sources to learn more about what their home communities were experiencing when groups of young people were sent to Carlisle. A teacher can have students use maps to trace where young people came from and the journey they embarked on that took them to Carlisle, or have them practice their skills in analyzing and interpreting Carlisle's infamous before-and-after photographs.

In addition to having teaching guides available online, the CISDRC project team also developed a physical teaching kit, with funding provided by the NHPRC to print and freely distribute copies to schools across the country. The teaching kits consist of 35 color facsimile reproductions of selected photographs, publications, printed programs, and even a student's personal autograph book. Numerous teachers have commented on the value of having some items to handle as a further aid to the learning process and as a way to more deeply engage students. Reading articles in a newspaper dating from 1888 and handling copies of photographs of students can aid with understanding the materiality of Carlisle and the public face it promoted. Even further, many teachers noted that not all schools have computing technology available for every student, or have consistently reliable internet, so physical teaching kits make it possible to study aspects of Carlisle and the Indian boarding school movement even when the website itself may not be accessible to all school audiences.

## Conclusion

In a way, the creation of physical teaching kits may seem to run counter to the theme of re-presentation. However, the balanced combination of administrative, ephemeral, and personal materials included in the kits echoes the digital re-presentation and amalgamation of materials presented via the CISDRC. Simply by presenting these disparate materials together, the colonial archive and its suppressive bias can be examined and undermined, much in the same way that artist Fred Wilson's landmark exhibit *Mining the Museum* recontextualized slavery-imbued collections through physical juxtaposition.[30]

Collective memory is a constructive and reconstructive process. In this way, historical narrative is not merely reproduced; we create and recreate narratives in response to ever-changing political and social circumstances.[31] As Trouillot asserts,

> Authenticity implies a relation with what is known that duplicates the two sides of historicity: it engages us as actors and as narrators... Whether it invokes, claims, or rejects The Past, authenticity obtains only in regard to current practices

that engages us as witnesses, actors, and commentators... Even in our relation to The Past our authenticity resides in the struggles of our present.[32]

We need to examine the critical relationship between memory (the past) and expectation (of the future) and why the stakes of social memory are so high.

The CISDRC, and digitization projects in general, are a key part of this societal reevaluation of our shared histories, both in terms of equalizing information access and elevating the prominence of traditionally suppressed perspectives. By re-presenting vital, albeit perpetually incomplete, information about those who were enrolled at Carlisle, and by adding to it through a robust framework of metadata, the CISDRC provides families, communities, and scholars alike with perhaps the most easily accessible collection of primary source materials related to the Indian Boarding School Movement. These documents now need to be interpreted and analyzed, the colonial archive mined for what it can reveal about Carlisle's students and staff, and about educational visions and experiments at the end of the 19th century. Through this continued analysis, and through the interjection of long-neglected indigenous perspectives into the colonial archive, critical questions can be raised about historic and modern-day oppression and injustice.

## Notes

1 "Carlisle Indian School Digital Resource Center," Dickinson College Archives and Special Collections. Accessed July 15, 2019. http://carlisleindian.dickinson.edu.
2 Bell, Genevieve. "Telling Stories Out of School: Remembering the Carlisle Indian Industrial School, 1879–1918." Ph.D. diss., Stanford University, 1998: 12–13.
3 Records of the Carlisle Indian Industrial School, National Archives and Records Administration, Record Group 75, Series 1322-1349C.
4 Central Classified Files – Letters Received by the Bureau of Indian Affairs, 1907–1939, National Archives and Records Administration, Record Group 75, Series 121.
5 Letters Received by the Bureau of Indian Affairs, 1881–1907, National Archives and Records Administration, Record Group 75, Series 91.
6 Letters Received by the Bureau of Indian Affairs, 1824–1880s, National Archives and Records Administration, Record Group 75, Series 79.
7 Brief inventories, indexes, and descriptions of holdings available online at Carlisle Indian School History, Cumberland County Historical Society. Accessed July 15, 2019. http://carlisleindian.historicalsociety.com.
8 Richard Henry Pratt papers, WA MSS S-1174, Beinecke Library, Yale University.
9 Carlisle Indian Industrial School photograph collection, RG99s, Army Heritage and Education Center.
10 The Carlisle Barracks Collection, the U.S. Army Military History Institute, Carlisle Barracks, PA.
11 Tough, Alistair, "Oral Culture, Written Records and Understanding the Twentieth-Century Colonial Archive: The Significance of Understanding from Within." *Archival Science* 12 (2012): 249–254.
12 For example, see the discussion of methodology in Bell. "Telling Stories Out of School," 12–13.
13 Fear-Segal, Jacqueline. *White Man's Club: Schools, Race, and the Struggle for Indian Acculturation*. Lincoln: University of Nebraska Press, 2007: 206–230.
14 See, for example, Obituary of John Renville, *Eadle Keatah Toh* 1, no. 5, August 1880s: 3; and Menaul, John, Bellville, PA, to H.H.S., Carlisle, PA, 9 August 1881, reprinted in *The School News* 2, no. 3, August 1881: 4.
15 Pratt, Richard Henry, Carlisle, PA, to Commissioner of Indian Affairs Ezra A. Hayt, Washington, DC, January 27, 1880s, in National Archives and Records Administration,

Record Group 75, Series 79, Letters Received by the Bureau of Indian Affairs, box 574, Miscellaneous-1880-P-#0132.

16 Friedman Moses, Carlisle, PA, to Charles W. Goodman, Phoenix, AZ, October 4, 1913, in National Archives and Records Administration, Record Group 75, Series 121, Central Classified Files, Letters Received by the Bureau of Indian Affairs, 1907–1939, #118318-1913-Carlisle-175.

17 See Risling Baldy, Cutcha. *We Are Dancing For You.* Seattle: University of Washington Press, 2019: 7–10, who speaks of Indigenous methods and language in terms of (re) claiming, (re)writing, re(righting), and re(riting) as practices of decolonizing research.

18 Vilcan, Mamie, Carlisle, PA, to Commissioner of Indian Affairs Cato Sells, Washington, DC, November 2, 1913, in National Archives and Records Administration, Record Group 75, Series 121, Central Classified Files, Letters Received by the Bureau of Indian Affairs, 1907–1939, #129343-1913-Carlisle-820.

19 See, for example, Robert Charboneau, Carlisle, PA, to Commissioner of Indian Affairs Francis E. Leupp, Washington, DC, September 17, 1907, in National Archives and Records Administration, Record Group 75, Series 121, Central Classified Files, Letters Received by the Bureau of Indian Affairs, 1907–1939, #77702-1907-Carlisle-820; "Fair Deal" [Anonymous], Carlisle, PA, to Commissioner of Indian Affairs Robert G. Valentine, Washington, DC, September 28, 1911, in National Archives and Records Administration, Record Group 75, Series 121, Central Classified Files, Letters Received by the Bureau of Indian Affairs, 1907–1939, #83567-1911-Carlisle-154; and Bebeaux, Anna, Carlisle, PA, to Commissioner of Indian Affairs Robert G. Valentine, Washington, DC, 18 November 1910s, in National Archives and Records Administration, Record Group 75, Series 121, Central Classified Files, Letters Received by the Bureau of Indian Affairs, 1907–1939, #92161-1910-Carlisle-154.

20 Student File of Wesley Two Moons (Smoking Bear), National Archives and Records Administration, Record Group 75, Series 1327, box 155, folder 6129.

21 "Fair Deal" [Anonymous] to Pratt, September 28, 1911.

22 The CISDRC has identified approximately 8,000 individuals who attended the school, using a combination of record types; however, around 6150 unique student files survive. Analysis suggests that those identified individuals with missing student files fall into two categories: those who transferred from Carlisle to another school (where their file was subsequently sent), and those who died before the school closed.

23 See, for example, Goshorn, Shan. *Resisting the Mission* (edited by Phillip Earenfight). Carlisle: The Trout Gallery of Dickinson College, 2018; Arthur, Emily, Marwin Begaye and John Hitchcock. *Re-riding History: From the Southern Plains to the Matanzas Bay* (edited by Phillip Earenfight). Carlisle: The Trout Gallery of Dickinson College, 2018; Chabitnoy, Abigail. *How to Dress a Fish.* Middletown: Wesleyan University Press, 2019; Thomason, Dovie. "The Spirit Survives." In Jacqueline Fear-Segal and Susan Rose (eds.). *Carlisle Indian Industrial School: Indigenous Histories, Memories, and Reclamations.* Lincoln: University of Nebraska Press, 2016, 315–332; and Thomason, Dovie. "How the Wild West Was Spun" (presentation, Clarke Forum for Contemporary Issues at Dickinson College, Carlisle, PA). November 1, 2018.

24 Standing Bear, Luther. *My People, the Sioux.* Boston: Houghton Mifflin, 1928; Betzinez, Jason. *I Fought with Geronimo.* Harrisburg: Stackpole, 1959; and Burns, Michael. *The Only One Living to Tell: The Autobiography of a Yavapai Indian* (edited by Gregory McNamee). Tucson: University of Arizona Press, 2012.

25 Blight, David W. "W.E.B. DuBois and the Struggle for American Historical Memory." In Genevieve Fabre and Robert O'Meally (eds.). *History and Memory in African-American Culture.* New York: Oxford University Press, 1994, 46.

26 Rose, Susan and Saralegui, Manuel. *The Lost Ones: The Long Journey Home, a Documentary Film,* presented by Jacqueline Fear-Segal and Barbara Landis. Carlisle: Dickinson College Community Studies Center, 2009, DVD.

27 Child, Brenda. "The Boarding School as Metaphor," *Journal of American Indian Education* 57, no. 1, 2018: 37–57.

28 Trouillot, Michel-Rolph. *Silencing the Past: The Power and Production of History.* Boston: Beacon Press, 1994, 27.

29 "Teaching Resources – Carlisle Indian School Digital Resource Center," Dickinson College Archives and Special Collections. Accessed July 15, 2019. http://.carlisleindian.dickinson.edu/teaching.

30 Wilson, Fred. *Mining the Museum: An Installation* (edited by Lisa G. Corrin). Baltimore: Maryland Historical Society, 1994; Frankel, Noralee. "Reviewed Work: *Mining the Museum* by Fred Wilson." *The Public Historian* 15, no. 3, 1993: 105–108.

31 Blight, "Struggle for American Historical Memory," 52.

32 Trouillot, *Silencing the Past*, 150–151.

## References

Arthur, Emily, Begaye, Marwin Begaye, and John Hitchcock, *Re-riding History: From the Southern Plains to the Matanzas Bay*. Edited by Phillip Earenfight. Carlisle: The Trout Gallery of Dickinson College, 2018.

Bebeaux, Anna, Carlisle, PA, to Commissioner of Indian Affairs Robert G. Valentine, Washington, DC. 18 November 1910. National Archives and Records Administration, Record Group 75, Series 121. Central Classified Files, Letters Received by the Bureau of Indian Affairs, 1907–1939, #92161-1910-Carlisle-154.

Bell, Genevieve. "*Telling Stories Out of School: Remembering the Carlisle Indian Industrial School, 1879–1918*." Ph.D. diss., Stanford University, 1998.

Betzinez, Jason. *I Fought with Geronimo*. Harrisburg: Stackpole, 1959.

Blight, David W. "W.E.B. DuBois and the Struggle for American Historical Memory." In *History and Memory in African-American Culture*. Edited by Genevieve Fabre and Robert O'Meally. New York: Oxford University Press, 1994.

Burns, Michael. *The Only One Living to Tell: The Autobiography of a Yavapai Indian*. Edited by Gregory McNamee. Tucson: University of Arizona Press, 2012.

Carlisle Indian Industrial School Photograph Collection. *RG99s. U.S. Army Heritage and Education Center*, PA, Carlisle Barracks.

Carlisle Indian School Digital Resource Center. "Dickinson College Archives and Special Collections." Accessed July15, 2019. http://carlisleindian.dickinson.edu.

Carlisle Indian School Digital Resource Center. Dickinson College Archives and Special Collections. "Teaching Resources – Carlisle Indian School Digital Resource Center." Accessed July15, 2019. http://.carlisleindian.dickinson.edu/teaching.

Central Classified Files - Letters Received by the Bureau of Indian Affairs, 1907–1939. National Archives and Records Administration, Record Group 75, Series 121.

Chabitnoy, Abigail. *How to Dress a Fish*. Middletown: Wesleyan University Press, 2019.

Charboneau, Robert, Carlisle, PA, to Commissioner of Indian Affairs Francis E. Leupp, Washington, D.C. 17 September 1907. National Archives and Records Administration, Record Group 75, Series 121. Central Classified Files, Letters Received by the Bureau of Indian Affairs, 1907-1939, #77702-1907-Carlisle-820.

Child, Brenda. "The Boarding School as Metaphor." *Journal of American Indian Education* 57, no. 1 (2018).

Cumberland County Historical Society. "Carlisle Indian School History." Accessed July15, 2019. http://carlisleindian.historicalsociety.com.

"Fair Deal" [Anonymous], Carlisle, PA, to Commissioner of Indian Affairs Robert G. Valentine, Washington, DC. September 28, 1911. National Archives and Records Administration, Record Group 75, Series 121. Central Classified Files, Letters Received by the Bureau of Indian Affairs, 1907-1939, #83567-1911-Carlisle-154.

Fear-Segal, Jacqueline. *White Man's Club: Schools, Race, and the Struggle for Indian Acculturation*. Lincoln: University of Nebraska Press, 2007.

Frankel, Noralee. "Reviewed Work: *Mining the Museum* by Fred Wilson." *The Public Historian 15*, no. 3 (1993).

Friedman, Moses, Carlisle, PA, to Charles W. Goodman, Phoenix, AZ. 4 October 1913. National Archives and Records Administration, Record Group 75, Series 121. Central Classified Files, Letters Received by the Bureau of Indian Affairs, 1907–1939, #118318-1913-Carlisle-175.

Goshorn, Shan. *Resisting the Mission*. Edited by Phillip Earenfight. Carlisle: The Trout Gallery of Dickinson College, 2018.

Letters Received by the Bureau of Indian Affairs, 1824–1880. National Archives and Records Administration, Record Group 75, Series 79.

Letters Received by the Bureau of Indian Affairs, 1881–1907. National Archives and Records Administration, Record Group 75, Series 91.

Menaul, John, Bellville, PA, to H.H.S., Carlisle, PA. 9 August 1881. Reprinted in *The School News* 2, no. 3, August 1881: 4.

Obituary of John Renville. "Eadle Keatah Toh" *1*, no. 5, August 1880, p. 3.

Pratt, Richard Henry, Carlisle, PA, to Commissioner of Indian Affairs Ezra A. Hayt, Washington, DC. January 27, 1880. National Archives and Records Administration, Record Group 75, Series 79. Letters Received by the Bureau of Indian Affairs, box 574, Miscellaneous-1880-P-#0132.

Records of the Carlisle Indian Industrial School. National Archives and Records Administration, Record Group 75, Series 1322-1349C.

Richard Henry Pratt Papers. WA MSS S-1174. Americana, Beinecke Rare Book and Manuscript Library, Yale University.

Risling Baldy, Cutcha. *We Are Dancing For You*. Seattle: University of Washington Press, 2019.

Rose, Susan and Manuel Saralegui. *The Lost Ones: The Long Journey Home, a Documentary Film*. Presented by Jacqueline Fear-Segal and Barbara Landis. Carlisle: Dickinson College Community Studies Center, 2009. DVD.

Standing Bear, Luther. *My People, the Sioux*. Boston: Houghton Mifflin, 1928.

Student File of Wesley Two Moons (Smoking Bear). National Archives and Records Administration, Record Group 75, Series 1327, box 155, folder 6129.

The Carlisle Barracks Collection. U.S. Army Heritage and Education Center, Carlisle Barracks, PA.

Thomason, Dovie. "The Spirit Survives." In *Carlisle Indian Industrial School: Indigenous Histories, Memories, and Reclamations*. Edited by Jacqueline Fear-Segal and Susan Rose. Lincoln: University of Nebraska Press, 2016: 315–332.

Thomason, Dovie. "*How the Wild West Was Spun*." Presentation. Clarke Forum for Contemporary Issues at Dickinson College, Carlisle, PA. November 1, 2018.

Tough, Alistair. "Oral Culture, Written Records and Understanding the Twentieth-Century Colonial Archive: The Significance of Understanding from Within." *Archival Science 12* (2012).

Trouillot, Michel-Rolph. *Silencing the Past: The Power and Production of History*. Boston: Beacon Press, 1994.

Vilcan, Mamie, Carlisle, PA, to Commissioner of Indian Affairs Cato Sells, Washington, DC, November 2, 1913. National Archives and Records Administration, Record Group 75, Series 121. Central Classified Files, Letters Received by the Bureau of Indian Affairs, 1907-1939, #129343-1913-Carlisle-820.

Wilson, Fred. *Mining the Museum: An Installation*. Edited by Lisa G. Corrin. Baltimore: Maryland Historical Society, 1994.

# 3 Access to Truth, Healing, and Justice

## Digitizing the Records of U.S. Indian Boarding Schools

*Christine Diindiisi McCleave, and Rose Miron*

### Brief Overview of Boarding School History and Context

The U.S. Indian boarding school policies lasted more than 150 years over the 19th and 20th centuries and removed hundreds of thousands of American Indian and Alaskan Native children from their parents and communities. Structured through policies[1] that aimed to "civilize" Indigenous peoples, the schools sought to destroy Indian languages and cultures and "Kill the Indian, Save the Man."[2] This policy ultimately attempted to dismantle Indian nations and enable the U.S. government to obtain more Indian land. As Brenda Child writes about in "Boarding School Seasons," assimilation was not the true purpose of the boarding schools. If so, why segregate American Indians and Alaskan Natives? "The reality is that properties and assets were still at stake, and a campaign for land and resources was waged every single day of the boarding school era."[3]

Physical, mental, and sexual abuse were all too common at the schools, and this trauma continues to affect American Indian and Alaskan Native communities today.[4] Yet, despite their far-reaching impacts, the history of these schools remains largely missing from U.S. education curriculum, and this history is still understudied. Numerous academic texts have examined periods of boarding school history[5] or the histories of single schools,[6] but few if any works have studied historical Indian boarding schools comprehensively or comparatively. This is because the records of these schools are scattered across the nation in hundreds of different federal, state, and church archives. Some records are beginning to be made accessible through single-school digital archives, but accessing large portions of these documents is still difficult, and there is no place to search all boarding school records, making comprehensive research on this subject nearly impossible.[7] Still, these materials constitute an important source on American, American Indian, and Alaskan Native history and U.S. policy regarding child removal and Indian child welfare.

The National Native American Boarding School (NABS) Healing Coalition was formed to address this lack of public awareness about the truth in history of U.S. Indian boarding schools. Our aim is to understand and address the profound trauma experienced by Native individuals and cultivate community-led healing for Native families and individuals. Research and data on boarding schools is an essential part of this mission. In spite of the academic literature cited above and a Freedom of Information Act (FOIA) request to the Bureau of Indian Affairs (BIA) in 2016, we still do not know how many Native children were taken from their families and placed in boarding schools. Though our independent research has identified more than 357

Indian boarding schools across the nation, the number of schools continues to grow as we seek out additional information from countless archives.[8] We are working to make these records more accessible and searchable to scholars who can initiate new paths of inquiry and to Native individuals and communities who are seeking information about their relatives. To do so, NABS is harnessing technology through a forthcoming digital archive project that aims to bring new knowledge and research about Indian boarding schools to the public. Compiling the scattered records of boarding schools and making information about each institution available digitally will make analysis, understanding, and teaching of this history significantly easier. A single portal that links to other existing digital archive collections will be the first time that multiple digitized collections of boarding school materials will be made searchable through a single point of access. In this chapter, we discuss how our digital platform will promote new research on the Boarding School Era, how we are using technology to teach history that is often "difficult," and how tribal consultation and collaboration are essential parts of our digitization process. Ultimately, our project contributes to larger trends in digitization, while calling attention to the important considerations that must be taken when digitizing and sharing information related to Indigenous peoples. Our work enables educators to raise awareness about this history, incorporate it into historical narratives, and inform community-led healing.

## The Significance of U.S. Indian Boarding Schools

Between 1819, when the Indian Civilization Fund Act was passed, and the shift in boarding school policy in the 1960s, when the federal government began transferring ownership of most of the schools to Indian nations, the political, economic, and cultural status of American Indian and Alaskan Native nations changed dramatically. The records of these schools reveal the motivations behind U.S. assimilation techniques, the conditions that Native students lived in when they attended these schools, and the experiences of these children within and beyond the institutions. While at these schools, many Native children were subjected to physical, mental, sexual, and spiritual abuse, which has led to wide-ranging intergenerational impacts on survivors and their descendants. Boarding school trauma significantly impacted survivors' abilities to maintain their own health and provide healthy environments for their descendants, and recent research in epigenetics has shown that the trauma from these experiences may actually passed down through our DNA.[9] Today, Native children are disproportionately affected by childhood trauma, and Native youth experience PTSD (post-traumatic stress syndrome) and suicide at rates which are three times that of the general public—the same rate as veterans of the war in Iraq.[10] There is value in understanding the roots of this trauma and the environment that survivors were steeped in during their years at boarding school. The documents help us reconstruct the environments in the schools, as well as create a picture of what life was like for Native children across the United States in the 19th and mid-20th centuries. Through these lines of inquiry, we can further understand the subsequent effects on Native communities so that future research can provide recommendations for culturally relevant, community-led healing. These records reveal stories and information that is too often absent from the public record, but that are increasingly sought by academics in the humanities and historians of American Indian, Alaskan Native, and American history as they construct new narratives that examine child removal, assimilation, and race-based discrimination throughout the past.

The value of boarding school collections has been demonstrated by the scholarly work that has begun to analyze this history. In his important work *Education for Extinction: American Indians and the Boarding School Experience, 1875–1928,* David Wallace Adams relies on records from the Beinecke Library at Yale University (Connecticut), the Hampton Archives (Virginia), and the records of the Bureau of Indian Affairs in the National Archives in Washington, D.C.[11] Likewise, in her book *Boarding School Seasons: American Indian Families, 1900–1940,* Brenda Child relies mostly on National Archive Records at the Kansas City Branch.[12] Other scholars who have worked on regional histories or the narratives of a single school like Matthew Sakiestewa Gilbert, Clyde Ellis, K. Tsianinia Lomawaima, Harvey Markowitz, Jaqueline Fear-Segal, and others have used records from the National Archive Records at Riverside, the National Archive Records at Fort Worth, the Oklahoma Historical Society, the Cumberland County Historical Society (Pennsylvania), Dickinson College (Pennsylvania), and the U.S. Army War College (Pennsylvania). Others like John Gram have used records from the University of New Mexico Library and the National Archive Records at Denver to construct geographically specific histories of boarding schools within the scope of empire, while scholars like Kevin Whalen have used the Bancroft Library (California) and the National Archive Records at Riverside, California, to construct a history of labor at boarding schools. In other words, the records of these schools have already been used to create a number of rich narratives, but the sources used to create these new secondary sources are quite geographically widespread.

## Canadian TRC Records Versus U.S. Records—Access to Justice and the Right to Truth

In boarding schools, Indian children were forcibly abducted by government agents, sent to schools hundreds of miles away, and beaten, starved, or otherwise abused when they spoke their Native languages. Canada, the U.S., New Zealand, and Australia all had boarding schools (or residential schools) aimed exclusively at assimilating Indigenous/Aboriginal children in their countries. All those countries' governments have acknowledged these policies and their impacts, except the United States. In 2010, a Truth and Reconciliation Commission (TRC) was established in Canada as part of a class action lawsuit for Indian Residential School abuses. The settlement included a court order for all the schools' records to be brought forth for review and analysis under the commission. The five-year TRC resulted in a seven-volume report, 94 calls to action, and a national apology for cultural genocide.[13] Here in the U.S., the statute of limitations (40 years at the federal level) prohibits us from pursuing any court actions; consequently, there is no access to justice for boarding school survivors and their descendants in the U.S.[14]

The United Nations Office on Human Rights Rule of Law (E/CN.4/2006/91)[15] states that Moreover, the United Nations Declaration on the Rights of Indigenous Peoples, Article 8, states that we have the right "not to be subjected to forced assimilation or destruction of [our] culture."[16] However, the United States has never acknowledged or made reparations for its Indian Boarding School Policy. To date, survivors and descendants of boarding school experiences have not had access to justice for harms inflicted or access to the truth and information about relatives who died or went missing at these federal-run boarding schools. Yet the fate of the many Indigenous children who never returned home after forced removal by the U.S. to boarding schools, including those in the many unmarked graves at former boarding school sites, remains an ongoing human rights violation under international law. The NABS programmatic work aims for truth, healing, and justice for boarding school survivors and their descendants. The time for truth and healing is now, and digital projects can be a key part of this effort. In April 2019, NABS and a coalition of tribes, organizations, and independent researchers filed a submission with the United Nations Working Group on Enforced and Involuntary Disappearances (UNWGEID) detailing a number of children who were taken into federal custody and whose fate and whereabouts remain unknown to this day. The filing outlined how the U.S. had not acknowledged, accepted responsibility for, nor shown accountability for the many children that did not return home from federal Indian boarding schools. We also noted that the U.S. has not provided any evidence that they systematically notified families or tribes when the children passed away or went missing from schools, despite attempts by the coalition to obtain this information through the Freedom of Information Act (FOIA) request process.

XXXX

The 2016 FOIA appeal[17] requested identification of children taken to boarding schools under government supervision. To date, the only response has been that there are too many files to respond to the request without significantly refining the request. However, there isn't any refinement that would focus on the information needed from the files. In 2017, NABS was informed, via a third party, that our FOIA request had been closed.[18] No written communication of that status has been received by our attorneys at the Native American Rights Fund or NABS to date. Therefore, we have been left to do independent research at our own cost and have filed a submission to the UNWGEID to get the answers we seek.

The filing to the UNWGEID outlined that the United States enforced attendance at federal and church-run boarding schools by withholding rations from families or by incarceration of family members and independent research was cited regarding the many children who died at the schools or went missing once taken into federal custody. The coalition urged the UNWGEID to call on the United States to provide a full accounting of American Indian and Alaskan Native children who were taken into government custody under the United States' Indian Boarding School Policy. The coalition that filed the UNWGEID submission includes the National Native American Boarding School (NABS) Healing Coalition, the International Indian Treaty Council (IITC), the Native American Rights Fund (NARF), the National Indian Child Welfare Association (NICWA), the Rosebud Sioux Tribe, the Saginaw Chippewa Indian Tribe of Michigan, and independent researchers Preston McBride, Marsha Small, and Eleanor Hadden.

Over the past several years, NABS has also contracted with independent researchers to find boarding school records and statistics. In 2017, Dr. Denise Lajimodiere (Anishinaabe), co-founder of NABS, completed six years of research that identified 357 Indian boarding schools across 29 states in the U.S. (64 of those schools are still open today and are largely run with tribal influence, which promotes language and culture rather than prohibiting it). No other comprehensive list of boarding schools exists to date. Another NABS researcher, UCLA PhD candidate

Preston McBride (Comanche descent), researched death rates at four schools: Carlisle, Haskell, Chemawa, and Sherman from 1900–1939. He found that many of the children died from preventable diseases such as tuberculosis. He found the names of nine children who went missing and dozens of others with irregularities in notification to families about their death. He also found approximately 100 students who ran away from the outing program (forced summer labor in private residences and neighboring communities) at Carlisle, and the whereabouts of some are still unknown. McBride's research looked at only four of the 357 schools in the U.S. **How many children died or went missing all together?** In *Education for Extinction: American Indians and the Boarding School Experience*, David Wallace Adams writes that by 1900 there were more than 20,000 children in boarding schools and by 1925 this number had tripled to 60,889.[19] Since these years only make up a fraction of the boarding school era, we estimate that more than 100,000 children attended these institutions. This estimate is also informed by the Canadian Truth and Reconciliation Commission's finding that over 150,000 children attended 150 residential schools in Canada. The U.S. had more than 357 boarding schools, so we can only estimate that potentially hundreds of thousands of children attended these institutions.

The search for accurate data about Indian boarding schools in the U.S. continues today with independent research. Native American Boarding School researchers Anna Naruta-Moya and Daniel Moya (Pojoaque Pueblo) found unmarked cemeteries in Santa Fe and Albuquerque. Researcher Marsha Small (Tsistsistas) found evidence of mass graves at the boarding school cemetery in Chemawa. The Saginaw Chippewa Tribe noted that the federal records indicated five deaths at Mt. Pleasant Indian Boarding School; however, there were more than 250 deaths reported at the school under state death records. Initial independent research has uncovered many heinous facts that help Native people understand the extent of the genocide that has been inflicted on us. However, access to the full scope of federal and church-run boarding school records has been an obstacle to fact-finding and truth-telling efforts.

In searching for the location of the Indian boarding school records, we have assessed that the U.S. National Archives and Records Administration (NARA) has only 61 of 357 schools located within their records groups. That is less than 20% of the records and even those that are held within NARA collections have limited access up to 1944 due to U.S. privacy law (75 years from 2019). The NABS has determined that the rest of the boarding school records must still be held within church archives or private collections. We have worked with the Catholic Indian Boarding School Network, the University of Notre Dame (Indiana), and Marquette University (Wisconsin) to identify the location of Catholic-run Indian boarding school records. We have also reached out to the Presbyterian, Episcopal, and Methodist General Assemblies to pass resolutions in support of church-led research and identification of their Indian boarding school records. These documents, in tandem with testimonies from boarding school survivors, help us reconstruct the environments in the schools and further understand the subsequent effects on Native communities, so that we can know the full scope of impact from boarding school experiences—the truth in history about this era—and inform programs and policies around addressing historical trauma in Native communities as a result of cultural genocide and child removal. As such, the truth about these institutions remains largely unknown because the records of boarding schools are scattered across multiple archives that are mostly inaccessible to independent researchers and Tribal Nations.

Why do we need access to boarding school records? We need that data to be able to research and understand the whole truth about Indian boarding schools. Records and testimonies can give us more details about what happened in these schools, allowing us to further understand how these institutions impacted future generations. Truth is also the foundation of healing. The history of what happened in these schools has been covered up and erased from common historical narratives. In order for survivors and descendants to heal, everyone needs to understand this history and recognize how it continues to impact American Indian and Alaskan Native families and communities. These stories need to be told by Indigenous peoples.

In spite of the trauma boarding schools inflicted on our communities, there is still so much that we don't know about what happened, and only a small portion of U.S. residents even know that Indian boarding schools existed. How many children attended these institutions in total? How many died at schools and never returned home? How many went missing after they were taken into government custody? How are the health, family, and other challenges our communities face connected to this history? As NABS continues to pursue new research, the number of schools we discover continues to grow. More research will only reveal more information about the environments our ancestors and relatives experienced and help us heal our Nations.

It is urgent that we locate and protect these records before any more of them disappear. Records hold the key to really understanding what happened to our children and to healing. Right now, even the records we do know of are difficult to access. The records of each school are typically grouped together, but many ended up spread across multiple different archives, like the Chemawa Indian Industrial School, which has records at the National Archives and Records Administration (NARA) in Seattle and in San Francisco, Marquette University, Wisconsin, and Pacific University Archives. Some collections are dedicated to specific schools, while others have materials from a larger Indian agency and multiple schools. For these collections, boarding school records must be manually mined from the larger record group. Some of the collections include detailed descriptions, while others have none at all. Some are organized chronologically, some alphabetically, and some are randomly compiled.

Current research on U.S. Indian boarding schools requires traveling to multiple archives, even if you only want to research the history of one school. Then, further analysis requires time-consuming, manual evaluation of often unorganized collections. The state of these records has created many barriers for both Native and non-Native researchers as well as community members. It's no coincidence that very little comprehensive or comparative research about boarding schools has been completed. Most researchers have their hands full with just researching the history of one school.

Technology can aid the process of searching for more information about these schools. Digitizing the scattered records allows boarding school survivors and descendants to access information and records online rather than having to spend time and money traveling to potentially multiple different archives. Bringing together previously separated records from the same schools helps researchers better understand the full scope of these institutions. Moreover, linking newly digitized records with existing digital collections puts similar materials within the context of other similar materials, facilitating new research and paths of inquiry that can, for the first time, examine this history comprehensively and comparatively across years and institutions. And yet, making the records of these schools publicly available online raises important questions about privacy, access, and Indigenous sovereignty.

## Data Sovereignty, Tribal Consultations, and Prioritizing Indigenous Research and Perspectives over Academic Colonization

As an organization whose board and executive staff is 100% Native American, NABS operates on the belief that healing must be directed by boarding school survivors and their descendants and that healing starts with the truth. We also believe that research and narratives about Native Americans should come from Native Americans. So much of modern-day research on Indian boarding schools comes from non-Native researchers looking in on Native communities from an outside perspective, much like the anthropologists of the 18th century who didn't understand Indigenous worldviews or cultures and wrote about them as if they were going extinct. In 2018, a study titled *Reclaiming Native Truth: A Project to Dispel America's Myths and Misconceptions about Native Americans*, stated that among non-Native peoples, "Contradictory stereotypes coexist [as] people comfortably accept and maintain conflicting narratives about Native Americans."[20] The same study also stated that "Most Americans likely form most of their perceptions from the news media, entertainment media and popular culture, including sports teams with Native-themed mascots." It is for these and many other reasons that any research about Native Americans should include Native American perspectives, voices, and worldviews if not actually originate from Native Americans themselves. Ultimately, NABS's vision aims to restore the cultural sovereignty that was taken away through boarding schools and that sovereignty includes the right to set the terms of research, healing, and reconciliation. Therefore, NABS advocates for the decolonization and re-indigenization of American Indian or Alaskan Native (AI/AN) representation in academic research, film, and media by supporting Natives in charge of our own narrative; we ask all researchers and allies to comply. Reindiginization is the return to Indigenous values and worldview.

Definitions:

- **Tribal Sovereignty** is the inherent authority of Indigenous/Tribal Nations to govern themselves.
- **Cultural Sovereignty** is the right to practice Indigenous language, culture, and lifeways and not to be subjected to forced assimilation or destruction of culture.
- **Indigenous Data Sovereignty** is the right of a Nation to govern the collection, ownership, and application of their own data.[21]
- **Academic Colonization:** the perpetuation of the paternalistic settler-colonial mindset and erasure of Indigenous interpretations by appropriating Indigenous narratives that absents Indigenous voices, perspectives, and worldview from Indigenous histories especially when seeking to profit off those histories by non-Indigenous peoples.

The NABS has also begun to locate records and conduct more research in spite of not having support or co-operation from the federal government and churches. In fall 2017, we issued eight research contracts to almost entirely Native researchers to locate, scan, and analyze records of U.S. Indian boarding schools. As we do this work, we've found that more and more non-Native scholars are building their careers on boarding school research that does not include the voices, perspectives, or interests of Native people themselves. Colonization is truly extending to academia as these scholars co-opt and profit from Indigenous histories and knowledge with more benefit to themselves than to Indigenous communities. While many are well-intentioned, they should know better than anyone that boarding schools were also "well-intentioned" institutions. Being well-intentioned without significant tribal consultation and involvement, and without prioritizing the interests of Native communities, is not enough.

The NABS fights persistently against this continued type of academic colonization, prioritizing the voices and perspectives of Native people in every part of our work. As a Native-led organization, NABS's research, programs, and community relationships are centered around supporting tribal sovereignty, cultural sovereignty, and data sovereignty, and we vet the researchers that we work with to ensure that they are prioritizing the needs of Native communities and doing work that truly benefits, and is driven by, boarding school survivors and their descendants. As outlined later in this chapter, we also have specific policies around tribal consultation and data sovereignty that govern our ongoing digitization efforts.

## Trauma-Informed Research in the Digital Humanities

As digitization projects becomes more common across institutions like NABS, the potential to make history, particularly histories that have long been silenced, more accessible is growing. However, there are a number of specific cultural sensitivities that must be considered when digitizing information related to traumatic histories like boarding schools. Trauma-trigger and cultural-sensitivity warnings must be included on digital sites so that individuals have the opportunity to not view information that may be triggering. The NABS site contains the following as an example:[22]

> Advisory: The content contained herein may trigger secondary trauma or trauma responses; we encourage individuals to seek counseling or healing if you experience any stress related to boarding school history or experiences. Indigenous peoples are warned that this database contains images, names, and references to deceased persons

These considerations are especially important when conducting oral histories, which have continued to grow in popularity as there has been a demonstrated need to record the first-hand accounts of survivors and digital projects make sharing these stories more possible than ever. Yet significant responsibility comes with these opportunities. For many students of U.S. Indian boarding schools and their descendants, trauma lingers from their varied experiences of family separation, institutionalization, language and culture prohibition, forced labor, as well as sexual, physical, mental, emotional, and spiritual abuse. Therefore, NABS uses and suggests following an Indigenous research protocol. One such protocol was developed by Dr. Lajimodiere when she interviewed boarding school survivors for her book, *Stringing Rosaries.*[23]

We never know how a situation or setting might affect people or when a trauma response may be triggered, but we do know that researching or discussing boarding school content and experiences can be difficult and can trigger historical trauma. Research tells us how people react to trauma triggers (a spectrum between dissociation to hyper-reactivity). It is important that all researchers know that counselors and traditional healers have this knowledge, know how to handle those reactions that they expect to see as responses to triggers, and provide trauma resources and support for people when engaging in the work of boarding school research and discussions with boarding school survivors and descendants. While it may not be the intention to activate one's trauma, it should be the intention to provide a safe space for healing around this historical narrative. Both in Canada and the U.S., boarding school survivors have documented testimonies of their experiences. While in most cases sharing one's story can bring healing, in some rare cases they have experienced extreme trauma responses afterward leading to depression, isolation, and anxiety. Therefore, we take collecting testimony and oral histories very seriously and proceed with extreme caution.

In addition to the above considerations about providing safe spaces and warnings for people with past trauma, NABS has the following guidelines and recommendations for conducting boarding school research or dialogs and collecting oral histories in Native communities:

- Work with and comply with Tribal Internal Review Board (IRB) or research protocol processes if conducting events/interviews on tribal land to abide by data sovereignty principles and protect tribal citizens.
- Research projects collecting American Indian or Alaskan Native (AI/AN) oral histories should be led by AI/AN researchers to understand cultural sensitivities and cultural protocols.
- Documentary films focused on AI/AN oral histories should have AI/AN leadership on the film project with Native producers, directors, and film crew and should also go through a Tribal IRB or relevant tribal research protocols.
- Procure spiritual advisors and mental health professionals to support your event or interviews.
- If anyone feels upset during any of the presentations or interviews at any time, let them know they can stop the interview or leave the group to see a designated elder or counselor (who are procured by the facilitator/interviewer) to provide the subject with emotional and spiritual support.
- Provide a private, separate space for counseling/praying to take place for trauma survivors.
- Provide a consent form and include trigger warnings for trauma survivors when sensitive content is going to be presented or asked about.
- Always provide a list of resources to use in the days and weeks following the event should participants have lingering emotions or trauma responses.
- Follow up with participants in the days and weeks following the event/interview to inquire about lingering emotions or trauma responses—refer to professional support if symptoms present.

We encourage survivors and descendants to carefully consider the following when sharing their own oral history:

1. Listening to content about trauma or sharing about one's own trauma may trigger a trauma response or post-traumatic stress. Trauma responses can include feeling disassociation, stress, panic, anxiety, depression, hopelessness, exhaustion, or overwhelm.

- Sharing about traumatic experiences may also cause second-hand (or vicarious) trauma in others who are listening.
- We encourage individuals to seek counseling or traditional healing if you experience any stress related to boarding school history.
- If you've never talked about your boarding school experiences, we ask that you speak with a counselor or spiritual advisor before participating in interviews or sharing your oral history.

## National Indian Boarding School Digital Archive with Aggregated Search and Digital Map

As part of our mission to understand and address the impacts of the U.S. Indian boarding school policy and boarding school experiences, NABS seeks to locate, copy, and analyze the records of the boarding schools across the country and to make this information available to the tribal nations whose children attended those schools. In 2017, NABS created an online database of boarding school research, references, and archive locations. In 2018, NABS began development of the National Indian Boarding School Digital Archive (NIBSDA) that includes digital image files of government records, church records, personal family records, and other documents related to federally funded U.S. Indian boarding schools that operated between 1819 and 1970. The site will also include a digital map of Indian boarding schools in the U.S similar to the one created by the Canadian TRC that will allow users to interact with the compiled records in a more visual way.[24] We have established a Research Advisory Council, a Digitization Policy that includes Tribal Consultation and Data Sovereignty, and a Digitization Plan to outline materials that are acceptable to be digitized and published based on the specified parameters.

In an effort to create one central comprehensive digital archive for all the boarding school records in the U.S., NABS has begun to analyze the collections at various locations to digitize boarding school records that have never been digitized before. Using an aggregated search, NIBSDA will also directly link to existing one-school digital archives, such as those of the Carlisle Indian School Digital Resource Center,[25] the Genoa Indian School Digital Reconciliation Project,[26] the Sherman Indian Museum Digitization Project,[27] the Indigenous Digital Archive,[28] and the Ziibiwing Center of Anishinabe Culture and Lifeways Digital Archive,[29] making these existing records searchable within NIBSDA. Our use of Elevator, a digital content management system developed at and hosted by the University of Minnesota, will allow us to bulk-import the metadata from these other sites, creating entries for their digitized materials so that all digitized content on boarding schools is searchable in one location. Elevator can support images, audio, video, 3D objects, PDF files, and more, and it allows for the mixing and matching of metadata in order to make this type of bulk import possible. To view the records themselves, users will be re-directed to the original site. This means, for example, that a descendant searching for their relative's records will no longer need to go to multiple sites to address their query. Users will be able enter the name of a tribe, school, or individual and view records for multiple schools held in separate repositories. For the first time, comparative and comprehensive research that would have previously required traveling to multiple different physical archives and searching multiple different digital archives will be possible.

We will also work with these archives and others to revise a Shared Protocols on the Management of Boarding School Records, which will shape the cataloging, transcribing, and creation of metadata for boarding school records so that they are more consistently organized and easily searchable across these platforms. The scans of records included in this digital archive project will make all boarding school records accessible in a single location so that scholars, boarding school survivors, and survivors' descendants can more readily access this information. Making these records readily available to Tribal Nations allows them to shape the healing programs and resources they need within their communities.

A central, digital repository and portal will make these collections more accessible and enable new narratives of both American Indian/Alaskan Native and American history. The NIBSDA will be the first time that multiple digitized collections of boarding school materials will be made searchable through a single point of access, and thus will become the largest collection of boarding school materials available online. In the long-term, NABS hopes to digitize or create links to all boarding school records through the creation of a digital map. This will allow users to search for information using keywords or begin their query by exploring records associated with a specific school or a geographic area. The digital map will summarize any basic information about the schools such as what years it was open, what tribal nations had children attend, and who ran the school. It will also include information on any known cemetery or burial sites associated with the school, any remaining buildings associated with the school, and what type of land the cemetery and/or remaining buildings are on in preparation for potential future repatriation efforts. The use of ArcGIS software and other digital mapping technologies will be leveraged. This will give Tribal Nations, boarding school survivors, their descendants, scholars, and the general public summary information about each individual boarding school. It will promote the preservation of these historical landmark buildings and may also help inform repatriation efforts from boarding school cemeteries.

These materials constitute an important source on American Indian history and policy and shed light on a longer history of removing children from their families throughout American history, a topic that will certainly be studied at increasing rates given current political events. After the launch of our digital archive, we anticipate a significant growth in scholarship that initiates new paths of inquiry about the history of U.S. Indian boarding schools across the United States and in research in broader fields like American history, psychology, anthropology, education, and Indigenous studies. Due to our planned outreach and consultation, we especially anticipate an increase in scholarship and community-led healing initiatives that are created and/or shaped by Native communities directly.

We expect that this project will also yield new paths of inquiry such as tracking patterns of changes in Indigenous languages over time, enabling comparisons of environments at different boarding schools, and examining the ways in which U.S. schools influenced the creation of residential schools internationally in other countries like Canada, New Zealand, and Australia. Boarding school history and Indigenous history continue to be major areas of research within the humanities, and in light of continued Indian child removal and arguments about the Indian Child Welfare Act, as well as our current immigration crisis, histories of assimilation and patterns of child removal throughout U.S. history will undoubtedly continue to be important topics for scholarly discussion. The increased use of these materials by both Native and non-Native scholars and students and non-academic Native researchers is a crucial step in raising awareness of the truth in boarding school history and ultimately influencing policy decisions regarding child removal and healing within Native communities.

However, in spite of the positive opportunities this project holds for new research on Indian boarding schools in the U.S., we do not digitize these materials without significant consideration. Many boarding school records contain private cultural or family information, and in line with NABS's policy on data sovereignty, it's important that tribal nations have the right to determine what information about their citizens is made available online, and in what way. To address this, NABS has created a clear Digitization Policy that outlines definitive protocols for tribal consultation. NABS has proposed to make all relevant digital records available to tribal representatives prior to their publication online, which will give tribes the opportunity to request that access to certain materials be restricted, or that the way items are described be changed. Prior to making these materials available to tribes, NABS staff will specifically flag materials that either certainly or possibly contain culturally or other sensitive information.

**Definitions of Terms for Digital Protocols:**

Culturally sensitive information is defined as "tangible and intangible property and knowledge which pertains to the distinct values, beliefs, and ways of living for a culture. It often includes property and knowledge that is not intended to be shared outside the community of origin or outside of specific groups within a community."[30]

Examples of culturally sensitive information related to the records of U.S. Indian boarding schools may include:[31]

- Descriptions or photographs of burials or funerals;
- Descriptions or photographs of cemeteries;
- Archaeological data related to cemeteries;
- Descriptions, recordings, transcripts, or photographs of ceremonies, songs, religious practices, healing or medicine of any kind;
- Community histories or oral histories.

Other sensitive information will be identified on a case by case basis using the judgementjudgment of NABS staff with input from the NABS Research Advisory Council and Tribal Nations. We loosely define this as any data that relays certain medical information, information about violent behaviors, or other private family matters that may impact living descendants.

Materials that certainly contain culturally or other sensitive information will require tribal permission before publication, but we will still flag materials that potentially contain culturally, or other sensitive information and encourage tribal representatives to review these records first. Following review, or at any point after the records are made public in the NABS digital archive, NABS plans to comply with all requests regarding tribal data, which could include that:

- Certain materials not to be published or made available to anyone.
- Certain materials be made available on a restricted basis to only tribal members, or to only certain groups of tribal members based on cultural protocols.
- Certain information within materials be redacted.
- The digital entry for the source includes a TK (Traditional Knowledge) label, so that items that remain accessible to the public or to tribal members are labeled as containing cultural or sensitive information.[32]
- That the description of materials be changed to better reflect tribal knowledge or include tribal perspectives.
- That information about cultural protocols associated with an item (i.e. that a story should only be read during a certain time of year or a song must be played from start to finish) be added to an item's description and be made available to users prior to accessing material.

These important considerations around culture and other sensitive information is another reason that we chose Elevator as the digital platform for this site. Elevator's capabilities include setting different permission levels for each record or for groups of records, and the site also allows us to make these materials accessible to only designated tribal community leaders before we determine if they are appropriate for publication. This will allow us to follow the protocols we've outlined earlier, which include requiring tribal permission prior to the publication of some materials and flagging additional materials for tribal review. Tribal leaders will also be able to write metadata descriptions or add information about cultural protocols surrounding materials if they so choose, and this content will be reviewed, then published by NABS staff.

Elevator's flexibility around metadata also allows us to add multiple tags to items, and create categories for people, tribal community, school, and more, in the hopes of giving users the best chance of finding information about their relatives and tribes. The image below gives an example of what this could look like. However, it's important to note that since our metadata schema is being created with significant input from tribal partners, it will not be complete until we are ready to launch the project. Until then, we are still conducting sharing and listening sessions with AI/AN people and tribal communities across the country, so that we can create item descriptions and metadata categories that are most useful to boarding school survivors and their descendants.

We also plan to use the Shared Protocols for the Management of Boarding School materials that we will be developing with other digital archives to define specific vocabulary around the records of these schools, as well as guidelines for how to organize and make them digitally accessible. For too long the research of Indian boarding schools and the sharing of Native American materials more broadly (whether in museum exhibits or online) has been led by non-Native individuals and

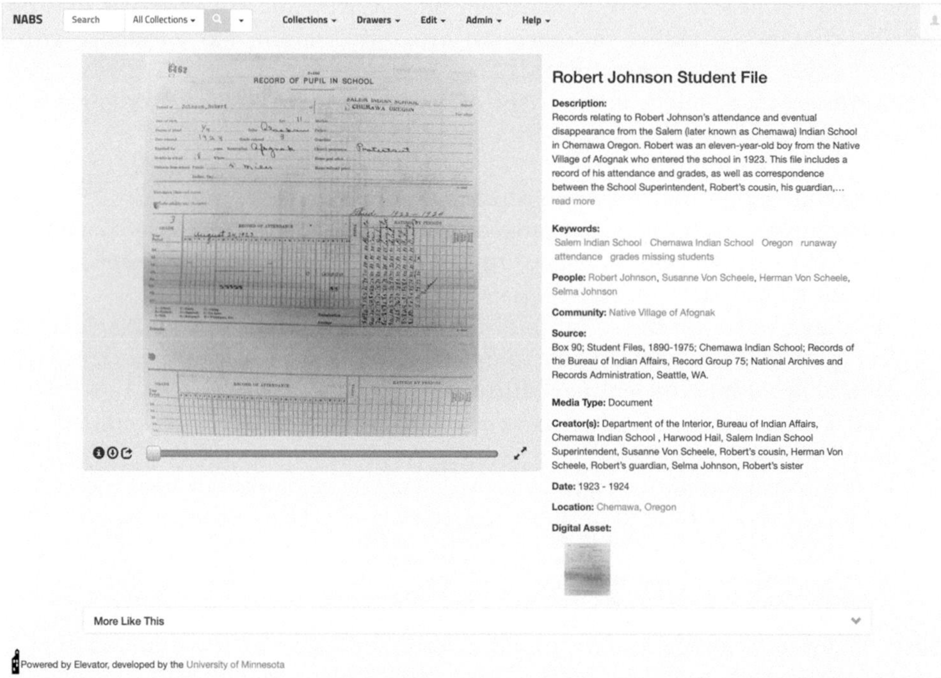


*Figure 3.1* An example of what a student record may look like in the National Indian Boarding School Digital Archive (categories are subject to change based on our ongoing consultation with Tribal Nations). This is the record of a student who went missing from Chemawa Indian School.

institutions without collaboration or consultation with Native peoples and Tribal Nations. The NIBSDA aims to reverse this trend by creating one of the first digital archives that will be compiled and managed by a Native-led organization and built in close collaboration with tribal nations.

Our desire to maintain control over this new digital collection, which has multiple cultural and other sensitive issues that must be considered, was another reason that we chose Elevator as our digital platform and are thus partnering with the University of Minnesota (UMN). As a small non-profit organization, NABS lacks the institutional capacity to create and maintain the technical side of such a large digital project. By partnering with UMN's College of Liberal Arts, we are able to back up and store the significant data associated with this site through the College's existing Amazon S3 account, and use Elevator's existing checksums and digital preservation processes to ensure our files are properly preserved and protected. Elevator staff will also be providing technical support for the site. However, the partnership agreement spelled out with UMN's College of Liberal Arts is specific in noting that NABS still owns the data and controls the content on the site. Though our data on Amazon's S3 server is only available to NABS Staff and Elevator's technical support team, our agreement also outlines a clear process for pulling data off of Amazon's S3 server should we need to do so. We hope that our work will solidify the importance of Native people directing research about their histories and cultures, and that more non-Native institutions will create similar digitization policies before cultural information is made freely available online without the consent of Native people.

## Conclusion: Acts of Justice

The history of Indian boarding schools in the U.S. remains largely missing from U.S. education curriculum, this time period is still understudied, and the truth about what really happened in these institutions remains elusive. We still don't know how many children were taken to these schools, how many died while in the custody of the U.S. federal government, and how many remain unaccounted for. Until the survivors of these schools, their descendants, and broader society knows the truth of what happened, our Nations cannot fully heal. Digital projects such as ours hold the potential to learn more about what happened to our children in these institutions and to raise awareness about Native history on a broader scale. However, we must emphasize the importance that any digital project concerning American Indian or Alaskan Native individuals adhere to standard data sovereignty protocols that ensure Native voices and views are centered. Only by ensuring Native people are able to restrict access to cultural or sensitive information can we prevent the further desecration of Native cultures in public institutions like museums, archives and digital projects. Only by reversing the trend of non-Native institutions and individuals having control over Native histories and cultures can we truly move closer to truth and healing. These are acts of justice.

## Notes

1 The Indian Civilization Fund Act of 1819 authorized up to $10,000 a year to support the efforts of religious groups and interested individuals willing to live among and teach Indians. Under the Grant Peace Policy of 1869, the Bureau of Indian Affairs gave Christian denominations power over Indian programs on reservations, leading the way to Indian boarding schools. See: Stat. 516, March 3, 1819.

2 Pratt, Richard H., Official Report of the Nineteenth Annual Conference of Charities and Correction. 1892, 46–59. Reprinted in Pratt, Richard H. "The Advantages of Mingling Indians with Whites" In *Americanizing the American Indians: Writings by the "Friends of the Indian" 1880–1900*, Harvard University Press, 1973, 260.

3 Child, Brenda J. *Boarding School Seasons: American Indian Families, 1900–1940*. Lincoln: University of Nebraska Press, 2000.

4 Brave Heart, Maria Yellow Horse and Lemyra M. DeBruyn, "The American Indian Holocaust: Healing Historical Unresolved Grief." *American Indian and Alaska Native Mental Health Research* 8, no. 2 (1998): 60–82; Center for Disease Control and Prevention and Kaiser Permanente, "The Adverse Childhood Experiences (ACE) Study," Accessed May 11, 2018. https://www.cdc.gov/violenceprevention/acestudy/about.html; Yehuda, Rachel, et al., "Holocaust Exposure Induced Intergenerational Effects on FKBP5 Methylation." *Biological Psychiatry* 80, (2016): 372–380.

5 For example, see Adams, David Wallace. *Education for Extinction: American Indians and the Boarding School Experience*. Lawrence: University Press of Kansas, 1995.

6 For example, see Fear-Segal, Jaqueline. *Carlisle Indian Industrial School: Indigenous Histories, Memories, and Reclamations*. Lincoln: University of Nebraska Press, 2016; Lomawaima, K. Tsianina. *They Called It Prairie Light: The Story of Chilocco Indian School*. Lincoln: University of Nebraska Press, 1994; Gram, John R. *Education at the Edge of Empire: Negotiating Pueblo Identity in New Mexico's Indian Boarding Schools*. Seattle: University of Washington Press, 2015; Gilbert, Matthew Sakiestewa. *Education Beyond*

*the Mesas: Hopi Students at Sherman Institute, 1920–1929*. Lincoln: University of Nebraska Press, 2010; Trennert Jr., Robert A. *The Phoenix Indian School: Forced Assimilation in Arizona, 1891–1935*. Norman: University of Oklahoma Press, 1988; Vuckovic, Myriam. *Voices from Haskell: Indian Students Between Two Worlds, 1884–1928*. Lawrence: University Press of Kansas, 2008.

7 See Carlisle Indian School Digital Resource Center. http://carlisleindian.dickinson.edu/ and Genoa Indian School Digital Reconciliation Project. https://genoaindianschool.org/.

8 Lajimodiere, Denise, "American Indian Boarding Schools By State." In *Healing Voices Volume 1: A Primer on American Indian and Alaska Native Boarding Schools in the U.S.*, September 2018, 8.

9 See: Brown-Rice, Kathleen, "Examining the Theory of Historical Trauma Among Native Americans." *The Professional Counselor* 3 no. 3 (2013): 117–130; Daniels, Judy and Michael D'Andea. "Trauma and the Soul Wound: A Multicultural-Social Justice Perspective." *Counseling Today*. Accessed May 22, 2018. https://ct.counseling.org/2007/06/dignity-development-diversity-6/; Max-Planck-Gesellschaft. *Childhood Trauma Leaves Mark on DNA of Some Victims*. Munchen: Max-Planck-Gesellschaft Institute of Psychiatry, 2012. https://www.mpg.de/6643282/childhood-trauma-dna; Yehuda, et al., "Holocaust Intergenerational Effects."

10 Executive Office of the President, "2014 Native Youth Report." *Obama White House Archives* (December 2014), https://obamawhitehouse.archives.gov/sites/default/files/docs/20141129nativeyouthreport_final.pdf.

11 Adams, David Wallace. *Education for Extinction: American Indians and the Boarding School Experience, 1875–1928*. Lawrence: University of Kansas, 1995.

12 Child, Brenda. *Boarding School Seasons*.

13 The Truth and Reconciliation Commission of Canada, Honouring the Truth, Reconciling for the Future: Summary of the final report of the Truth and Reconciliation of Canada (2015). http://nctr.ca/reports.php.

14 Stamatopoulou, Elsa and Wilton Littlechild. *Indigenous Peoples' Access to Justice, Including Truth and Reconciliation Processes*. New York: Institute for the Study on Human Rights at Columbia University, 2014, 255–261.

15 The United Nations. "Access to Justice." United Nations and the Rule of Law. https://www.un.org/ruleoflaw/thematic-areas/access-to-justice-and-rule-of-law-institutions/access-to-justice/

16 The United Nations. "United Nations Declaration on the Rights of Indigenous Peoples." United Nations Department of Economic and Social Affairs. https://www.un.org/development/desa/indigenouspeoples/declaration-on-the-rights-of-indigenous-peoples.html.

17 FOIA Control number BIA-2016-01054.

18 Hilleary, Cecily. "Native Americans No Closer to Learning Fates of Boarding School Ancestors." Voice of America (October 20, 2017). https://www.voanews.com/usa/native-americans-no-closer-learning-fates-boarding-school-ancestors.

19 Adams. *Education for Extinction*.

20 First Nations Development Institute. "Research Findings: Compilation of All Research," Reclaiming Native Truth: A Project to Dispel America's Myths and Misconceptions (June 2018), https://rnt.firstnations.org/wp-content/uploads/2018/06/FullFindingsReport-screen.pdf.

21 US Indigenous Data Sovereignty Network. "About Us." Native Nations Institute at the University of Arizona. https://usindigenousdata.arizona.edu/.

22 The National Native American Boarding School Healing Coalition. "Resource Database Center.". https://boardingschoolhealing.org/resource-database-center/.

23 Lajimodiere, Denise, *Stringing Rosaries: The History, the Unforgivable, and the Healing of Northern Plains American Indian Boarding School Survivors*. Fargo: North Dakota State University Press, 2019.

24 National Centre for Truth and Reconciliation. "Interactive Map." University of Manitoba. https://nctr.ca/map.php.

25 Carlisle Indian School Digital Resource Center.

26 Genoa Indian School Digital Reconciliation Project.

27 The Sherman Indian Museum. http://www.shermanindianmuseum.org/.

28 The Indigenous Digital Archive, the Museum of Indian Arts & Culture, New Mexico. https://omeka.dlcs-ida.org/s/ida/page/home.
29 Ziibiwing Center of Anishinabe Lifeways & Culture. Saginaw Chippewa Indian Tribe of Michigan. http://www.sagchip.org/ziibiwing/.
30 First Archivist Circle. "Protocols for Native American Archival Materials." http://www2.nau.edu/libnap-p/protocols.html.
31 Adapted from First Archivist Circle, "Protocols."
32 Local Contexts, "Traditional Knowledge (TK) Labels," Local Contexts. http://localcontexts.org/. "The TK Labels are a tool for Indigenous communities to add existing local protocols for access and use to recorded cultural heritage that is digitally circulating outside community contexts. The TK Labels offer an educative and informational strategy to help non-community users of this cultural heritage understand its importance and significance to the communities from where it derives and continues to have meaning. TK Labeling is designed to identify and clarify which material has community-specific restrictions regarding access and use. This is especially with respect to sacred and/or ceremonial material, material that has gender restrictions, seasonal conditions of use and/or materials specifically designed for outreach purposes. The TK Labels also can be used to add information that might be considered 'missing,' including the name of the community who remains the creator or cultural custodian of the material, and how to contact the relevant family, clan or community to arrange appropriate permissions."

## References

Adams, David Wallace. *Education for Extinction: American Indians and the Boarding School Experience, 1875–1928*. Lawrence: University Press of Kansas, 1995.

Brown-Rice, Kathleen. "Examining the Theory of Historical Trauma Among Native Americans." *The Professional Counselor 3* no. 3 (2013): 117–130.

Carlisle Indian School Digital Resource Center. http://carlisleindian.dickinson.edu/.

Center for Disease Control and Prevention and Kaiser Permanente. "The Adverse Childhood Experiences (ACE) Study." Center for Disease Control and Prevention. Accessed May 11, 2018. https://www.cdc.gov/violenceprevention/acestudy/about.html.

Child, Brenda J. *Boarding School Seasons: American Indian Families, 1900–1940*. Lincoln, University of Nebraska Press, 2000.

Daniels, Judy and Michael D'Andea. "Trauma and the Soul Wound: A Multicultural-Social Justice Perspective." *Counseling Today*, Accessed May 22, 2018, https://ct.counseling.org/2007/06/dignity-development-diversity-6/.

Executive Office of the President, "2014 Native Youth Report." *Obama White House Archives*, December 2014, https://obamawhitehouse.archives.gov/sites/default/files/docs/20141129nativeyouthreport_final.pdf.

Fear-Segal, Jaqueline. *Carlisle Indian Industrial School: Indigenous Histories, Memories, and Reclamations*. Lincoln: University of Nebraska Press, 2016.

First Archivist Circle. "Protocols for Native American Archival Materials." http://www2.nau.edu/libnap-p/protocols.html.

First Nations Development Institute. "Research Findings: Compilation of All Research." *Reclaiming Native Truth: A Project to Dispel America's Myths and Misconceptions*, June 2018. https://rnt.firstnations.org/wp-content/uploads/2018/06/FullFindingsReport-screen.pdf.

Genoa Indian School Digital Reconciliation Project. https://genoaindianschool.org/.

Gilbert, Matthew Sakiestewa. *Education Beyond the Mesas: Hopi Students at Sherman Institute, 1920–1929*. Lincoln: University of Nebraska Press, 2010.

Gram, John R. *Education at the Edge of Empire: Negotiating Pueblo Identity in New Mexico's Indian Boarding Schools*. Seattle: University of Washington Press, 2015.

Hilleary, Cecily. "Native Americans No Closer to Learning Fates of Boarding School Ancestors." *Voice of America*, October 20, 2017. https://www.voanews.com/usa/native-americans-no-closer-learning-fates-boarding-school-ancestors. Indian Civilization Fund Act, Stat. 516, March 3, 1819.

"The Indigenous Digital Archive." The Museum of Indian Arts & Culture, New Mexico. https://omeka.dlcs-ida.org/s/ida/page/home.

Lajimodiere, Denise. *Stringing Rosaries: The History, the Unforgivable, and the Healing of Northern Plains American Indian Boarding School Survivors*. Fargo: North Dakota State University Press, 2019.

Local Contexts, "Traditional Knowledge (TK) Labels," *Local Contexts*, http://localcontexts.org/

Lomawaima, K. Tsianina. *They Called It Prairie Light: The Story of Chilocco Indian School*. Lincoln: University of Nebraska Press, 1994.

Max-Planck-Gesellschaft, "*Childhood Trauma Leaves Mark on DNA of Some Victims.*" Munchen: Max-Planck-Gesellschaft Institute of Psychiatry, 2012. https://www.mpg.de/6643282/childhood-trauma-dna.

National Centre for Truth and Reconciliation. "*Interactive Map.*" University of Manitoba. https://nctr.ca/map.php.

The National Native American Boarding School Healing Coalition. "Resource Database Center." https://boardingschoolhealing.org/resource-database-center/.

Pratt, Richard H. *Official Report of the Nineteenth Annual Conference of Charities and Correction* (1892). Reprinted in Richard H. Pratt, "*The Advantages of Mingling Indians with Whites,*" *Americanizing the American Indians: Writings by the "Friends of the Indian" 1880–1900*. Cambridge: Harvard University Press, 1973.

The Sherman Indian Museum. http://www.shermanindianmuseum.org/.

Stamatopoulou, Elsa and Wilton Littlechild, *Indigenous Peoples' Access to Justice, Including Truth and Reconciliation Processes*. New York: Institute for the Study on Human Rights at Columbia University, 2014.

Trennert Jr., Robert A. *The Phoenix Indian School: Forced Assimilation in Arizona, 1891–1935*. Norman: University of Oklahoma Press, 1988.

The Truth and Reconciliation Commission of Canada. *Honouring the Truth, Reconciling for the Future: Summary of the Final Report of the Truth and Reconciliation of Canada* 2015, http://nctr.ca/reports.php.

The United Nations, "Access to Justice." *United Nations and the Rule of Law*. https://www.un.org/ruleoflaw/thematic-areas/access-to-justice-and-rule-of-law-institutions/access-to-justice/.

The United Nations, "United Nations Declaration on the Rights of Indigenous Peoples." *United Nations Department of Economic and Social Affairs*. https://www.un.org/development/desa/indigenouspeoples/declaration-on-the-rights-of-indigenous-peoples.html.

US Indigenous Data Sovereignty Network. "About Us." *Native Nations Institute at the University of Arizona*, https://usindigenousdata.arizona.edu/.

Vuckovic, Myriam. *Voices from Haskell: Indian Students Between Two Worlds, 1884–1928*. Lawrence: University Press of Kansas, 2008.

Yehuda, Rachel, Nikolaos P. Daskalakis, Linda M. Bierer, Heather N. Bader, Torsten Klengel, Florian Holsboer, Elisabeth B. Binder, "Holocaust Exposure Induced Intergenerational Effects on FKBP5 Methylation," *Biological Psychiatry 80* (2016): 372–380.

Yellow Horse Brave HeartMaria and Lemyra M. DeBruyn. "The American Indian Holocaust: Healing Historical Unresolved Grief." *American Indian and Alaska Native Mental Health Research8*, no. 2 (1998): 60–82.

Ziibiwing Center of Anishinabe Lifeways & Culture. "*Saginaw Chippewa Indian Tribe of Michigan.*" http://www.sagchip.org/ziibiwing/.

# 4 The Indigenous Digital Archive

## Creating Effective Access to and Collaboration with Government Records

*Anna Naruta-Moya*

### Introduction

The Indigenous Digital Archive (IDA), a project of the Museum of Indian Arts and Culture in partnership with the New Mexico State Library, the Indian Pueblo Cultural Center, and the New Mexico History Museum, plays a special role in digital archives in a number of important ways. First, the project arose from Tribal constituents' articulations of community needs during a museum strategic planning process and continues with Native direction and participation throughout. Second, the project has developed and enacts a practice of respectful online access to be responsive to concerns about what is appropriate to share publicly online. Third, the project leverages new developments technology, particularly a novel and useful combination of the Open Annotation and International Image Interoperability Framework (IIIF) W3C standards and computer-assisted indexing through natural language processing (NLP) techniques, to create an open source platform that makes the content of mass digitized historical documents more accessible to people whose communities they relate and to other researchers.

Fourth, the IDA creates tools to facilitate collaborative research with and making meaning of the online documents, recognizing the social nature of research while sharing authority for describing documents. We recognize the need to make meaning together and over time by tagging within pages of documents, adding comments or annotations, and even creating counter narratives to what's in the records. Importantly, these tools are created in a framework of system independence and digital preservation, so the content that people contribute is not dependent on, and inadvertently made obsolete by, any particular digital system.

Finally, through support and mentoring, including an IDA Fellows pilot project initially open to members of the 23 tribes of New Mexico, plus Hopi (geographically separated while culturally and genealogically related), the IDA has begun enacting an additional element ethically required of digital and other institutional projects related to historically minoritized people; that financial, technical, and mentoring support must necessarily be provided to enable members of historically minoritized communities to make full use of digital knowledge tools.

### Community Origin to Meet Community Needs

The Museum of Indian Arts and Culture (MIAC) has its origins in the founding of the Museum of New Mexico in 1909 and the Rockefeller funded Laboratory of

Anthropology of the 1920s. A separate building for MIAC opened in 1987. In 2013, the Museum of Indian Arts and Culture gained its first-ever Native American director, Della Warrior (Otoe-Missouria), a Harvard educated administrator. The MIAC began strategic planning that included a process to hear from New Mexico's Native American constituents and identify ways in which MIAC could forge closer and meaningful relationships between the resources of the museum and Native peoples throughout New Mexico. At a 2014 MIAC World Café planning session, Native American constituents expressed 1) wanting MIAC to provide online access to documents relating to their history, and 2) for MIAC to provide them opportunities to gain experience with archives, as they gain skills in connecting with archival information as well as develop and operate their own governmental and cultural archives.

These expressions became the seed of the Indigenous Digital Archive project, which was further refined through continual conversations, pilot studies, participation of, and direction from Native scholars, educators, community leaders, and community members, and Native and Native-serving library and archives practitioners.

The MIAC was able to begin the Indigenous Digital Archive project with a 2016 IMLS National Leadership grant and a Knight Foundation prototyping grant to create an open source software toolkit and a digital repository use case. We began by creating user stories,[1] descriptions of people who would be using the IDA and their anticipated uses and goals, to guide the user experience and technological development.

MIAC's use of the IDA project addresses the lack of access to public government records related to the build up and operation of U.S. government boarding and day schools. These records include the period of the "Indian Wars" and reforms related to the "Indian New Deal" in the 1930s. The State Co-ordinator of Tribal Libraries, who often receives reference requests related to information the documents the project will make available, notes that now having even just a pile of documents of student names online filled a need no one has been able to respond to before in connecting people affected by these government policies generations onward with the information they're seeking.

Creating effective access locally is seen as particularly important at this time, as this is a window of opportunity where Native people have the benefit of understanding the records with the input of those who are Elders today (who were young children at the time of the creation of the later records), and others who still have first-hand stories from their parents or grandparents in the 1920s–1930s and even earlier.

The MIAC additionally partnered with the New Mexico State Library Tribal Libraries Program and the Indian Pueblo Cultural Center in training in archives and archival management for Tribal librarians and other Tribal and Tribal serving practitioners. This has many wide-reaching effects for archive development and digital literacy: as Tribal libraries play a key role in establishing digital inclusion in many Native American communities,[2] helping strengthen local connections to Tribal libraries makes all online resources more accessible.

## Respectful Online Access

From the outset, the project realized that there is a difference between material being publicly available in a repository and what it can mean for individuals, families, and communities when that material is scanned and made available online. On the one hand, there is a need for digitization and online access to mitigate the significant geographical and financial barriers to access material in archival repositories. At the same time, it is important to carefully select what material will be broadcast online. This can be especially important in areas like the Southwest where Native communities feel keenly about having historically been subjected to outside individuals, extracting information and physical objects in colonial encounters, with effects that continue today.[3] The Indigenous Digital Archive provides explanation and direction for creating respectful online access.[4]

## Technological Frameworks to Empower Community Researchers

The Indigenous Digital Archive toolkit is a set of open source software components built over the Omeka-S content management system widely popular with libraries, museums, and digital humanities projects, but designed as modular so it is adaptable to other digital repository systems. The IDA toolkit was designed to leverage technological advances of the last half decade to meet community needs and challenges articulated in the library and archives professions.[5]

Since the beginning of social media and Web 2.0, those in charge of providing access to archives and special collections have talked about the need to create interfaces to bring in diverse voices and share the authority for describing material, particularly for detailed descriptions.[6] However, even what was long a model experiment in archives for adding user-driven information to records—the Polar Bear Expedition Digital Collections[7] —is no longer available, leaving almost no trace in the current online information; as without another major project, it was unable to be integrated with tools serving needs in preservation and access.[8] The need for an effective toolkit compatible with the forward migration of content management systems and digital preservation has not flagged, however. In Kate Theimer's 2011 volume collecting examples and plans of digital engagement and crowdsourcing in repositories, Yakel notes "an undiscovered power of the social web for cultural institutions" that can be engendered when "authority for the description and representation of [online materials] is shared."[9]

In recent years, the impetus for allowing people to contribute to knowledge about collections online has given rise to projects like the Citizen Science projects of *Zooniverse*, *Zooniverse*'s new AnnoTate transcription project, and the Smithsonian Transcription Center. These, and similar recent projects, have developed strong tools for repositories allowing people to conduct very specific tasks online, such as full-text transcription, correction of automated optical character recognition (OCR), and transcription of portions of regular records, such as weather logs, an institution's register of bird specimens, or logs of troop movements, into structured data. In each case by design there is a defined and specific task to accomplish, rather than being an interface to aid exploration of digitized material. In a recent analysis of existing tools facilitating crowdsourced tasks, Ben Brumfield and Mia Ridge[10] noted the "line-at-a-time, queue-oriented, multi-track transcription workflow" characteristic of the *Zooniverse* interfaces, don't allow for users to return to something they've worked on, or see what others have done and discuss among themselves.

These interfaces are designed to collect a limited range of structured data according to a particular research design, and are becoming increasingly sophisticated at directing the workflow. However, these are not tools for exploring collections beyond a specific kind of encounter. Sometimes, the narrow task orientation and gamification of some interfaces might mean that a user is presented with a single image of text that is completely interesting to them, only to see it whisked away after they've completed a transcription task, with no clue of where it came from, or ability to see the whole item.

As a tool allowing more flexible interactions with content, *FromThePage* has made great advances in user interactions around transcriptions and translations, OCR correction, and exploring content based on indexed key terms. *FromThePage* also integrates and allows use of content stored in the Internet Archive or Omeka, an opensource content management system adopted by many GLAMs (Galleries/Libraries/Archives/Museums), and allows users to create articles on indexed key terms. Most recently, *FromThePage* also made the advance of incorporating the system's focused International Image Interoperability Format (IIIF) standard. Being able to work with images in an IIIF compatible environment brings many advantages in user interface, such as the ability for smooth deep zooming, and being able to work with and compare material from multiple IIIF-compatible repositories in one browser interface.

However, tools are still needed to enable work with mass digitized documents. As Brumfield and Ridge[11] note, *FromThePage* is designed for work with small collections, and its limited discover interface means it would be prohibitive to use on collections of hundreds or thousands of documents. Additionally, indexing in *FromThePage* depends on a transcription in which the subjects are identified and hyper-linked. For mass digitized archival documents, access needs are not always met by transcription. This is not only because full transcription or OCR correction is usually much more time consuming than selecting a tag (a name, event, concept, or place) that would be meaningful for someone looking for the content, but also because what often would be used as a keyword does not actually appear in that text. (For example, a derivative, alternate, or misspelled form of a name is used, or what would receive a keyword tag of "boarding school deaths" appears in euphemistic language.)

The need to create online access to allow collections to reach a wider group of users means that repositories do continue to look to mass digitization as part of their strategies.[12] Usability studies on the posting of a large backlog of tens of thousands of archival photographs with minimal processing found that users preferred to be able to access the material even with the most minimal metadata, and in some cases supplied information about the items that staff then used to augment the catalog record.[13] At the Hoover Institution Archives at Stanford University, California, which holds over 27 linear miles of mainly 20th century material, Miller[14] has suggested mass digitizing unprocessed archival collections, with searchability to be provided through OCR, with the idea that this would give users a level of accessibility that would be familiar to them from Google searches.

The need for more user-directed tagging while exploring content, as opposed to straight transcription or the narrow task orientation of *Zooniverse*, have led to some projects finding a solution in outsourcing the content (and contributions) to commercial ventures such as Flickr. While people are eager to interact with digitized historic cultural content online in a collaborative environment, and using Web 2.0 tools can enhance a person's ability to find the digitized documents they're looking for and provide their input into what is presented with the material,[15] using these commercial content services introduces its own problems, such as user privacy from commercial data-mining now or in the future, the large expenditure of effort on creation of material that is "locked" inside the company's system, the need to conduct additional activities for digital preservation purposes (web harvesting or web archiving of the site), and limitation of formats (e.g. single images). Additionally, there is not yet a match between such services and potentially culturally sensitive material, or where participating community members don't feel comfortable with the contributions they are making becoming part of a commercial venture rather than being curated by a museum or archive.

With respect to user-directed exploration and tagging, the Mukurtu content management system, with an emphasis on responding to the needs of cultural sensitivity issues in a digital repository, has built an effective system designed for gating and providing access to material appropriate to one's tribal affiliation, clan, gender, age group, and other Indigenous community-defined considerations. Users may create tags and add commentary, at the level of the entire digital object and catalog record. For now, this constraint limits the ability of Mukurtu to provide access with the same kind of interactivity to mass digitized material, or allow someone to show exactly where in a passage or multi-page document their comment applies.

Talking about the potentials for crowdsourcing in increasing access to collections at the May 2015 Crowd Consortium workshop (as part of the IMLS-funded National Forum in Crowdsourcing for Libraries and Archives), HathiTrust's Jeremy York noted that while we hold collections for the benefit of our communities, "yet many in the community do not know what is in our collections, [and] they can't find the useful materials; we are not being as effective as we could in fulfilling our vision."[16] This challenge is still urgent to address.

The IDA's toolkits that are based on international standards help bridge that gap. The IDA's software tools that interface with the Omeka-S digital content management system for cultural heritage institutions create an online access and collaboration layer that enables creating effective access to mass-digitized archival documents, including typescript and print documents as well as other images often highly resistant to automated processes such as OCR, through enabling a suite of interactive features based on the system's focused International Image Interoperability Format standard (IIIF, http://iiif.io) and Open Annotation.

Use of the IIIF format enables a suite of abilities, such as the ability for a user to quote all or part of an image using just a URL, seamless deep zoom, the ability to add keyword tags[17] or annotations to a portion of or even a range of images; and even, as seen in IIIF clients such as the Mirador viewer, being able to view and work with objects from multiple IIIF-enabled repositories in one browser interface.[18] Use of the Open Annotation format, the standard used in the web annotation software Hypothes.is (maintained by the W3C, the international organization responsible for managing the standards that make the internet able to communicate across different languages and applications), addresses the need to be able to create online collaborations including crowdsourcing applications where the data will stay linked to the source images. It would also be maintained in an internationally agreed standard format that allows it to be sustained and useful outside of any one particular software application, aiding long-term digital stewardship.

The IIIF standard has been adopted by U.S. libraries and institutions such as ARTStor, the libraries of Stanford, Yale, Princeton, and Harvard universities, and now the Smithsonian Institution and others. Digital Public Library of America (DPLA) encourages it as a way for contributing repositories to have better representation of their content. Internationally, IIIF has been adopted by La Bibliothèque Nationale de France; the National Libraries of Wales, Austria, Denmark, Israel, Poland, Serbia, Norway, Australia, and New Zealand; Oxford University's Digital Bodleian; and others of an expanding list of participating organizations. Shims or patches were made for existing repository systems such as Content DM and the Internet Archive ahead of formal adoption.

The IDA project also builds on the Universal Viewer, an open source project to enable cultural heritage institutions to present their digital artifacts in an IIIF-compliant and highly customizable user interface. The project was initially developed to provide an interface to the content of the many different subunits of the British Library, and the Wellcome Library's emerging Digital Library Cloud Services (DLCS), a service to provide IIIF-compliant image hosting with additional services for cultural heritage digital projects, including OCR indexing and searching, annotation storage, and easy-to-use APIs.

Tools the IDA has developed greatly complement new tools developed for Omeka-S, a rewrite of Omeka code that came from an IMLS National Leadership grant project that began one year earlier than ours. For example, the new LOD (linked open data) tools Omeka-S are implementing to aid standardization in the collection and object level catalog metadata are complemented by the IDA tool allowing semantic tagging by users within the documents. Additionally, others could take parts of the semantic tagging toolkit that best apply to their applications.

We also began an experiment using natural language processing (NLP) techniques for computer-assisted indexing to build starting points for people to access the records. Our NLP toolkit identifies dates, geographic places, names of individuals, organizations, and tribes, and other topics.

Further, as the ultimate goal of this project is to build an interface that is really focused on creating the best user experience possible, this IDA builds on the emerging work in developing what generous interfaces, user interfaces that rather than present the user with just an empty search box, instead organize and present preliminary faceted data to provide the most information to a person in the most apprehendable way, with the most effective possibilities for next steps for effective interaction.

We matched these national needs for tools among libraries, archives, and museums with the needs articulated among our local communities to refine the toolkits.

## Support and Mentoring

We have been able to conduct an IDA Fellows pilot project initially open to members of the 23 tribes of New Mexico plus Hopi (geographically separated while culturally and genealogically related), to provide stipend support to a cohort of Native people interested in conducting researching with the digital records. In doing so, the IDA has begun enacting an additional element ethically required of digital and other institutional projects related to historically minoritized people: that financial, technical, and mentoring support must necessarily be provided to enable members of historically minoritized communities to make full use of digital knowledge tools.[19]

In addition to having positive contributions to ethics, such supported opportunities are well positioned to follow a plan of meeting needs for fostering digital engagement in museums articulated by Ridge.[20] Participants can also help create model contributions, as recommended from museum participation research that finds high quality model contributions to open participation opportunities can encourage other people to take contributing seriously, and "can inspire and energize less-skilled visitors without making them feel inferior."[21]

Members of our pilot cohort of IDA Fellows conducted research on topics of interest to them and their communities. They presented their projects at the Museum of Indian Arts and Culture and in their home communities, and took part in evaluation activities. We used what we learned from the pilot project to craft an IDA Technical and Research Fellow supported position open to Native emerging professionals for our next phases of project implementation. We also hope to partner to expand our IDA Fellows pilot project to create residency research and interpretation opportunities that can be supported at a higher level. We would also like to add support for research mentor positions drawn from Native scholars and other practitioners nationwide.

## Recursive Design and Research

Our initial experiments with natural language processing for computer-assisted indexing showed great promise and provided beginning tags for generating a user interface design that could give people some sense of the content of the records and ways to find it. Clearly, computer-assisted indexing through NLP—a range of techniques for analyzing texts that identify things like mentions of people, places, dates, and other topics—has a productive role to play in helping people find information of interest. The question remains how to configure the possible steps in NLP to be best suited for scanned archival material, and what of those results are the most useful for people seeking information, and what arrangements and features of an online repository interface informed by this are most useful and satisfying for people to use?

From these beginnings we designed and, in 2019, began to implement a series of NLP experiments in concert with usability testing to determine three things: 1) what sorts of NLP provide the most useful results in connecting people with the content of records, 2) how is it usefully combined with traditional archival description like series information, and 3) how can that information be most effectively presented to people. With a focus on meeting the needs of people seeking information that is so far unprecedented in NLP work, our three year project will: 1) follow established standards for statistical significance to select a range of scanned archival documents and add topic tags by experts to about 1000 pages to create a much needed sample of well tagged archival documents (a "testing corpus," in computer science terms) for benchmarking potential NLP techniques; and 2) run a series of empirical tests on the data to develop a set of NLP and data prep techniques that optimize the information returned. Once the initial experiments with the data and data processing are available, we will 3) add various configurations of our NLP computer indexing results, series information, and different content, and conduct evaluations of the online repository interface for usability, including evaluations using the Indigenous Evaluation Framework (NSF grant REC-0438720), until the final iteration achieves at least 90% usability and satisfaction. For inclusiveness, we will also validate the interface against accessibility audits and one on one tests to ensure usability by people with disabilities. We will make the open source software code and testing corpus available on Github,

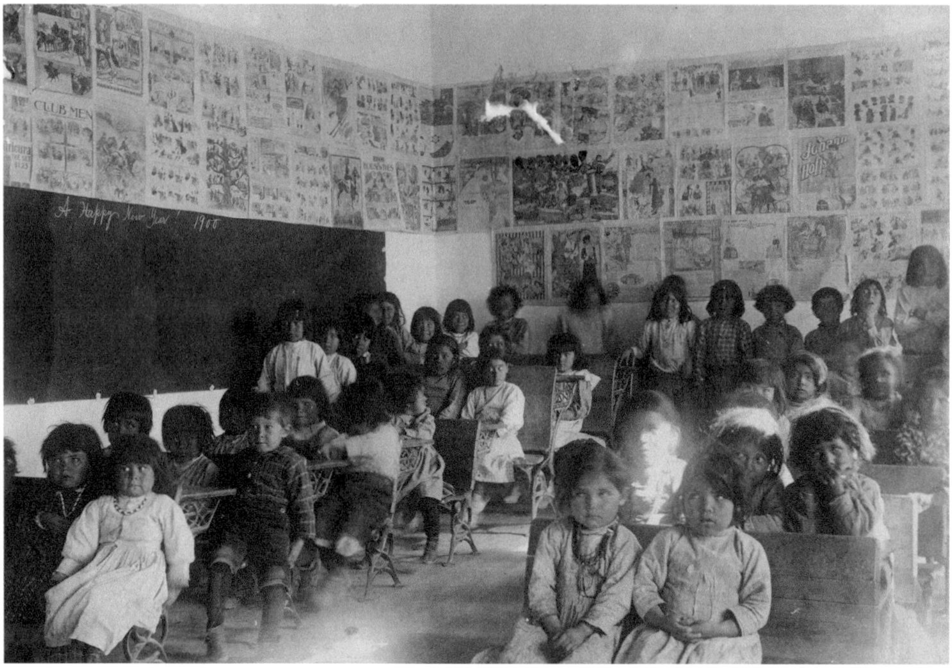


*Figure 4.1* Young Students at the United States Indian School, Santa Fe, New Mexico, 1900, Courtesy New Mexico History Museum Palace of the Governors Photo Archives (NMHM/DCA) 1036.

and disseminate results and receive further shaping feedback through presentations at conferences connecting with diverse audiences of Native and non-Native library, archives, and museum practitioners, researchers, and community members.

The results of this project will allow libraries, museums, and archives to make more effective choices in how to provide access to mass digitized or born digital content, and increase experience and confidence with online repositories among historically unserved communities. By making available our expertly tagged sample of records, this project can also encourage computer scientists and others to work to make further refinements in connecting people with information in masses of records like the complex collections that can come from multiple jurisdiction government repositories, and make that information more readily available to the people and communities to which it relates.

## Treaties Explorer Partnership

In 2019, we also began a different kind of partnership with a mainstream institution looking to more effectively engage with Native American constituents. Our Treaties Project partnership with the U.S. National Archives Office of Innovation nationally expands our collaboration with Native scholars and builds a new approach to allowing people to explore the treaties between the U.S. and Native nations. In this project, we are synthesizing scholarship and creating a Treaties Explorer interface that people can use for research and casual exploration all the way from treaty boundaries on the map to the contextual information that currently requires a scholar to piece together and interpret. The Treaties Explorer web portal allows someone to start with a state or zip code or location on their phone, and see the treaties that relate to their area.[22] Or someone may decide to look up the name of a tribe to see related treaties. Launching at the national Indigenous Pop X conference in 2019, the Treaties Explorer will also provide lesson plans ready for K–12 classroom use and video units on treaties, their historical context, and research you can do using the Treaties Explorer at DigiTreaties.org.

A focus on the U.S. allows us to do a deeper context dive, and beyond treaties involving land cessation, we start with the original copies of the 374 Ratified Indian Treaties since 1784, held by the U.S. National Archives, thanks to an anonymous donation to the National Archives Foundation that is providing conservation, imaging, and this rare partnership to connect people with the documents.

The Treaties Explorer project continues our practice of integral involvement of Native scholars and practitioners from planning through implementation. Since the Treaties Explorer web portal is built on the IDA and its interoperability principles, as for the IDA toolkit and its contents, the treaties and surrounding context information are available as reusable resources. We look forward to people benefiting from this work to augment other map, database, research, or exhibition projects.

## Notes

1 Our initial guiding user stories are available at http://blog.indigenousdigitalarchive.org/post/2017/10/10/Technology-behind-the-IDA-toolkit-a-roundup-of-links-and-videos or https://is.gd/UserStories2016.

2 Jorgensen, Miriam, Traci Morris, and Susan Feller. *Digital Inclusion in Native Communities: The Role of Tribal Libraries*. Oklahoma City: Association of Tribal Archives, Libraries, and Museums, 2014.

3 Phippen, J. Weston. "The Auction of Native American Artifacts – Next Week an Auction House in France Will Sell Hundreds of Native American Items, Some of Which are Considered Sacred." *The Atlantic* (27 May 2016). https://www.theatlantic.com/international/archive/2016/05/native-american-auction/484316/; Reuters, "Hopi Sacred Masks Auctioned in Paris Despite Protests." June 11, 2015. https://www.reuters.com/article/us-france-auction-masks/hopi-sacred-masks-auctioned-in-paris-despite-protests-idUSKBN0OR1DG20150611; Lozada, Lucas Iberico. "The Professor and the Pueblo: Was the Disclosure of Acoma Traditions Exploitation or Scholarship?" *Santa Fe Reporter* (26 January 2016). https://www.sfreporter.com/news/coverstories/2016/01/26/the-professor-and-the-pueblo/.

4 See https://omeka.dlcs-ida.org/s/ida/page/respect.

5 Naruta-Moya, Anna. *The Indigenous Digital Archive: Effective Access to and Collaboration with Mass Digitized Historic Documents*. Washington, DC: Society of American Archivists, Research Forum, August 14, 2018.

6 Yakel, Elizabeth. "Balancing Archival Authority with Encouraging Authentic Voices to Engage with Records." In Kate Theimer (ed.). *A Different Kind of Web: New Connections Between Archives and Our Users*. Society of American Archivists, 2011; Krause, Magia and Elizabeth Yakel, "Interaction in Virtual Archives: The Polar Bear Expedition Digital Collections Next Generation Finding Aid." *The American Archivist* 70, no. 2 (Fall/Winter) (September 2007): 282–314.

7 Krause and Yakel. "Interaction in Virtual Archives." 2007.

8 Bentley Historical Library, University of Michigan. "About the Polar Bear Expedition Digital Collections," n.d. http://quod.lib.umich.edu/p/polaread/about.html.

9 Yakel, "Balancing Archival Authority." In Theimer, 2011.

10 Brumfield, Ben and Mia Ridge. *Wellcome Library Transcribing Recipes Final Report.* London, UK: Wellcome Library, 2015, 17. http://www.slideshare.net/Wellcome/wellcome-library-transcribing-recipes-report.
11 Ibid., 28.
12 Rocke, Michael and Lukas Klic. "Revealing Renaissance Art: Mass Digitization of the Berenson Library Photo Archive," Berenson Library, Harvard University, 2015. http://library.harvard.edu/revealing-renaissance-art-mass-digitization-berenson-library-photo-archive.
13 Kelly, Elizabeth, "Processing Through Digitization: University Photographs at Loyola University New Orlean." *Archival Practice* 1, no. 1 (2014). http://libjournal.uncg.edu/ap/article/view/865/563.
14 Miller, Larisa K., "All Text Considered: A Perspective on Mass Digitizing and Archival Processing." *The American Archivist* 76, no. 2 (2013): 536–537.
15 Zinckham and Springer, 2010: 107.
16 Flanagan, Mary, Neil Fraistat and Andrea Wiggins. "Engaging the Public: Best Practices for Crowdsourcing Across the Disciplines," A report of Dartmouth's 2014 IMLS-funded National Forum in Crowdsourcing for Libraries and Archives (CCLA), 2015: 132. http://crowdconsortium.org.
17 Appleby, Mike. "IIIF Tagging and Discovery. Sharing Images of Global Culture." *National Gallery of Art* (5 May 2015). http://bit.ly/1Vtl2AV.
18 Snydman, Stuart, Rob Sanderson and Tom Cramer. "The International Image Interoperability Framework (IIIF): A Community & Technology Approach for Web-Based Images." *Archiving 2015*, Proceedings of the Society for Imaging Science and Technology Conference, Los Angeles, CA, May 19–22, 2015, http://purl.stanford.edu/df650pk4327.
19 Naruta-Moya, Anna, "The Indigenous Digital Archive: Effective Access to and Collaboration with Mass Digitized Historic Documents," Digital Library Federation DLF 2017 Forum, Pittsburgh, PA, October 24, 2017.
20 Ridge, Mia, "Digital Participation, Engagement and Crowdsourcing in Museums," *London Museums Group* (15 August 2013). http://www.londonmuseumsgroup.org/2013/08/15/digital-participation-engagement-and-crowdsourcing-in-museums/.
21 Simon, Nina. *The Participatory Museum*. Santa Cruz, California: Museum 2.0, 2010.
22 Native-land.ca has made a great start in this kind of exploration, for treaties involving land cessation, over multiple continents.

## References


Appleby, Mike. "IIIF Tagging and Discovery. Sharing Images of Global Culture." *National Gallery of Art*, May 5, 2015, http://bit.ly/1Vtl2AV.

Bentley Historical Library, University of Michigan. "About the Polar Bear Expedition Digital Collections". n.d. http://quod.lib.umich.edu/p/polaread/about.html.

Brumfield, Ben and Mia Ridge. "*Wellcome Library Transcribing Recipes Final Report.*" London, UK: Wellcome Library, September 2015, http://www.slideshare.net/Wellcome/wellcome-library-transcribing-recipes-report.

Duff, Wendy M., et al. "The Development, Testing, and Evaluation of the Archival Metrics Toolkits." *The American Archivist 73.2* (2010): 569–599.

Flanagan, Mary, Neil Fraistat, and Andrea Wiggins. "Engaging the Public: Best Practices for Crowdsourcing Across the Disciplines." *A Report of Dartmouth's 2014 IMLS-funded National Forum in Crowdsourcing for Libraries and Archives (CCLA)*, 2015, http://crowdconsortium.org.

Jorgensen, Miriam, Traci Morris, and Susan Feller. *Digital Inclusion in Native Communities: The Role of Tribal Libraries*. Oklahoma City: Association of Tribal Archives, Libraries, and Museums, 2014.

Kelly, Elizabeth. "Processing Through Digitization: University Photographs at Loyola University New Orleans." *Archival Practice 1*, no. 1, 2014, http://libjournal.uncg.edu/ap/article/view/865/563.

Krause, Magia and Elizabeth Yakel. "Interaction in Virtual Archives: The Polar Bear Expedition Digital Collections Next Generation Finding Aid." *The American Archivist 70*, no. 2 (Fall/Winter, 2007) pp. 282–314.

Lozada, Lucas Iberico. "The Professor and the Pueblo: Was the Disclosure of Acoma Traditions Exploitation or Scholarship?" *Santa Fe Reporter* January 26, 2016, https://www.sfreporter.com/news/coverstories/2016/01/26/the-professor-and-the-pueblo/.

Miller, Larisa K, "All Text Considered: A Perspective on Mass Digitizing and Archival Processing." *The American Archivist* 76, no. 2 (2013): 536–537.

Naruta-Moya, Anna, "The Indigenous Digital Archive: Effective Access to and Collaboration with Mass Digitized Historic Documents," *Digital Library Federation DLF 2017 Forum, Pittsburgh, PA*, October 24, 2017.

Naruta-Moya, Anna, *"The Indigenous Digital Archive: Effective Access to and Collaboration with Mass Digitized Historic Documents."* Society of American Archivists, Research Forum, Washington, DC, August 14, 2018.

Phippen, J. Weston. "The Auction of Native American Artifacts – Next Week an Auction House in France Will Sell Hundreds of Native American Items, Some of Which Are Considered Sacred." *The Atlantic*, May 27, 2016, https://www.theatlantic.com/international/archive/2016/05/native-american-auction/484316/.

Reuters, "Hopi Sacred Masks Auctioned in Paris Despite Protests." June 11, 2015, https://www.reuters.com/article/us-france-auction-masks/hopi-sacred-masks-auctioned-in-paris-despite-protests-idUSKBN0OR1DG20150611.

Reynolds, Julie. "Digital Participation, Engagement and Crowdsourcing in Museums." *London Museums Group*. August 15, 2013, http://www.londonmuseumsgroup.org/2013/08/15/digital-participation-engagement-and-crowdsourcing-in-museums/.

Rocke, Michael, and Lukas Klic. "Revealing Renaissance Art: Mass Digitization of the Berenson Library Photo Archive." *Berenson Library, Harvard University*. 2015, http://library.harvard.edu/revealing-renaissance-art-mass-digitization-berenson-library-photo-archive.

Simon, Nina. *The Participatory Museum*. Santa Cruz, CA: Museum 2.0, 2010.

Snydman, Stuart, Rob Sanderson and Tom Cramer. "The International Image Interoperability Framework (IIIF): A community & technology approach for web-based images." *Archiving 2015, Proceedings of the Society for Imaging Science and Technology Conference*, Los Angeles, CA, May 19–22, 2015, http://purl.stanford.edu/df650pk4327.

Whitelaw, Michael. Generous Interfaces for Digital Cultural Collections. *Digital Humanities Quarterly* 2015 9,no. 1, http://www.digitalhumanities.org/dhq/vol/9/1/000205/000205.html.

Yakel, Elizabeth, "Balancing Archival Authority with Encouraging Authentic Voices to Engage with Records" in Theimer, Kate, (Ed.) *A Different Kind of Web: New Connections between Archives and Our Users*. Chicago: Society of American Archivists, 2011.

# 5 Myaamiaataweenki Eekincikoonihkiinki Eeyoonki Aapisaataweenki

## A Miami Language Digital Tool for Language Reclamation

*Daryl Baldwin, David J. Costa, and Douglas Troy*

### Introduction

The Myaamiaki (Miami People) are an aboriginal people historically located around the southern Great Lakes region of North America with primary village locations along the Wabash River Valley in what is today north-central Indiana. The language spoken by this group is referred to as myaamiaataweenki (Miami Language). The Myaamiaki spoke a dialect of the same language spoken by the various subtribes of the Illinois (Inoka), in particular the Peoria and the Kaskaskia. Today, linguists group together these different dialects as the Miami-Illinois language.[1]

The Myaamiaki signed 13 treaties that were ratified with the United States Federal Government between the years of 1795 and 1867.[2] The 1840 treaty included provisions for the removal of the Myaamiaki from their traditional land base in Indiana. The Miami Tribe, defined by these treaties as a legal entity, was militarily forced in October 1846 to move to a reservation in the unorganized territory, which later became Kansas. Several families were exempt from that removal and remained in the homelands of Indiana as absentee members of the tribal nation for a period of time. In 1937 these Indiana descendants organized themselves separately under a State 501(c)(3) non-profit organization known today as the Miami Nation of Indians of the State of Indiana, Inc.

From 1846 to 1873 the Miami Nation remained on its new reservation lands in Kansas until a second removal to Indian Territory (what would become the state of Oklahoma in 1907) occurred between the years of 1873 and 1884, again leaving some individuals and families behind in Kansas. The fragmentation of the Myaamiaki through forced removal, the government's allotment of tribal lands, and the boarding school era all hindered the ability of the Myaamiaki to maintain communally based activities, including language use. By the early 20th century, myaamiaataweenki fell into disuse, and by the mid-20th century few if any fluent speakers could be found. It should also be noted that the descendants of what have historically been called the Illinois people are today represented by the Peoria Tribe of Indians of Oklahoma. The Peoria dialect of the Miami-Illinois language appears to have lost its last fluent speakers at roughly the same time as the Myaamia dialect, in the mid 1900s.

During the late 1980s, graduate student David J. Costa began inquiring about myaamiaataweenki as a potential dissertation topic in linguistics at the University of California at Berkeley. Costa's initial inquiries showed that little was known about the language and that few if any speakers were still living. However, prompted by curiosity to describe this unknown Algonquian language, Costa's search would uncover two and a half centuries of documentation, which became the basis for his doctoral dissertation.[3]

By the mid 1990s the Miami Tribe of Oklahoma became interested in supporting community efforts toward language revitalization as a result of several families initiating home learning efforts. This was a significant step, since for several years the language had been labeled "extinct" or "dead," and it took a great deal of work to reverse this unfortunate perception among tribal citizens. The first tribally supported community effort came in 1997 with an unpublished Myaamia phrasebook, the result of an Administration for Native American (ANA) Language grant awarded to the Miami Tribe of Oklahoma.[4] During this period of time Costa continued his work reconstructing the grammar, phonology and lexicon of Miami-Illinois, leading to his completed dissertation in 1994.[5]

The Miami Tribe and Miami University, located at Oxford, Ohio, have shared a long-standing relationship that dates back to the 1970s. Miami Tribe students began attending Miami University as a result of a scholarship program created in 1991 called the Miami Indian Heritage Award.[6] Through this mutually supportive relationship, tribal and university leaders sensed an ideal opportunity to advance much-needed language research and development. In the fall of 2000, the Miami Tribe of Oklahoma approached their allies at Miami University to develop a language research project. In 2001, both the Tribe and University agreed to support the development of the Myaamia Project, a teaching and research unit within the University with emphasis on language research and cultural education. The project has evolved and expanded over time to become the Miami Tribe of Oklahoma's "research arm," and now collaborates with multiple departments and programs on campus, including the Department of Computer Science and Software Engineering, our collaborator on MIDA. In recognition of this growth, the Myaamia Project transitioned into the Myaamia Center within the University.[7]

This background and history is all significant in understanding the evolutionary process that has led to current efforts to digitize, analyze, store and make accessible online the vast linguistic archives available for the myaamiaataweenki. Relationship building is always at the core of our work, including any research or technologies we develop.

## The Myaamiaataweenki Sources

Miami-Illinois may be unique among Native North American languages for not having been natively spoken for at least half a century, yet still having extremely extensive written documentation spanning almost 250 years, most of which exists as unpublished manuscripts in archives and libraries. Purely in terms of written records, Miami-Illinois is one of the best documented Algonquian languages, far more extensively recorded than many other Native American languages which still have speakers.

The earliest documentation of Miami-Illinois consists of three Illinois dictionary manuscripts compiled by French Jesuit missionaries from the 1690s through the 1720s.[8] Taken together, these manuscripts contain tens of thousands of words, collected at a time when the language was in daily use by large, monolingual communities still living in a traditional manner.

In the early colonial period, the language was again documented in several vocabularies, some quite extensive, from the 1790s through the 1860s. Starting in the

1890s, the language received its first attention from the Bureau of American Ethnology, by the Swiss-born linguist Albert Gatschet, who recorded extensive field notes, several native texts and thousands of vocabulary cards, which reside at the National Anthropological Archives at Suitland, Maryland.[9] Not long after Gatschet's work, the Miami-Illinois language was extensively documented for more than ten years by the Indianapolis lawyer and avocational linguist Jacob P. Dunn, who re-elicited much of Gatschet's data, as well as collecting several new texts and a huge amount of new vocabulary from speakers in both Indiana and Oklahoma. A small portion of Dunn's data was poorly redacted and published by linguist Carl Voegelin in the late 1930s,[10] but the bulk of Dunn's data has never been published, and remains at the Indiana State Library and the National Anthropological Archives to this day.[11] The last substantial documentation of Miami-Illinois was undertaken by the Bureau of American Ethnology linguist Truman Michelson, who, in one week's worth of fieldwork on the Peoria dialect in Oklahoma in 1916, collected three native texts, a full schedule of kinship terms, numerous verb paradigms and a fair amount of vocabulary. Truman Michelson was the closest thing to a trained Algonquianist and linguist who ever conducted fieldwork with fluent speakers of Miami-Illinois, and so his records are quite valuable, though nowhere near as extensive as one might wish. Again, none of Michelson's notes were ever published, and they too are now preserved at the National Anthropological Archives.[12]

From the 1930s through the early 1960s, the last generation of Miami-Illinois speakers was recorded in a handful of brief vocabularies, some recorded by linguists but most by hobbyists. The most valuable of these vocabularies was recorded by linguist Charles Hockett during two days' worth of fieldwork in 1938 with Myaamia and Peoria semi-speakers living in Oklahoma.[13] After Hockett's work, the Miami-Illinois language was documented in a scattering of small wordlists, the importance of which is greatly diminished by the lack of any kind of linguistic training by the people who collected them as well as the greatly decreased fluency of the semi-speakers and members who were still alive in the 1940s and thereafter. Although people with ancestral Native knowledge of Miami-Illinois survived until the late 1970s in Indiana, and to a lesser extent in Oklahoma, no trained linguists worked with these people. Very unfortunately, no significant sound recordings were ever made of any speakers of Miami-Illinois.

Despite being documented for more than two centuries, all the records of Miami-Illinois are problematic, as none of the data was recorded with fully modern standards of phonetic accuracy, and no sources consistently mark all the contrastive sounds of the language. Data from the most fluent speakers tends to have been written down by the least skilled transcribers, while the more accurate later records of the language, transcribed by the first generation of trained linguists, are from a time when speakers were less fluent. Thus, even though there is a massive amount of data on Miami-Illinois, little of it was competently recorded from fluent speakers. As a result, none of the recorded corpus of Miami-Illinois data can be taken at face value, and careful philological analysis must be brought to bear on all of it.

The two most problematic phonological features in determining the correct pronunciation of Miami-Illinois words are vowel length and preaspiration of consonants. Vowel length and preaspiration are crucial features in the phonology of Miami-Illinois, are both fully contrastive, carrying a high functional load, yet they are seldom indicated in most recordings of the language. In the extensive French Jesuit records, vowel length is never marked and preaspiration is only infrequently marked; in the late-19th and early-20th century records of the language, both length and preaspiration are marked somewhat more often, though still not dependably. There is no source on Miami-Illinois that marks both features consistently, so these contrastive phonological features must be filled in and all the data phonemicized in order to make materials usable for either linguistic or pedagogical purposes.

There are two primary methods by which phonological details can be filled in for Miami-Illinois data. One is by comparing all the varying original transcriptions for the words, and the other is by comparing the Miami-Illinois words with cognate data from its closely related sister languages. Both vowel length and preaspiration are found in essentially the same places in Miami-Illinois as in neighboring Algonquian languages such as Meskwaki, Ojibwe, Shawnee, and Kickapoo, and so comparing Miami-Illinois transcriptions to cognate words from these languages is extremely helpful in determining the true phonological shape of Miami-Illinois words.

## Challenges in Working with the Sources

Costa's primary means of organizing the massive amount of data needed for his dissertation analysis consisted of several large alphabetized Word documents that he created containing annotated data from all the different manuscripts. These Word documents, exceeding 1,300 pages, served as the organized data source for Costa's work for more than 25 years. Also at this time linguistic databases were being created, most notably the Summer Institute of Linguistics' (SIL) Shoebox program.[14] An early Shoebox database for myaamiaataweenki was constructed during the late 1980s by Myaamia Center researchers, but this database was later abandoned after SIL discontinued support for the Macintosh platform. To store, translate, and analyze the extensive corpus of documentation, it was obvious that a much more capable database tool was needed.

The unique challenges of working with the Miami-Illinois corpus have directly dictated the form that any database for storing the data must take. Virtually all preexisting dictionary database programs are designed for the cataloging of phonemic data taken directly from speakers. Examples are LingSync, WeSay, Lexique Pro, Language Explorer, Online Linguistic Database, and Miromaa. The needs of a database for Miami-Illinois are far different. In a Miami-Illinois database, one cannot simply include glossed phonemic data with no commentary, since almost none of the primary data in the language is in phonemic form. In support of the phonemicized data, one must also include the original primary data and glosses, precisely as they appear in the sources, as well as whatever additional data is deemed necessary to further support the phonemicizations and glosses. For example, it is not enough to give the phonemically spelled Miami-Illinois word for "no, not," *moohci*; it is also necessary to give the original forms, such as the one given in the early-18th century French Jesuit sources, ⟨m8tchi⟩ , as well as its translations given there, "seulement" ("only") and "même" ("even"). Additionally, one needs to include later transcriptions of this word such as Gatschet's ⟨mú'htchi⟩ and Truman Michelson's ⟨mṓ'tci⟩ , which both support the phonemic reconstruction *moohci*, as well as the fact that in all sources from the late-1700s onwards this word is translated as simply "not" or "no." In further support of the phonemicizations, it is often helpful to include cognates from other Algonquian languages, so one must also have a dedicated field to include related words, such as Meskwaki *mo·hči* and Shawnee *mohči* "even." Essentially the entries for the individual words must include not only all the data needed to interpret the words for language instruction, but they must also include all the primary evidence gathered to support how the translations, grammatical analysis, and corrected spellings were arrived at. For example, in many verbs, the only phonological feature distinguishing the first person singular and the second person singular is vowel length: for example, compare *mee-naani* "I drink" versus *meenani* "you drink."[15]

## Toward a Technical Solution

Processing the older Jesuit-era materials proved challenging for many years. In 1999, myaamiaataweenki revivalists, who were interested in gaining access to these early sources, initiated the Illinois Project.[16] The primary goal of this early project was simply to develop a process for systematically transcribing and translating the Jesuit era manuscripts in order to gain access to new language materials for reclamation efforts. The Illinois Project was initially overseen through the joint effort of members of the language committees of the Miami Tribe of Oklahoma and the Miami Nation of Indians of the State of Indiana, Inc., under the guidance of a ten-year compact agreement signed in 1997. This agreement was intended to create a collaborative environment for each entity to work together and provide volunteer resources for the project and other language reclamation efforts. The agreement reached its end in 2007 and was not renewed. From this point in 2007, the Myaamia Project (now the Myaamia Center) assumed the responsibility of moving the Illinois Project's goals forward. During this time period we relied heavily on volunteer transcriptionists and project organizers as there were no funds available to support full or part-time research staff. Many hours were dedicated by volunteers over a span of several years who organized this early project and created a great deal of transcription materials that would be used in a later phase of this work. Transcription work to properly prepare and process the vast amount of early data was especially challenging. For these reasons, it became very difficult to move the project forward and so it eventually fell dormant for several years. Reflecting back on this early struggle, the challenges became clear: How do we organize, store, and retrieve massive amounts of data as needed? What would an ideal database system look like? And what functionality would be available to us that would make linguistic analysis more efficient and easier to perform?

In a second attempt to address these problems, the Illinois Project was reinvigorated in 2012 when the Myaamia Center received an award from the National Endowment for the Humanities (NEH).[17] The NEH award, along with technological advances, allowed the Illinois Project (now referred to as the Ilaatawaakani Project) to take a significant leap forward by allowing us to re-examine ways of accessing Jesuit-era source materials, as well as to reassess technological advances in order to determine what type of digital archive and digital research tools were possible. To test the waters, we began working with one of the three primary Jesuit manuscripts.[18] This document was selected as the first to be transcribed due to its numerous example sentences, its well-organized format, and the fact that it had never been edited before. The translated, annotated and analyzed redaction of the LeBoullenger manuscript, with its vast amount of data, would constitute an invaluable source of readily accessible data and properly test our ideas for archival access and development. During the three years of the NEH supported grant, the Ilaatawaakani Project would create the first ever Miami-Illinois Digital Archive (MIDA).[19]

Before looking any further at developmental concepts, it should be noted that MIDA was never intended to be a language learning tool. The purpose for creating MIDA was to eliminate, to a large degree, the cumbersome need to work directly with original source materials as well as to create what current linguistic databases were not providing. The MIDA has brought order to the large corpus of language data and allows us to filter out specific kinds of information. As an example, it is not uncommon for useful and important language data to be embedded within the manuscripts in such a way that a typical visual search of the physical pages (in whatever order the manuscript was written) does not easily allow a user to find the entry. For example, the interesting word *alaamatayi* (spelled in the original ⟨aramataye⟩ ) is glossed by LeBoullenger "avant que de naitre dans le ventre de sa mere," which translated literally into English is "before being born in his/her mother's womb." Upon further analysis, this word turns out to be an adverb, basically meaning "in the womb, in utero." However, this word is not listed under any word for "womb," nor in any kind of list of terms having to do with childbirth or even body part terms, but is instead hidden under LeBoullenger's keyword "avant," which in English means "before." This is a fairly typical example of how hundreds of interesting vocabulary items in this manuscript often lurk in places where they cannot be "looked up" in any way until the manuscript is available in a searchable database.

To initiate development of the NEH-supported Ilaatawaakani Project, we pulled together a team of linguists, tribal researchers, computer programmers, and communication specialists to design and build a new software application that would address our identifiable needs. Initial design requirements for MIDA included the following stipulations:

- The database must be designed to function as simply as possible and to not do more than what was necessary to meet the Myaamia Center's research needs.
- It must have a robust search function. Finding something within and among several manuscripts would hinge on a well-designed search application.
- It must be online and accessible to anyone interested for research purposes.
- Some language content created through MIDA, such as stem and morpheme lists, would be shared with the online Myaamia Dictionary, a separate online resource serving as a community language learning tool.[20]

Over the initial three years of developing the Ilaatawaakani Project, our understanding has evolved tremendously. This has led to the development of MIDA, which has already become one of the most significant research tools we have developed to date. Future versions of MIDA will include all known linguistic source materials. In the following sections we describe in greater detail project organization including the digitization of the manuscript, the process for building the archive (including search and administrative tools), and initial research uses in different disciplines.

## Steps in Processing a Manuscript

The MIDA requires a four-step process (Figure 5.1) starting with the preparation of the manuscript and ending with searchable data accessible through the MIDA search engine and the Myaamia online dictionary.

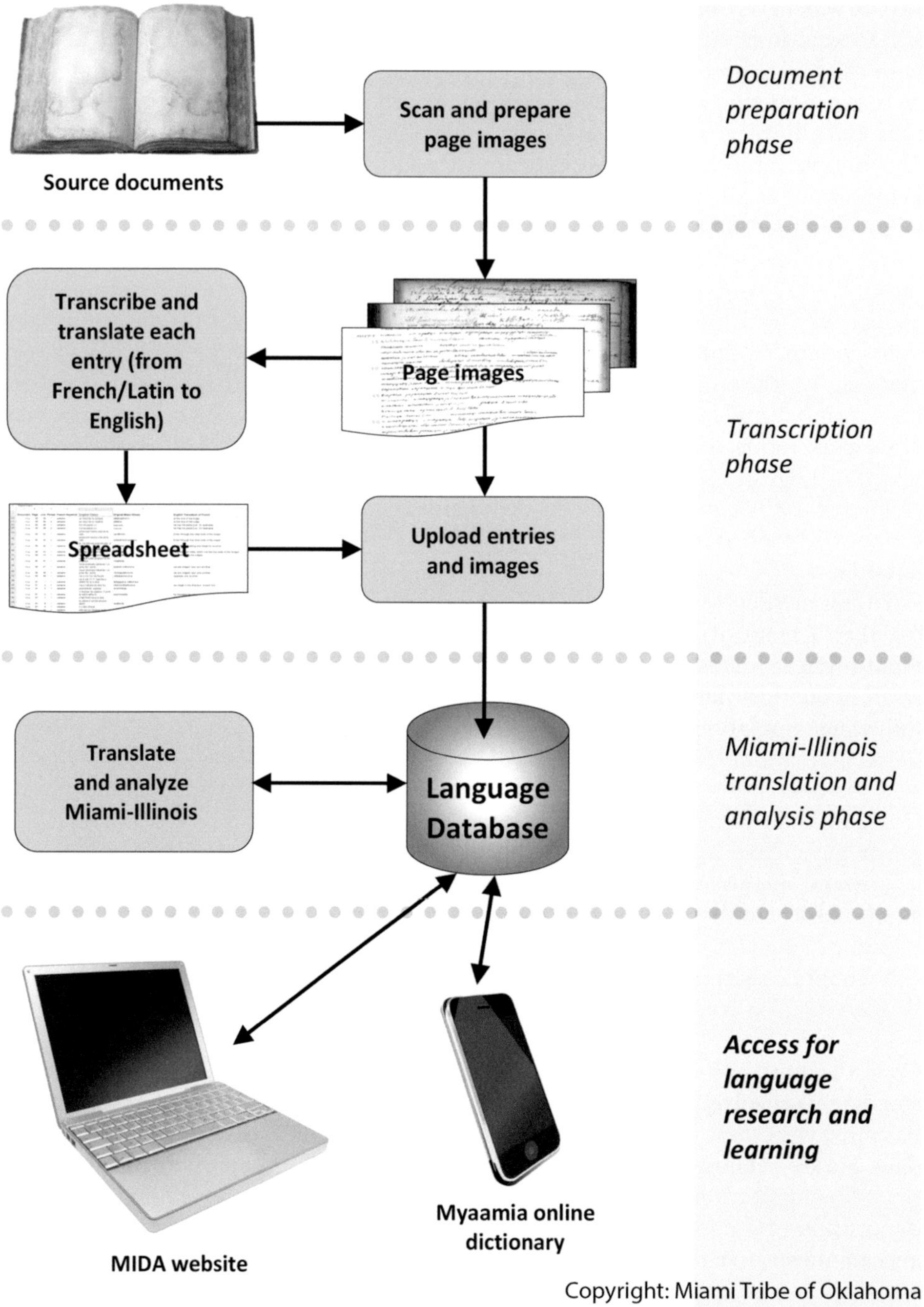


*Figure 5.1* Four (4) step process in building data for MIDA (Miami-Illinois Digital Archive).

To begin the process, the transcriptionist receives, from the archival library, high resolution (300–600dpi) digital scans for each page in a document and adds page and line numbers for reference directly to the image in preparation for upload (Figure 5.2). Each document page, including blank pages, are numbered and the text is numbered every ten (10) lines.

A spreadsheet is provided for the transcriptionist to begin transcribing based on the organization of the original document (Figure 5.3), which in this case includes the keyword (if available), original French or English, and original Miami-Illinois entries into the spreadsheet. Each entry has its corresponding page, line, and phrase number for reference to the original document. A new spreadsheet was required approximately every 15 pages to maintain usability of the spreadsheets, as they grew quite large.

The spreadsheets were created in a web-based service allowing access to documents from anywhere, as well as the critical feature of allowing multiple users to edit the same spreadsheet simultaneously. This was important so that collaborators who are not always in the same location can not only do their work, but assist each other with reading original text and translations.

The transcription and French translation steps are always done in spreadsheet form for reasons of efficiency. Figure 5.4 illustrates a few lines from a partially completed spreadsheet. Final transcription work from the LeBoullenger document alone resulted in approximately 25,000 spreadsheet lines of data.

Once the transcription and translation of the French text is complete, the spreadsheet data is then uploaded into the MIDA database for further work. The process begins by providing English translations and analyses of the original Miami-Illinois data transcribed from the LeBoullenger document on the MIDA website. There are three primary parts to this task; the first step is filling in the contemporary spelling of the Miami-Illinois words. As mentioned above, the data in the Illinois dictionaries is recorded in an inconsistent writing system, which fails to mark all the phonemic contrasts of the language. Thus, it is necessary to re-transcribe the Miami-Illinois data into the modern, phonemic orthography. We also provide a place for supporting evidence to show how these phonemicizations are decided upon. There is also a cognate field, discussed earlier, where one enters cognate words drawn from the sister languages as well as original forms of the words drawn from other Miami-Illinois sources.

The second step consists of filling in corrected English glosses to the Illinois data. While it is essential to provide the literal English translations of the French glosses, often the original glosses are just as imprecise as the original transcriptions of the Illinois words. Thus, it is necessary to provide revised English translations, informed by our actual knowledge of Miami-Illinois grammar and data elsewhere in the corpus. For example, LeBoullenger's form ⟨ac8eng8sa⟩ is glossed as "petite s [ouris]," literally "little mouse," though it is clear from the modern records that this word in fact means "chipmunk." As a more subtle example, LeBoullenger glosses the imperative ⟨nissahanto⟩ as "abats cette perche," or "knock down that pole," though it is clear from the structure of this verb, and from related forms recorded elsewhere, that its actual meaning is more like "knock it down! (by instrument)"; that is, that this verb explicitly indicates that the action is accomplished by some kind of tool, and that it can refer to any kind of standing object being knocked down, not just poles. And finally, the third step is to provide the grammatical analysis of the

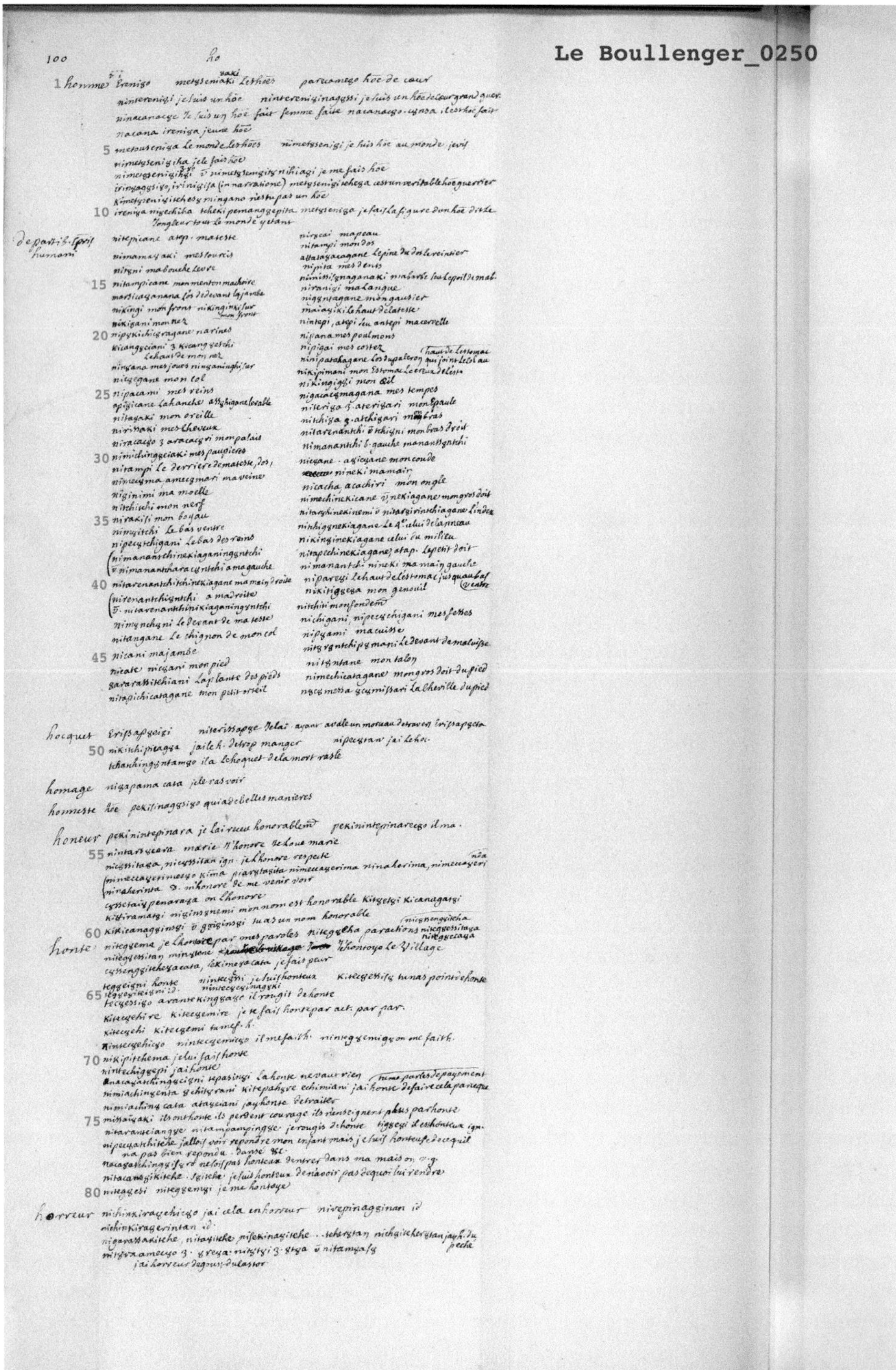


*Figure 5.2* LeBoullenger manuscript page with page and line numbers.

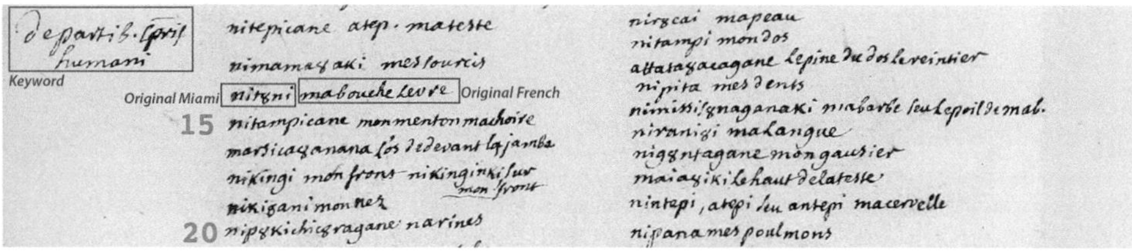


*Figure 5.3* Organization of the entries in LeBoullenger.

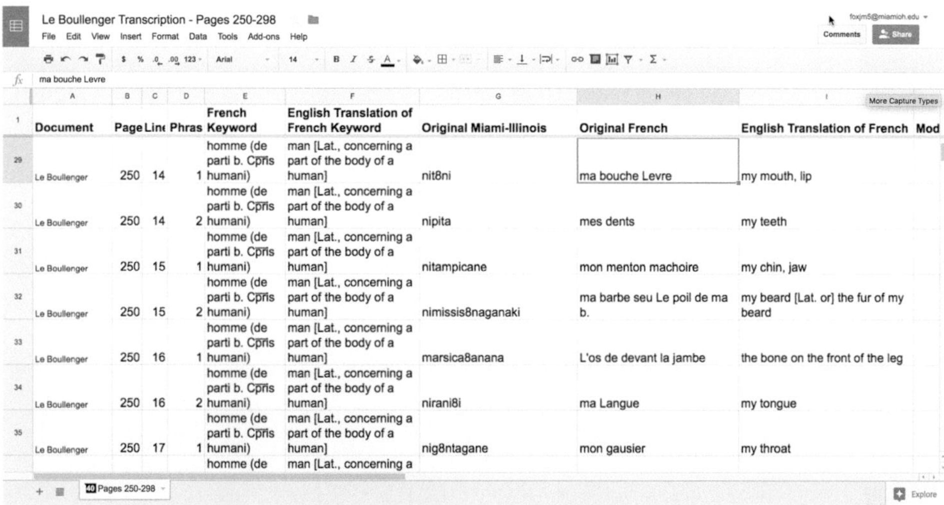
Le Boullenger Transcription - Pages 250-298

| Document | Page | Line | Phrase | French Keyword | English Translation of French Keyword | Original Miami-Illinois | Original French | English Translation of French | Mod |
|---|---|---|---|---|---|---|---|---|---|
| Le Boullenger | 250 | 14 | 1 | homme (de parti b. Cpris humani) | man [Lat., concerning a part of the body of a human] | nit8ni | ma bouche Levre | my mouth, lip | |
| Le Boullenger | 250 | 14 | 2 | homme (de parti b. Cpris humani) | man [Lat., concerning a part of the body of a human] | nipita | mes dents | my teeth | |
| Le Boullenger | 250 | 15 | 1 | homme (de parti b. Cpris humani) | man [Lat., concerning a part of the body of a human] | nitampicane | mon menton machoire | my chin, jaw | |
| Le Boullenger | 250 | 15 | 2 | homme (de parti b. Cpris humani) | man [Lat., concerning a part of the body of a human] | nimissis8naganaki | ma barbe seu Le poil de ma b. | my beard [Lat. or] the fur of my beard | |
| Le Boullenger | 250 | 16 | 1 | homme (de parti b. Cpris humani) | man [Lat., concerning a part of the body of a human] | marsica8anana | L'os de devant la jambe | the bone on the front of the leg | |
| Le Boullenger | 250 | 16 | 2 | homme (de parti b. Cpris humani) | man [Lat., concerning a part of the body of a human] | nirani8i | ma Langue | my tongue | |
| Le Boullenger | 250 | 17 | 1 | homme (de parti b. Cpris humani) | man [Lat., concerning a part of the body of a human] | nig8ntagane | mon gausier | my throat | |
| | | | | homme (de | man [Lat., concerning a | | | | |

*Figure 5.4* Spreadsheet entries for transcription and French translation.

Miami-Illinois words, by breaking down all the Miami-Illinois words into their constituent parts including their stems and stem components. Such data enables users to search on all words in the manuscripts, which share certain morphemes, so as to compare their usage across dozens or even hundreds of different words. Figures 5.5 and 5.6 show screenshots of a typical search and results based on these last steps.

Needless to say, completion of all three of these categories for each entry in the manuscript is an extremely time-consuming project. Some words submit to a very simple, obvious analysis, while just as many words are of a more obscure origin and require extensive research before their entries can be even partially analyzed. A significant number of words resist analysis entirely. Given the size of the LeBoullenger manuscript as well as that of the other equally large or larger Jesuit dictionaries that will be added to MIDA in the course of time, it is clear that an exhaustive analysis of all the data in all of these manuscripts is a process that will span decades, long after the process of keying in the data and translating the French glosses is complete.

As noted above, the transcriptions and their French translations were originally recorded in spreadsheets. Translation and linguistic work required a more

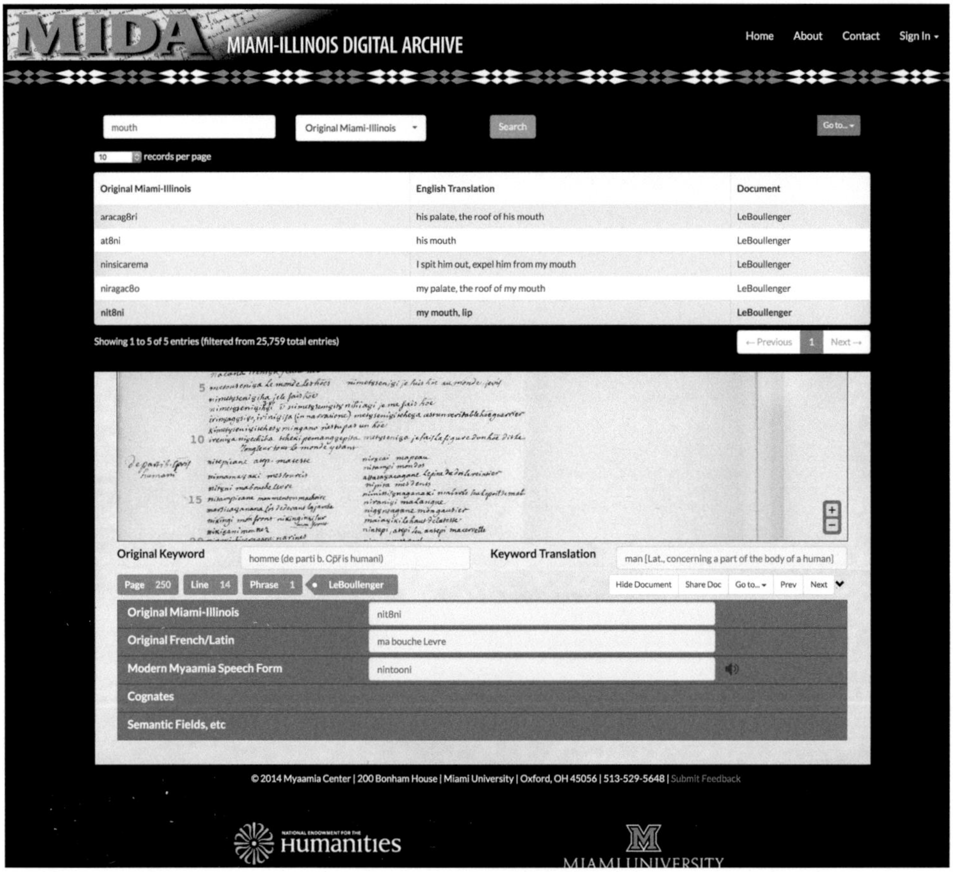


*Figure 5.5* Typical search result in the Miami-Illinois Digital Archive.

sophisticated database to support storage of stems, morphemes, and cognate information as well as comprehensive search capability to locate other entries in the corpus for cross-reference. These additional fields are added to MIDA after the spreadsheet data has been uploaded.

The MIDA contains an advanced search function allowing the user to search within or among manuscripts and by any data field. Figure 5.7 shows the field menu of the search function.

The MIDA also supports many additional features including:

- No login needed for general users (search only);
- Accounts and login required for editors (search and update);
- Comprehensive search engine for general users and researchers;
- A feature that logs changes made to records by editors;
- Administrative tools for spreadsheet data import/export, account management, and account creation.

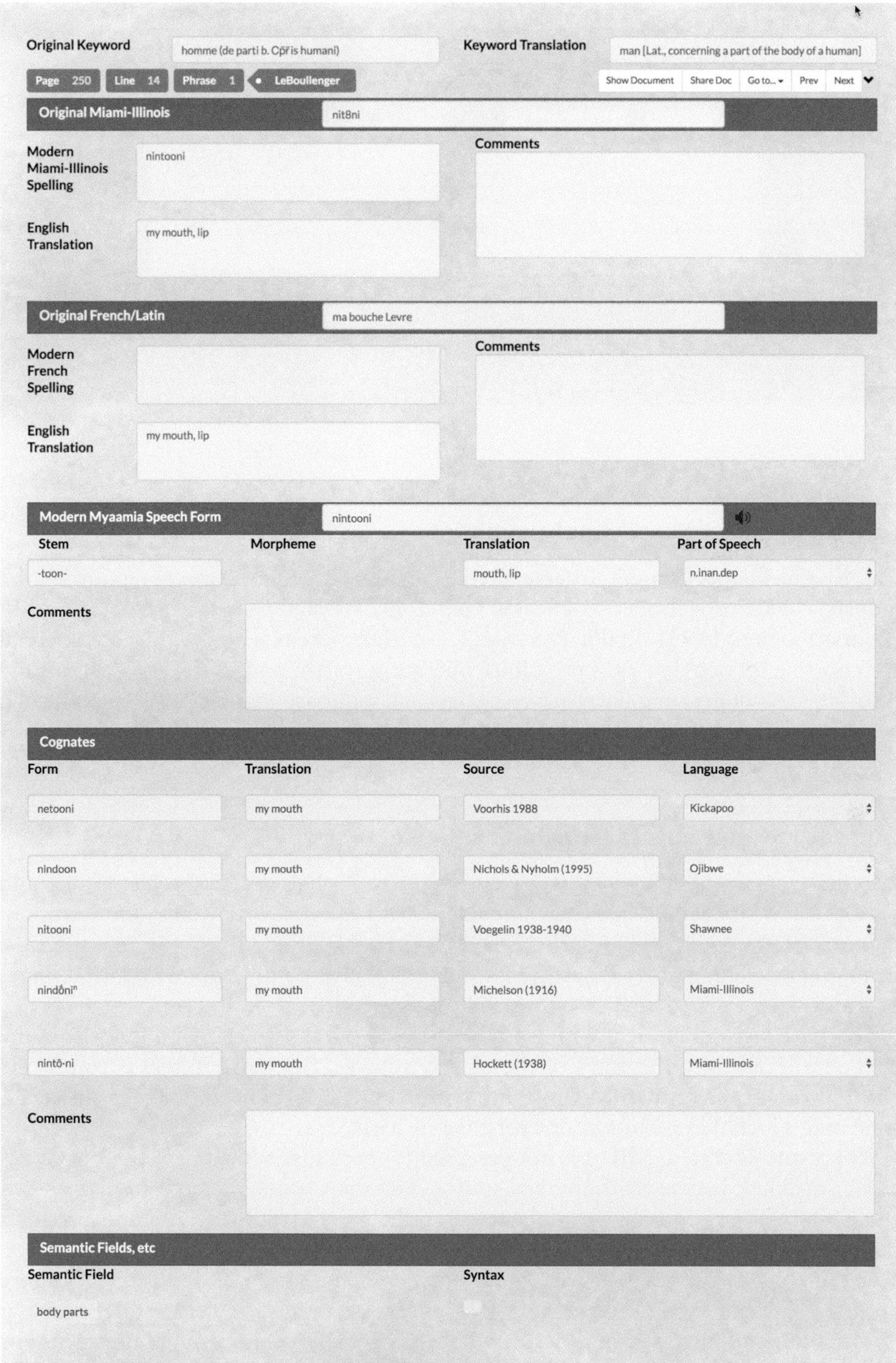


*Figure 5.6* Search result with expanded fields.

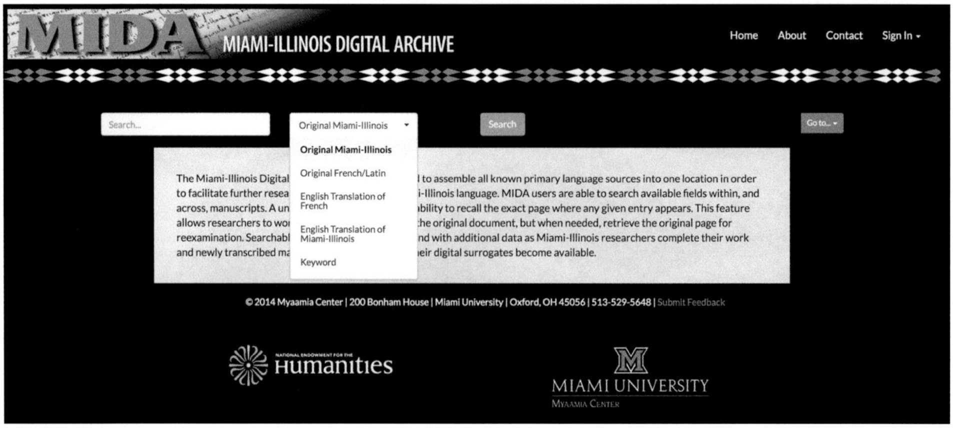


*Figure 5.7* Advanced search function menu of Miami-Illinois Digital Archive.

## MIDA Architecture and Implementation

The Miami-Illinois Digital Archive (MIDA) is implemented using a version of the industry-standard web service solution commonly referred to by the acronym LAMP (Linux-Apache-MySQL-PHP). The complete MIDA system is hosted on a commercial web hosting service that runs the Linux operating system and a web server (typically Apache). The underlying database is a relational database management system using MySQL. The web pages, search logic, and database loading code are programmed using a combination of the HTML, PHP and JavaScript programming languages.

## Current Use and Future Development

The first version of MIDA is in use by linguists, Tribal educators, historians, and ecologists. While MIDA has been populated with the data from only one Jesuit-era manuscript (LeBoullenger), at the time of this publication, two more manuscripts from the same era, namely Pinet and Largillier, are near completion with initial stages of transcription and their French translation finished.

The Largillier manuscript will likely be the next dataset to come online within the next year. Following the entry of these early sources, our current goal is to process and upload all the known Miami-Illinois source materials and to make them accessible and searchable within future versions of MIDA.

As previously stated, MIDA is not designed to teach myaamiaataweenki, but serves as a critical link between the archival sources and the various research specialists and educators who are looking for data to serve their individual interests and program needs. One of the most powerful features of this new software is its ability to retrieve a digital copy of the original page when further examination and interpretation are needed. This immediate access to copies of the original, coupled with contextual information and linguistic analysis, is what makes MIDA unique in its function. For communities having to reclaim their language from documentation, where documentation becomes the main source for language, a heavy research component is necessary and MIDA allows for that research to progress faster and more efficiently.

Easy access to linguistic, cultural, historical, and ecological information is invaluable in cultural revitalization work. The ability to search within and between documents at the same time for specific content is a luxury we have not had up to this time. Tribal educators who might be developing curriculum for something as simple as the reintroduction of traditional games can now search "game" in MIDA and obtain a wealth of information that can then be forwarded to linguists and cultural experts for help with translation and interpretation before they utilize the information in community culture and language vitalization programs. Ecologists are now able to search for a wide range of food plants in MIDA and can gain insight into traditional dietary information that has direct use and value for our ongoing "cooking with traditional foods" project.

## Conclusion

The Ilaatawaakani Project was essential in creating the Miami-Illinois Digital Archive (MIDA). The years of struggle, the numbers of individuals who contributed to ideas, and rapidly developing technologies all factored into our evolving understanding of what was possible. Just as there is no one-size-fits-all solution to revitalizing languages, there is no one tool that will serve every community's language archive needs. The Ilaatawaakani Project (formerly the Illinois Project) began in 1999, and 17 years later we have reached a milestone in the development of a long-term archival tool that specifically meets our linguistic research and cultural vitalization needs. Development was delayed by the need for more capacity building, secured funding, and other human resources. Bringing these multiple resources together in a way that supports further development takes time and much effort. The ground-breaking work of language reconstruction, the development of the Myaamia Center, and the relationship nurtured through Miami University's College of Engineering and Computing were also vital relationship building activities that laid the foundation for a collaborative project like this to be successful.

Despite the challenges, the Ilaatawaakani Project's process has allowed us to explore the extent to which we could develop a digital archive that specifically met the needs of community-based language research and development. Hopefully, our work in the reconstruction and reclamation of Myaamia language and culture will continue to afford us opportunities to establish new research tools and methodologies for archival work. The development of MIDA has not only opened up new possibilities for archiving and utilizing large amounts of linguistic sources, but has also impacted our ability to access critical cultural and ecological information for tribal educational programs. Many of us who work with the Miami Tribe of Oklahoma are honored to observe the true beneficiaries of all this work—the many Myaamia children who for the first time in over a 100 years are afforded the opportunity to hear and speak myaamiaataweenki and to learn many new and exciting aspects of their complex history, language, and culture.

The Miami-Illinois Digital Archive can be found at www.ilaatawaakani.org.

## Notes

1 Costa, David. *The Miami-Illinois Language*. Lincoln: University of Nebraska Press, 2003, 2.

2 Ironstrack, G. "Myaamiaki neehi Myaamionki: The Miami People and Their Homelands." In Curtis W. Ellison (ed.). *Miami University 1809–2009 Bicentennial Perspectives*. Ohio University Press, 2009.
3 Costa, *The Miami-Illinois Language*, 2003.
4 Miami Tribe of Oklahoma. *Iilaataweeyankwi. Our Language*. Miami, OK: Miami Tribe of Oklahoma, 1997.
5 Costa, *The Miami-Illinois Language*, 2003.
6 Burke, B. and Daryl W. Baldwin II. "A People With a Past, Not From the Past: Miami University and the Miami Tribe of Oklahoma." In Curtis W. Ellison, 2009.
7 Baldwin, Daryl. "Oowaaha Myaamiaataweenki: Miami Is Spoken Here." In Wiley, Terrence G., Joy Kreeft Peyton, Donna Christian, Sarah Catherine K. Moore and Na Liu (ed.), *Handbook of Heritage, Community, and Native American Languages in the United States*. New York: Routledge-Taylor and Francis Group, 2014.
8 LeBoullenger, Antoine-Robert, S.J., *French and Miami Dictionary* (Manuscript at the John Carter Brown Library, Brown University, Providence, Rhode Island), n.d. [ca. 1725]; Largillier, J., *Illinois-French Dictionary* (Manuscript in Watkinson Library, Trinity College, Hartford, CT), n.d. [ca. 1700]; Pinet, Pierre François. *French-Miami-Illinois Dictionary* (Manuscript at the Archives des Jesuites au Canada, Montreal, QC), n.d. [ca. 1702].
9 Gatshet, Albert. *Vocabulary and Text (Three Original Miami and Peoria Field Notebooks)* (Manuscript #236, National Anthropological Archives, Smithsonian Museum Support Center, Suitland, MD), n.d.
10 Voegelin, Carl F., *Shawnee Stems and the Jacob P. Dunn Miami Dictionary* (Indiana Historical Society Prehistory Research Series 1: 63–108, 135–167, 289–323, 345–406, 409–478). Indianapolis: Indiana Historical Society, 1938–1940.
11 Dunn, Jacob P., *Various Notes on Miami* (Manuscript at the Indiana State Library, Indianapolis), n.d.
12 Michelson, Truman. *Notes on Peoria* (Manuscript #2721, National Anthropological Archives, Smithsonian Museum Support Center, Suitland, MD), 1916.
13 Hockett, Charles F., "Notes on Peoria and Miami." *Algonquian & Iroquoian Linguistics* 10, no. 4 (1985): 29–41.
14 Shoebox (2016). http://www-01.sil.org/computing/shoebox/.
15 Wesay: https://www.lingsync.org/#/home; Lexiquepro: lexiquepro.com; Language Explorer (FLEx): http://fieldworks.sil.org/flex/; Online Linguistic Database (OLD): http://www.onlinelinguistic-database.org/; Language Database Software Miromaa: http://www.miromaa.org.au/.
16 MIDA (2016). http://ilaatawaakani.org.
17 Inokaatawaakani. *Illinois Dictionary Project* (# PD-50017-12), Washington: National Endowment for the Humanities, 2012.
18 LeBoullenger, Antoine-Robert S.J., *French and Miami Dictionary*, n.d.
19 MIDA (2016).
20 Myaamia Online Dictionary, 2016.

## References

Baldwin, Daryl. "Oowaaha Myaamiaataweenki: Miami Is Spoken Here." In Wiley, Terrence G., Joy Kreeft Peyton, Donna Christian, Sarah Catherine K. Moore & Na Liu (eds.), *Handbook of Heritage, Community, and Native American Languages in the United States*. New York: Routledge-Taylor & Francis Group, 2014.

Burke, B, & Daryl W. Baldwin II. "A People With a Past, Not From the Past: Miami University and the Miami Tribe of Oklahoma." In Curtis W. Ellison (ed.), *Miami University 1809–2009 Bicentennial Perspectives*. Athens, OH: Ohio University Press, 2009.

Costa, David. *The Miami-Illinois Language*. Lincoln: University of Nebraska Press, 2003.

Dunn, Jacob P. *Various Notes on Miami* (Manuscript at the Indiana State Library, Indianapolis), n.d.

Gatschet, Albert. *Vocabulary and Text (Three Original Miami and Peoria Field Notebooks)* (Manuscript #236, National Anthropological Archives, Smithsonian Museum Support Center, Suitland, MD), n.d.

Hockett, Charles F. "Notes on Peoria and Miami." *Algonquian & Iroquoian Linguistics 10*(4) (1985): 29–41.

Inokaatawaakani. *Illinois Dictionary Project (# PD-50017-12)*. Washington: National Endowment for the Humanities, 2012.

Ironstrack, George "Myaamiaki neehi Myaamionki: The Miami People and Their Homelands." In Curtis W. Ellison (ed.). *Miami University 1809–2009 Bicentennial Perspectives*. Athens, OH: Ohio University Press, 2009.

Largillier, Jacques. *Illinois-French Dictionary* (Manuscript in Watkinson Library, Trinity College, Hartford, CT), n.d. [ca. 1700].

LeBoullenger, Antoine-Robert S. J. *French and Miami Illinois Dictionary* (Manuscript at the John Carter Brown Library, Brown University, Providence, RI), n.d. [ca. 1725].

Miami Tribe of Oklahoma. *Iilaataweeyankwi. Our Language*. Miami, OK: Miami Tribe of Oklahoma, 1997.

Michelson, Truman. *Notes on Peoria* (Manuscript #2721, National Anthropological Archives, Smithsonian Museum Support Center, Suitland, MD), 1916.

MIDA. 2016. http://ilaatawaakani.org.

Myaamia Center. 2016. http://myaamiacenter.org.

Pinet, Pierre François. *French-Miami-Illinois Dictionary* (Manuscript at the Archives des Jesuites au Canada, Montreal, QC), n.d. [ca. 1702].

Shoebox. 2016. http://www-01.sil.org/computing/shoebox/.

Voegelin, Carl F. *Shawnee Stems and the Jacob P. Dunn Miami Dictionary*. Indiana Historical Society Prehistory Research Series 1: 63–108, 135–167, 289–323, 345–406, 409–478. Indianapolis, IN: Indiana Historical Society, 1938–1940.

# 6 A Cartographic History and Analyses of Indian-White Relations in the Great Plains

*Daniel G. Cole*

## Colonial Era

The Great Plains, accounting for much of the interior of the continent, began to be mapped in the 1600s by the French and Spanish. Typically, during that century and later, much guesswork was involved regarding peoples residing therein, the extent of the land, the topography, and the wildlife. Many maps were created based upon second- or third-hand knowledge from the field, previously published maps, guessing locations, and hearsay. Very often, villages and Native nations would be shown based in information from American Indian informants, while blank spaces were filled in with "Savages" and non-native animals.[1]

For instance, in 1673, Jacques Marquette and Louis Jolliet traveled down the Mississippi from Lake Michigan via Wisconsin, turned around near the mouth of the Arkansas River, and returned to Lake Michigan via the Illinois River. Jolliet and Marquette mapped Native villages along the Mississippi as well as along nearby tributaries. In 1682, Jean Baptiste Louis Franquelin mapped the Mississippi River with the Arctic and Pacific oceans perceived as being much closer than reality.[2] And on this map, west of the Mississippi was depicted with camels, ostriches, elk, and cattle to fill in space. Nonetheless, a number of named Native American villages are portrayed along the Mississippi and Missouri and a couple of other rivers trending west of the Mississippi. In 1685, Franquelin published another map with more details of native villages of the Ozages (Osage), Panai, Pani-maha (Pawnee), and others.[3]

Guillaume de L'Isle mapped the Mississippi in more detail in 1702. The top map sheet (of five) labels the Sioux de l'Ouest and the Sioux de l'Est on the west and east sides of the Mississippi in present-day Minnesota. On the second sheet are a couple of Native villages noted: Aiaouez ou des Paoute (Ioway), Village des Maha (Omaha), both south of the Nation of Tintons (Teton Sioux).[4] But no additional Native towns or nations are noted in the Plains in the lower three sheets. In 1718, De l'Isle produced the first detailed map of the Gulf Coast region and the Mississippi River, as well as the first printed map to show Texas (identified as "Mission de los Teijas etablie en 1716").[5] The map is also the first to identify New Orleans, founded in 1718 (includes an inset detail of the mouth of the St. Louis River). De l'Isle obtained most data from French explorers and fur traders traveling through North America. A close examination of the map reveals the land routes of early explorers in North America. Each route is mapped out and identified with the explorer's name and year of travel. The map represents the travels of Spanish explorer Hernando de Soto in Florida and the southeast in 1539–1542, Alonso de Leon in 1689, and French Canadian explorer

Saint Denis in 1713 and 1716, among others. De l'Isle accurately identified the location of many Native American tribal villages, marked by a small hut symbols and names; and he plotted the regions occupied by the same tribes noted in 1702, while adding the Panis (Pawnee), Padouca (Plains Apache), and Osage further south.

Meanwhile, the Englishman Herman Moll published two maps in 1715 and 1720 covering much of the continent. While he wasn't privy to the details already mapped by the French, he did include some Native American groups (with occasional villages) such as the Eokoros (Arikara), Tintin (Teton) Sioux, Panis (Pawnee), Maha (Omaha), the Plains Sioux, the Osage, and the Plains Apache.[6]

Later, in 1752, Philippe Buache compiled two maps on one sheet including a possible water route from Lake Superior to the Pacific from Ochagach (Cree) and a second map of the northern plains from two unpublished maps by French fur traders. These two maps illustrated territories of the Assiniboils (Assiniboine), Cristinaux or Cristinots (Cree), the Monsoni (Ojibwe) and the Sioux.[7] This map is the first printed record from the plains of an Indigenous map. Maps drawn by Native Americans were not common, but not unusual. While Indigenous peoples did not have paper to draft maps, many fur traders and explorers knew that the people they met often had good spatial cognition of their hunting grounds, neighboring tribes and villages, trade routes, rivers and lakes, mountains (as barriers) and passes, as well as the locations of various sacred sites. And when they were provided paper or something equivalent to draw a map, their spatial depictions were relatively accurate, especially given that the unstated scales were typically in terms of travel time. Moreover, as will be discussed further, Indigenous maps were rarely snapshots in time, but often entailed stories (or histories) of tribal activities over time.

Competition between the Euro-American colonies could be easily seen on the map by John Mitchell in 1755.[8] He provided defined colonial borders, irrespective of any tribe's existing territory, as well as negating French and Spanish territorial claims. Specifically, one may note the "sea-to-shining-sea" type boundaries given to Georgia, North and South Carolina, and Virginia. Regardless of the colonial land claims, at least the eastern and western Sioux, Padoucas (Comanche), Panis (Pawnee), and Osage were recognized. Even though this is a map made by the English, he borrowed information from the French concerning Louisiana.

The Spanish were also mapping the southern plains as part of northern New Spain as seen on Nicolas de Lafora's map of 1771.[9] While distortion on his map is clear, a lot of good data on southern plains and southwest tribes and villages can be ascertained. One can see Cumanches (Comanche) territory and villages along with three bands of Apaches, including the Xicarillas (Jicarilla), Pharaones (Faraon), Mescaleros and Natages (Mescalero). In 1787, a Comanche artist depicted on a large scale map,

> an engagement in the Spanish frontier province of New Mexico between the Comanche and their traditional enemy, the Mescalero Apache. According to an accompanying document signed by then governor Juan Bautista de Anza, the pictographic map documents illustrate the Battle of Sierra Blanca, July 30 1787.[10]

## Federal Era to 1900

Two maps were created in the 1790s (although both weren't published until several decades later) by the French of the central and northern plains depicting Indigenous territories and villages, water features, and stretching from Hudson Bay to the Pacific.

Tribal names included the C(h)ristanaux (Cree), Sioux, Chaguiene (Cheyenne), Arricaras and Richaare (Arikara), Mandanes (Mandan), Pitapahabe (Kiowa Apache), and others. Notably, on the 1795 map by Anthoine Soulard and Johann Kohl was indicated which Indian nations were migratory (nations embulantes) versus those which were sedentary (nations fixes). But the 1796 map by George Collot and P.F. Tardieu is more cartographically attractive; its treatment of the human landscape is a bit different than the earlier map, identifying more Native nations while not stating whether any were sedentary.[11] While either of these maps would have been helpful to the Lewis and Clark expedition, unfortunately neither was published until much after the expedition.

Another map that was likely unknown to Lewis and Clark was one of the northern plains drawn by a Blackfoot chief (Ackomakki or The Feathers) in 1801 and later redrawn at a smaller scale and labeled by Peter Fidler, along with notes added still later by Johann G. Kohl, who published it in 1850. Notably included were Blackfoot names of rivers plus 11 peaks in the Rocky Mountains were named (translated to English) as well. Kohl finishes his Notes with the following open-minded comment of praise:

> The map is, however, also interesting in this respect: It shows the field of geographical knowledge and perhaps in a certain degree the limits of the hunting excursions of a Blackfoot chief at about the year 1800. His knowledge of the Missouri sources was greater than the information of our geographers at that time.[12]

Maps that were studied and used as sources prior to the trip for a new expedition map included David Thomson's map of the Bend of the Missouri River. As noted on the Library of Congress website, Thomson's map "Gives number of warriors, houses and tents of six Mandan and Pawnee Indian villages in the vicinity of the junction of the Missouri and Knife Rivers."[13] And this map served as one of the sources for Nicholas King's map compiled in preparation for the journey. King not only incorporated Thomson's map data, but he also included additional information obtained from fur traders.[14]

Unfortunately, when William Clark extensively compiled spatial information from Native sources concerning other Indigenous groups' locations and populations for his first map in 1805, he seemed to have ignored, or was unaware of, a Native-made map by Too-Né, an Arikara chief, which was discovered by Christopher Steinke while conducting his doctoral research in the French National Library (via Thomas Jefferson). Steinke notes that it

> captures some of the Arikara history that Clark left unrecorded. Drawn sometime in 1805 or early 1806, it shows the course of the Missouri River, Lewis and Clark in council at the Arikara and Mandan villages, the locations of more than 30 different Indian groups, and significant places in the history of the Arikara people.[15]

The first map published from the Lewis and Clark expedition was a copy by Nicholas King (1805) of a sketch by Clark. This map included tribal village locations, as well as populations of warriors or men. The follow-up map to this was drafted by Samuel Lewis, who copied Clark's later drawing in 1814.[16] Like the earlier map, this map also depicted tribal village locations, but, in terms of populations, instead of merely noting warriors or men, now identified the numbers of souls, which included women and children. Nonetheless, both maps relied heavily upon Indigenous informants since many of the tribal locations and populations were located a long distance away from the expedition's outbound and return routes.

In 1823, Major Stephen H. Long published two maps of the Mississippi drainage as part of an account of an expedition from Pittsburg to the Rocky Mountains. The map of the western drainage covered the Great Plains and plotted not only numerous tribal villages, but also trails of the Pawnee and Osage. Regardless, he referred to the southern plains as the Great American Desert with the following note: "The Great Desert is frequented by roving bands of Indians who have no fixed places of residence but roam from place to place in quest of game." While these people may not have been sedentary, his failure to note the existence of more permanent residences farther east is questionable.[17]

Gero-Schunu-Wy-Ha (Oto) drew in 1825

> a map of events on the middle Missouri River between Council Bluffs and the Little Missouri River and also traces the route of an Oto war party that attacked the Arapahos in the area between the upper Arkansas and upper Cimarron rivers.

As Malcolm Lewis notes,

> The map illustrates the route of an Oto war party that attacked the Arapahos in the area between the upper Arkansas and upper Cimarron rivers. The events are depicted in typical Indian pictographic style against a network of rivers. The gross distortion of the network reflects the constraints imposed on the Indian by the rectangular sheet of paper. Even so, it is a remarkable map, covering about a third of a million square miles of the northern and Central American plains.[18]

George Catlin (1848) illustrated the Native American landscape in 1833. Most people think only about Catlin's portraits of American Indians and are not aware that he made this map. While he properly identifies tribal locations across the country, Catlin notes the "Hostile Ground" in the map's center (eastern Colorado and western Nebraska and Kansas). But no reason is given for that label.[19]

Prior to Texas becoming a republic, and later a state, it was relatively accurately mapped by David Burr in 1835, including adjacent areas of the United States and Mexico. This map's primary focus was to illustrate the expatriate American empresario land grants across the territory. Nonetheless, he also included the location of the short-lived Shawnee Reservation in Texas along the Red River.[20]

In 1836, several maps were printed that either looked back at former lands occupied by American Indians or looked forward to lands being ceded or assigned to tribes via treaties. The American Antiquarian Society published a map by Albert Gallatin, who drafted a continent-wide map of Indian territories, which was hand-colored to show the location of Indian tribes and 11 linguistic families. For the plains, he included most of the major Indian nations including the Assiniboine in the north to the Panis (Pawnee) in the south. While his ethnographic map depicted tribal lands as held in the east ca. 1600, for the rest of the continent, including the plains, territories were shown ca. 1800. Meanwhile, two other maps appeared from the Topographical Bureau detailing the lands ceded or set aside for the Indian nations. One was made by

William Featherstonhaugh, as a follow-up to an 1834 geological report of the areas around the present states of Iowa and Missouri, while the other was strictly created to illustrate the land cessions and assignments for the tribes in areas around Iowa, eastern Nebraska and Kansas, and Oklahoma as part of Col. Henry Dodge's expedition to the Rocky Mountains. This map included four tables of acreages of assigned lands to the eastern tribes as well as populations (with variable degrees of precision) of tribes both east and west of the Mississippi. These latter maps defined the governmental control of Native Americans in the central plains at that time.[21]

Non-Chi-Ning-Ga (Iowa) created a map in 1837 of tribal migrations in the Upper Mississippi and Missouri river basins. Malcolm Lewis reviewed this map in his 1984 paper with the earlier map by Gero-Schunu-Wy-Ha (Oto). For both maps, he redrafts the maps with locations and interprets and geo-references them against modern maps of the same areas showing how well the two Indigenous cartographers fared in their depiction of the hydrological landscape.[22]

During the 1840s, several maps of note were created. Joseph Nicollet was responsible for a hydrological map published in 1843 of the Upper Mississippi basin as part of the Fremont expedition as a project of the War Department in the 1830s. This map accurately illustrates the rivers and lakes, and landforms depicted with hachures. Importantly, he included both Indigenous and English names for many of the rivers and lakes while identifying more than a dozen resident Indian nation occupiers of the landscape with labels such as Ponka Indian Country, Yankton Country, Otoe Indian Country, and others.[23]

In 1844, Joshua Gregg published a book that included a map he drafted of Indian nation lands, reservations, hunting grounds, villages, pueblos, along with trade routes, migratory trails, military expeditions, forts and trading posts, missions, hydrology and topography. As Bernstein notes, Gregg heavily relied on Comanche Chief Tabba-quenna (Big Eagle), who drew him a map (that apparently was not preserved) which accurately delineated all of the primary rivers from Missouri to Santa Fe along with a number of Mexican settlements.[24]

Charles Preuss compiled a series of seven maps in 1846 from the Fremont expedition, at a scale of ten miles to the inch, along the Oregon Trail from the intersection of the Kansas and Missouri rivers to the Walla Walla and Columbia rivers. These maps could be used as a travelogue since they included meteorological observations taken at regular intervals along the trail, notes from Fremont's report, nearby Indian territories, and remarks about wildlife, timber, grass for livestock grazing, and Indian relations (which typically included warnings about attacks and theft).[25]

Pursuant to the Laramie Treaty, Father Pierre-Jean de Smet in 1851, delimited tribal territories of the central and northern plains, a portion of the Plateau, and identified some Great Basin tribes to the west as well. In the area north of Indian Territory includes a region with a number of tribes are listed: Pawnees in the west, and Ponkas, Omahas, Otoes, Iowas, and Kickapoos in the east along the Missouri River. As Cole and Sutton pointed out, "This map reflects, in many instances, a benign effort on the part of government to map with a pro-tribal perspective, although it was not very long before intrusions upon those mapped territories would take place."[26]

In 1853, an anonymous Assiniboine warrior drew a map of the northern Missouri River from Fort Union, North Dakota, to west of Fort Benton in north central Montana. The inscriptions by Edwin Denig on the map read:

> Map of the north side of the Missouri River from Fort Union, mouth of the Yellow Stone, to Fort Benton, mouth of the Maria, drawn by an Assiniboine warrior at Fort Union, December 27, 1853. The artist was not acquainted with the country on the south side of the Mo. The dotted line is their usual war path to the Blackfeet. Names of rivers & c. written under his direction and explanation.

This map is another example of the good spatial cognition held by Native Americans. And Warhus notes

> that the Assiniboine's map was not made to give a western oriented picture of the landscape; like much of the plains picture writing it was made to record the warpath of the anonymous brave making a raid on the Blackfeet in Montana.[27]

Near the end of that decade, the Lieutenant Gouverneur K. Warren of the Department of War drafted a comprehensive map of the military departments in the American West, including the departmental boundaries, colored flags to denote fort locations, Indian nation territories and reservations, routes and dates of expeditions and surveys, and known hydrology and topography (Figure 6.1). This map was created to prepare for the proposed trans-continental railroads yet to be built. As Bernstein states:

> Warren called it the 'Indian map' for obvious reasons. In what he considered the authoritative printing, a colorful patchwork covers the western United States, indicating which tribe(s) controlled which territory. With the exception of a few relocated groups along the Missouri [and in present-day Oklahoma], these regions did not correspond with treaties or any formal claims. Instead, they depicted the current geopolitical situation by acknowledging Indian control of the trans-Missouri west.[28]

After the Civil War, four maps appeared to further establish governmental and corporate control and religious influence in the western states. In 1866, the U.S. Army Corps of Topographical Engineers compiled and produced a map of Indian Territory and nearby areas illustrating the evolving nature or shrinkage of Indian country, reservations, and leased lands. In the following year, G. K. Warren of the Topographical Engineers drafted a map of Nebraska and Dakota region with Indian territories and reservations. The 1868 map by Joseph Gorlinski of the General Land office illustrated proposed trans-continental railroad routes, survey boundaries, mineral deposits, Indian reservations, states and territories. And in the same year, the G.W. & C.B. Colton Company published a national map of Indian tribes, reservations, and undefined territories marketed to mission groups that planned to administer to the tribes.[29]

Julius J. Durage, of the War Department, drafted a map of the Dakota Territory in 1878 depicting reservations, forts, roads, railroads, battle sites, and extensive hydrologic and topographic features. Foreshadowing coming events, the east side of the map shows the beginnings of the Public Lands Survey System into the area. One reservation that was notably included in the survey was the Sissiton and Wahpeton Reservation (now known as the Lake Traverse Res.), that became heavily checkerboarded during the allotment era.[30]

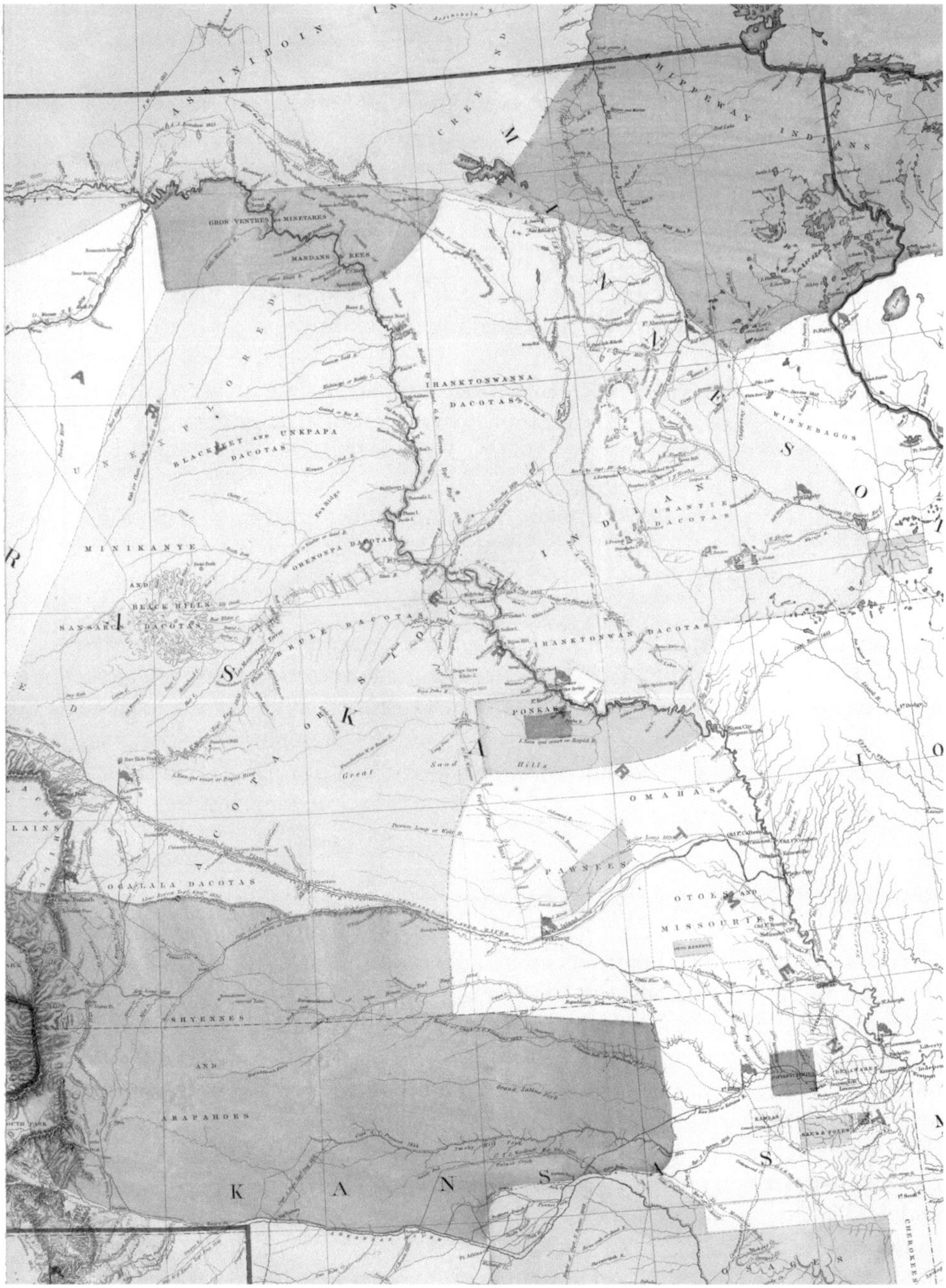


*Figure 6.1* Cropped portion of G. K. Warren's map (1858) of Indian lands in the northern and central plains.

Around 1880, John Crazy Mule, a Cheyenne scout, produced two maps commissioned by General Miles of the Army depicting his perspective of events in the upper Missouri basin. One map included two Indigenous attempts at survival: the Nez Perce and the Lakota, the former gleaned from imprisoned Nez Perce warriors regarding their escape route to Canada to north central Montana, and the latter concerning "Miles's invasion of Lakota leader Lame Deer's camp, he also documented the trail of abandoned camps the Lakotas left behind as they fled to North Dakota."[31] The second map covered Crazy Mule's personal experience with his fellow Cheyenne when they were forced from their homelands to Indian Territory (not shown) in 1878, moved again to Pine Ridge reservation in 1881, and finally to the Northern Cheyenne reservation during 1884 in Montana.

The dynamism of American Indian land tenure is seen throughout our country's history, and the 1880s were no exception. The Office of Indian Affairs (OIA) started making maps in that decade as part of its mission. Several nation-wide maps were published in 1883, 1885, and 1888.[32] While these maps show how Indian lands and reservations were being diminished, they also illustrated the Native American land base that existed prior to the Dawes Severalty Act of 1887. And in 1887 and 1889, the General Land Office and OIA, respectively, published two maps of Indian Territory. While these maps are stylistically different, the latter map certainly emphasizes Indian lands in the Territory both through its use of color and legend. After the passage of the Curtis Act of 1898, the five civilized Tribes were forced to be included in the allotment of their lands and regressive examples can be seen on maps of Creek lands in 1899 and 1902 by the Department of the Interior.[33]

The Smithsonian mapping efforts of plains Indians fell under the Bureau of American Ethnology. In 1892, John Wesley Powell, Director of the Bureau of American Ethnology (SI), published his map "Linguistic Stocks of American Indians North of Mexico" (The date is commonly given as 1891, the publication date on the title page). The culmination of a project to classify the Indians according to the relationships of their languages, this map provided a guidebook not only for linguistic study but also for the ethnographic description of the continent. Powell's classification recognized 58 language families, 26 of which consisted of but a single language, often poorly known, and these covered every square mile except for small areas along the coast of the Carolinas. In the following years, efforts were made to relate these families to each other. At the end of the decade, Charles C. Royce published his important work, *Indian Land Cessions,* which covered Indian land transfers including purchases, congressional actions, executive orders, as well as treaty and agreement cessions, from the period of 1784 to 1894. Nonetheless, several problems exist with the Royce maps: some polygons were drawn without survey input or from incompetent survey; he typically assumed that the public land survey systems consisted of straight township and range lines; edge-matching was often absent between state maps; he had poor geographic knowledge of drainage divides; he provided lack of acknowledgment of non-federal activities; and his work ended in 1894.[34]

Two Native Americans, Black Goose (Kiowa) and Amos Bad Heart Bull (Oglala Lakota), produced maps of their peoples' histories and interactions with the U.S. government. Black Goose's map is an interesting display covering both time and space (Figure 6.2). This pictographic map, probably made during the mid-1890s, contains camps, mountains, rivers, along with religious ceremonial and historical sites including locations of intertribal warfare, U.S. military forces, and subsistence

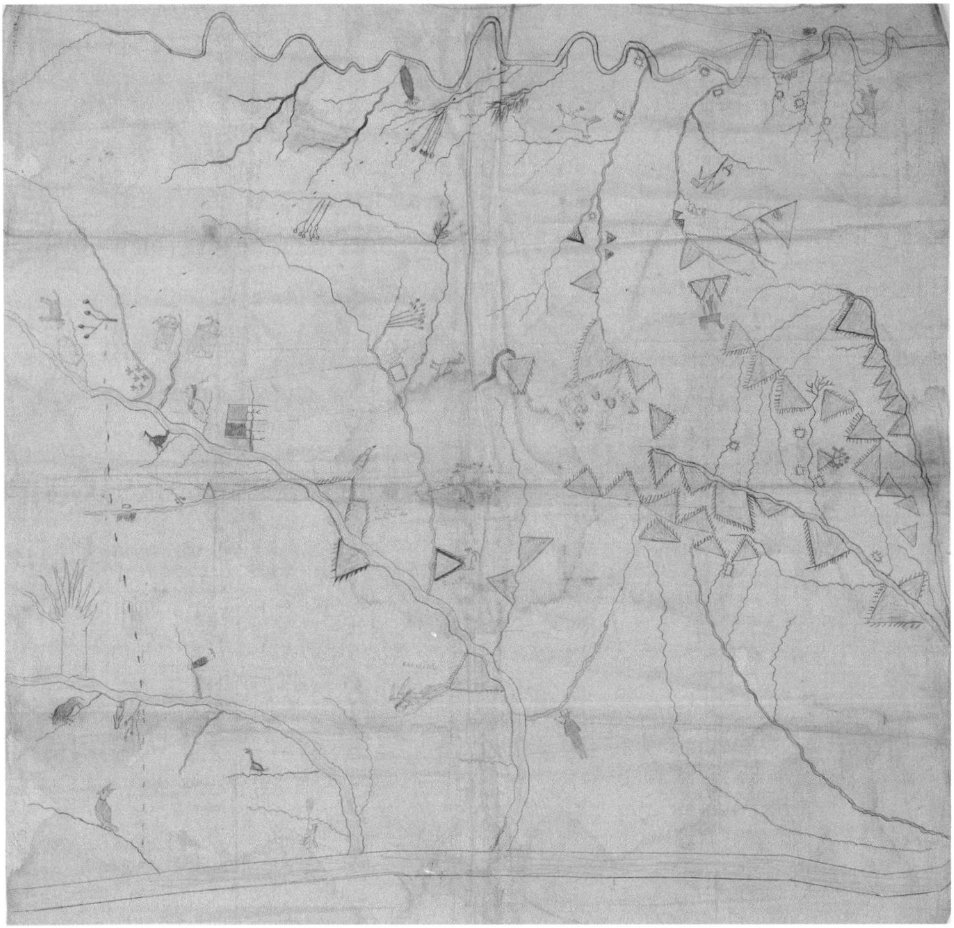

*Figure 6.2* Black Goose's pictographic map (ca. 1890s) of Kiowa political cartography (photograph by Don Hurlbert).

activity. Amos Bad Heart Bull was a prolific artist whose work in the years before and after 1900 was compiled and detailed by Helen Blish. Within that body of drawings included some pictographic maps of individual battle scenes at Little Big Horn, and fights with other Native nations such as the Crow and Cheyenne. Since some of these maps were of events before Bull was born, obviously, these were created out of descriptions that he acquired from oral traditions.[35]

## The Modern Era After 1900

An unsuccessful proposition to create the State of Sequoyah, mapped by D. W. Bolich, Muskogee, in 1905, that encompassed what was left of Indian Territory. Its significance is as a symbol of Indian resistance to statehood within Oklahoma Territory. Otherwise, the 20th century continued with the diminishment of Indian lands until the allotment era ended with the passage of the Indian Reorganization Act of 1934. In 1939, Sam Attahvich, of the OIA, drafted a nation-wide map of Indian tribes, lands, and settlements. On this map, he symbolized tribal lands in solid black, reservations allotted and open in cross-stitch, and reservations allotted in part in a diagonal line pattern. Since some reservations fall in both of the latter two categories, this can be spatially confusing to the reader. Additionally, his point symbology in the legend for colonies in Nevada and rancherias in California was significantly larger than what appeared on the map. Regardless, the map summarized the status of Indian lands at that time.[36]

In academia, Alfred Kroeber, an anthropologist at UC Berkeley, published in 1939, *Cultural and Natural Areas of Native North America*, within which he had mapped and delineated the Native American tribal groups that were frequently divided by watershed boundaries. And Sutton and Cole noted,

> Since Kroeber's focus was culture, he found that boundaries represented the weakest feature when mapping whole cultures. He spoke of tribes living along an 'interarea' boundary as having much in common, and he would have preferred a cultural map without boundaries. To be sure, many of the sources cited in Kroeber's work included references to Native informants, but ultimately his interpretations were those of the scholar.[37]

As more information was gathered about tribal territories in the following decades, William Sturtevant and Ives Goddard, of the *Handbook of North American Indians* office in the Anthropology department of the National Museum of Natural History, respectively published *Early Indian Tribes, Culture Areas, and Linguistic Stocks* (in 1967 as part of the National Atlas), and *Native Languages and Language Families of North America* (in 1999 as a revised insert to the *Languages* volume of the *Handbook*).[38]

On the government side, the Bureau of Indian Affairs (BIA) published a series of nation-wide maps titled *Indian Land Areas* starting in 1971. This map includes federal and state reservations, recognized tribes without trust lands, former reservations in Oklahoma, as well as terminated tribes between 1953 and 1970. This map was revised by me, and produced by the USGS (United States Geological Survey), with no terminated tribes included, in 1987 and 1989; and the BIA published a new version in 1992 with region boundaries. In 1998, the BIA's Geographic Data Services Center published *Indian Lands in the United States* map that was an example of a beautiful map that was hard to read: small reservations and off-reservation tribal trust lands are not much more than yellow speckles on the 3D background. In 2016, the BIA's Office of Trust Services, Branch of Geospatial Support (BOGS) produced a newer map of *Indian Lands in the United States*, with clearly visible off-reservation tribal trust lands depicted as tiny pink polygons. More recently, the BIA's BOGS published *U.S. Domestic Sovereign Nations: Land Areas of Federally Recognized Tribes*. But note in the plains, due to legal actions, the Lake Traverse Reservation has diminished borders, and the Osage Reservation no longer appears even though the Osage still maintain total subsurface rights under that reservation. While the BIA's online version is the same, it gives the user the ability to zoom, change base maps, and identify tribal lands with the BIA Land Area Representation (LAR). The LAR depicts the external extent of Federal Indian reservations, land held in "trust" by the U.S., "restricted fee" or "mixed ownership" tracts for Federally-recognized tribes and individual Indians.[39]

Other government agencies such as the Census Bureau and the Indian Health Service (IHS) have ongoing mapping efforts as well. The Census map, *American Indians and Alaskan Natives in the United States*, appeared after the 2010 census, and the agency plans to map the updated data of this topic after the 2020 census. In the last decade, the IHS had a printed map titled "A Culture of Caring" that illustrated IHS boundaries along with the locations of health centers, health stations, and hospitals throughout Indian and Alaskan Native country. That map has been replaced by another map produced by the Department of Veterans Affairs titled "Federal Health Care Facilities: Departments of Defense, Health and Human Services, Homeland Security and Veterans Affairs." But they also have an interactive map website with IHS boundaries, IHS headquarters, area headquarters, and IHS facilities.[40]

The U.S. Forest Service (USFS) has an interactive map depicting sessions from Royce (1899), present-day reservations, and National Forests and National Grasslands.[41] Unfortunately, the USFS map may be confusing to the average reader for two reasons: first, the present-day reservation boundaries don't match the Royce boundaries either because the reservations have been diminished or due to mapping errors by Royce, but the difference is not defined; and second, when a polygon is clicked on, the record that pops up gives the Royce tabular information, along with "Related American Indian Tribes" that at times includes some others that are not closely related while occasionally ignoring the present-day inhabitants. For instance, the Wind River Reservation in Wyoming only includes the Shoshone Tribe instead of the Eastern Shoshone and the Arapaho tribes, while rubric of "Related" tribes includes many Shoshone bands, tribes and communities in California, Idaho, Nevada and Oregon.

## Contemporary Indigenous Mapping and GIS

Indigenous mapping and GIS efforts during the past two decades has blossomed. Four editions of the *Tribal GIS* book have been published by Esri with a number of 1–2 page case studies on individual tribal activities.[42] For the plains, the most recent book includes discussions on community mapping at Fort Belknap, Fort Peck, and Northern Cheyenne reservations; mapping lead paint hazards in Standing Rock Reservation; spatial analysis of asthma in Fort Peck lands; GIS and emergency management services in the Chickasaw nation; and GIS education at Haskell Indian College.

An interactive map by Village Earth made allotments accessible to Pine Ridge Reservation residents. The Village Earth website notes that the Pine Ridge Land Information System "makes it possible for members of the Oglala Sioux tribe to: Search for individually allotted and Tribal owned trust lands using the Tract ID; View, print and share a web link for the boundaries of specific land tracts."[43]

## Conclusion

Overall, mapping of the Native American landscape, in the Great Plains, and elsewhere in North America, has been and continues to be dynamic. Indigenous people for the historic past have shown keen spatial knowledge of the landforms, rivers, lakes, mountains, sacred sites, hunting grounds, trade routes, the movement of wildlife, and so on, for much of their territories and adjacent areas. In their relations with other tribes, they were also aware of the contested territories involved. So there is little wonder that Euro-Americans relied heavily on American Indian input for their mapping efforts. And today, Native Americans continue to map and use GIS for their land and resource management, as well as planning for the future.

## Notes

1 See Jolliet's map of New France at http://www.encyclopedia.chicagohistory.org/pages/11574.html and Marquette's map (*Carte de la decouverte faite l'an 1673 dans l'Amerique Septentrionale*) at the Library of Congress (LoC). https://www.loc.gov/resource/g4042m.ct001908/.

2 Franquelin's 1682 map (*The Mississippi*) is from the LoC at https://www.loc.gov/resource/g4042m.ct000784/.

3 Franquelin's 1685 map (*Amérique septentrion.lle [i.e. septentrionale]: composée, corigée, et augmẽtée, sur les iournaux, mémoires, et observations les plus justes qui en ón'etes.tes en l'année 1685 & 1686, par plusieurs particuliés*) is from the LoC at https://www.loc.gov/resource/g3300.ct000667/.

4 De L'Isle's 1702 map (*Carte de la rivière de Mississipi: sur les mémoires de Mr. Le Sueur qui en a pris avec la boussole tous les tours et detours depuis la mer jusqu'à la rivière St. Pierre, et a pris la hauteur du pole en plusieurs endroits*) was drafted on five sheets (262×79 cm, sheets 57×83 cm). https://www.loc.gov/resource/g4042m.ct000665/.

5 De L'Isle's 1718 map (*Carte de la Louisiane et du cours du Mississippi…*) is from the LoC at https://www.loc.gov/resource/g3701s.ct003028/.

6 Moll's 1715 map (*This map of North America, according to ye newest and most exact observations is most humbly dedicated by your Lordship's most humble servant Herman Moll, geographer*) and 1720 map (A new map of the north parts of America claimed by France under ye names of Louisiana, Mississipi [i.e. Mississippi], Canada, and New France with ye adjoining territories of England and Spain: to Thomas Bromsall, esq., this map of Louisiana, Mississipi [i.e. Mississippi] & c. is most humbly dedicated, H. Moll, geographer) are from the LoC at: https://www.loc.gov/resource/g3300.ct007387/ and https://www.loc.gov/resource/g3300.ct000677/.

7 This sheet of two maps (*Carte physique des terreins les plus eleves de la partie occidentale du Canada*) by Ochagach and Buache is from the David Rumsey Map Collection (DRMC) at https://www.davidrumsey.com/luna/servlet/detail/RUMSEY~8~1~306194~90076573:Carte-physique-des-terreins-les-plu?sort=Pub_List_No_InitialSort&qvq=w4s:/who%2FBuache%25252C%2BPhilippe%25252C%2B1700-1773;sort:Pub_List_No_InitialSort;lc:RUMSEY~8~1&mi=13&trs=45. The original map by Ochagach is from The University of North Carolina Chapel Hill at http://rla.unc.edu/Mapfiles/BSH3/BSH%20B%204044-84.HMC.3.jpg.

8 Mitchell's 1755 map (*A map of the British and French dominions in North America, with the roads, distances, limits, and extent of the settlements, humbly inscribed to the Right Honourable the Earl of Halifax, and the other Right Honourable the Lords Commissioners for Trade & Plantations, by their Lordships most obliged and very humble servant, Jon. Mitchell*) is from the LoC at https://www.loc.gov/resource/g3300.np000009/.

9 Photographs of the east and west sides of De Lafora's 1771 map (*Mapa de la Frontera del Vireinato de Nueva Espana nuevamente construido por el Ingeniero*) are from the University of North Carolina Chapel Hill at http://rla.unc.edu/Mapfiles/HMC5/Guerra%20LM%208a-1a-a,51.HMC.5a.jpg and http://rla.unc.edu/Mapfiles/HMC5/Guerra%20LM%208a-1a-a,51.HMC.5b.jpg.

10 The Comanche map is from LoC at https://www.loc.gov/resource/g4322s.ct006428/ and the quote is from Viola, Herman J., "Smithsonian Expert Discusses Comanche Map." *Friends of the Geography and Map Division* 12, no. 4 (2014): 3. Translation of the Spanish

legend includes: A. Chief Ysampampi; B. 95 Comanche towards the anterior; C. 5 dead Apaches; D. 35 prisoners of the former of both sexes; E. 16 horses captured; F. 1 Comanche dead as a result of wounds; G. 6 wounded Comanche; H. 8 horses also wounded.

11 The 1795 map by Soulard and Kohl (*Idee topographique des hauts de Mississipi et du Missouri*) is from the LoC at https://www.loc.gov/resource/g4125.ct000683/ and the 1796 map by Collot and Tardieu is from the DRMC at https://www.davidrumsey.com/luna/servlet/detail/RUMSEY~8~1~1007~100110:Map-of-the-Missouri--of-the-higher-.
12 A copy of this map (*An Indian map of the Upper-Missouri, 1801*) with accompanying notes is from the LoC at https://www.loc.gov/resource/g4127m.ct000579/.
13 Thompson's map (*Bend of the Missouri River*) and the quote are from the LoC at https://www.loc.gov/resource/g4127m.ct000151/.
14 King's map (*Lewis and Clark map, with annotations in brown ink by Meriwether Lewis, tracing showing the Mississippi, the Missouri for a short distance above Kansas, Lakes Michigan, Superior, and Winnipeg, and the country onwards to the Pacific*) is from the LoC at https://www.loc.gov/resource/g4126s.ct000071/.
15 Steinke, Christopher. "'Here is my country': Too Né's Map of Lewis and Clark in the Great Plains." *William and Mary Quarterly* 71, no. 4 (2014): 589. Allegedly, Jefferson gave this map to the French.
16 A copy of this map (*A map of part of the continent of North America: between the 35th and 51st degrees of north latitude, and extending from 89 degrees of west longitude to the Pacific Ocean*) with accompanying notes is from the LoC at https://www.loc.gov/resource/g3300.ct000586/. The second map associated with the expedition (*A map of Lewis and Clark's track, across the western portion of North America from the Mississippi to the Pacific Ocean: by order of the executive of the United States in 1804, 5 & 6*) is from the LoC at https://www.loc.gov/resource/g4126s.ct000028/.
17 Long's map (*Country drained by the Mississippi: Western Section*) is from the LoC at https://www.loc.gov/resource/g4042m.ct002090/. See David Bernstein's discussion of this map in his book, *How the West Was Drawn: Mapping, Indians, and the Construction of the Trans-Mississippi West*. Lincoln: University of Nebraska Press, 2018, 138–160.
18 G. Malcolm Lewis discusses this map on pp. 98–100 in "Indian Maps: Their Place in the History of Cartography." *Great Plains Quarterly* (Spring 1984): 91–108. The map's source is the Cartographic and Architectural Branch, Record Group 75, Map 931, National Archives, College Park, MD.
19 Catlin's map (*Outline Map of Indian Localities in 1833*) is from the University of Cincinnati digital library at http://digital.libraries.uc.edu/luna/servlet/detail/univcincin~25~25~53995~102958:Outline-Map-of-Indian-Localities-in.
20 Burr's map (*Texas*) is from the DRMC at https://www.davidrumsey.com/luna/servlet/detail/RUMSEY~8~1~214~20053:Texas,-By-David-H-Burr-?sort=Pub_List_No_InitialSort&qvq=w4s:/who%2FBurr%25252C%2BDavid%2BH.%25252C%2B1803–1875;sort:Pub_List_No_InitialSort;lc:RUMSEY~8~1&mi=278&trs=281.
21 Gallatin's map (*Map of the Indian Tribes of North America about 1600 A.D. along the Atlantic & about 1800 A.D. westwardly*) is from the LoC at https://www.loc.gov/item/2002622260. Featherstonhaugh's map (A map of a portion of the Indian country lying east and west of the Mississippi River to the 46th degree of north latitude from personal observation made in the autumn of 1835 and recent authentic documents) is from the LoC at https://www.loc.gov/item/96683128/. The third map (*Map showing the lands assigned to Emigrant Indians west of Arkansas and Missouri*) is from the LoC at https://www.loc.gov/item/99446197/.
22 G. Malcolm Lewis, op. cit. The map's source is the Cartographic and Architectural Branch, Record Group 75, Map 821, National Archives, College Park, MD.
23 Nicollet's map (*Hydrographical Basin of the Upper Mississippi River from Astronomical and Barometrical Observations, Surveys, and Information*) is from the LoC at https://www.loc.gov/resource/g4042m.ct001419/. See the discussion of this map and related matters by Bernstein (2018: 172–193) concerning Nicollet and Fremont's work.
24 Gregg's map (*A Map of the Indian Territory: Northern Texas and New Mexico, Showing the Great Western Prairies*) is from the DRMC at https://www.davidrumsey.com/luna/

servlet/detail/RUMSEY~8~1~261609~5523519:A-map-of-the-Indian-territory---Nor?sort=Pub_List_No_InitialSort&qvq=w4s:/who%2FGregg%25252C%2BJosiah;sort:Pub_List_No_InitialSort;lc:RUMSEY~8~1&mi=0&trs=10. Gregg, Joshua. *Commerce of the Prairies or the Journal of a Santa Fe Trader during Eight Expeditions across The Great Western Prairies and Residence of nearly Nine Years in Northern Mexico*. New York: Langley, 1844. The quote is from Bernstein, David. *How the West Was Drawn: Mapping, Indians, and the Construction of the Trans-Mississippi West*. Lincoln: University of Nebraska Press, 2018, 9–10.

25 Preuss' maps (*Topographical map of the road from Missouri to Oregon, commencing at the mouth of the Kansas in the Missouri River and ending at the mouth of the Wallah-Wallah in the Columbia*) are from LoC at https://www.loc.gov/item/99446202/.

26 Cole, Daniel G. and Imre Sutton, "A Cartographic History of Indian-White Government Relations during the Past 400 Years." *American Indian Culture and Research Journal*, 37, no. 1 (2013): 32–34. De Smet's map (*Map of the upper Great Plains and Rocky Mountains region*) is from the LoC at https://www.loc.gov/item/2005630226/.

27 The Assiniboine map is from the National Anthropological Archives (NAA) at http://collections.si.edu/search/detail/ead_component:sova-naa-ms2600b1-ref19?q=map+of+the+north+side+of+teh+missouri+river&record=2&hlterm=map%2Bof%2Bthe%2Bnorth%2Bside%2Bof%2Bteh%2Bmissouri%2Briver&inline=true. See also Warhus, Mark. *Native American Maps and the History of Our Land*. New York: St. Martin's Press, 1997.

28 Warren's map of 1858 (*Map of the Territory of the United States from the Mississippi to the Pacific Ocean; ordered by Jeff'n Davis, Secretary of War to accompany the Reports of the Explorations for a Railroad Route*) is from the LoC at https://www.loc.gov/item/76695826/. The quote is from Bernstein, op. cit., 198.

29 The Army's map (*Indian Territory, with part of the adjoining state of Kansas, &c.*) is from the LoC at https://www.loc.gov/item/2011590003/. Warren's 1867 map (*Map of Nebraska and Dakota and portions of the states and territories bordering thereon*) is from the LoC at https://www.loc.gov/item/2006459247/. The General Land Office map (*Map of the United States and Territories, Showing the extent of Public Surveys and other details*) is from LoC at https://www.loc.gov/item/83691753/, and the Colton map (*Map showing the location of the Indian Tribes within the United States: Prepared to accompany the Manual of Missions*) is from LoC at https://www.loc.gov/item/98685156/. The quote is from Warhus, op. cit., 181.

30 Durage's map (*Dakota Territory*) is from the LoC at https://www.loc.gov/item/2007626713/. The quote is from Bernstein, op. cit., 198.

31 John Crazy Mule's maps are housed at the Joslyn Art Museum, Omaha, Nebraska, and the quote is from Lucchesi, Annita Hetoevehotohke'e. "'Indians Don't Make Maps': Indigenous Cartographic Traditions and Innovations." *American Indian Culture and Research Journal* 42, no. 3 (2018): 20. A more in-depth discussion of Crazy Mule's maps is found in Fredlund, Glen, Linea Sundstrom and Rebecca Armstrong, "Crazy Mule's Maps of the Upper Missouri, 1877–1880." *Plains Anthropologist* 41 (1996): 155.

32 The OIA's 1883 map (*Map showing Indian reservations with the limits of the United States: 1883*) by Paul T. Brodie, which actually only covered about ¾ of the country is from the LoC at https://www.loc.gov/item/2009579475/; the 1885 map (*Map showing the location of the Indian reservations within the limits of the United States and territories*) by Paul T. Brodie and Norris Peters is from the American Geographical Society Library Digital Map Collection at University of Wisconsin, Madison Library at https://collections.lib.uwm.edu/digital/collection/agdm/id/492/; and the 1888 map (*Map showing the location of the Indian reservations within the limits of the United States and territories*) by John H. Oberly is from the LoC at https://www.loc.gov/item/2009579476/.

33 The GLO's 1887 map (*Indian Territory: compiled from the official records of the records of the General Land Office and other sources under supervision of Geo. U. Mayo*) is from the LoC at https://www.loc.gov/item/98687105/, and the OIA's 1889 map (*Indian Territory: compiled under the direction of the Hon. John H. Oberly, Commissioner of Indian Affairs, by C. A. Maxwell*) is from the LoC at https://www.loc.gov/item/98687111/. The 1899 and 1902 maps (*Map showing progress of allotment in Creek Nation*) from the

Department of the Interior are from the LoC at https://www.loc.gov/item/2007627492/ and https://www.loc.gov/item/2007627513/. Quote is from Lucchesi, op. cit., 2018.

34 Powell's ethnographic map is from LoC at https://www.loc.gov/item/2001620496/. The entire 67 map series of Royce's Indian Land Sessions maps are from the LoC at https://www.loc.gov/resource/g3701em.gct00002 and one may browse by date, Tribe, and state/territory at https://memory.loc.gov/ammem/amlaw/lwss-ilc.html.

35 Black Goose's map is from the National Anthropological Archives at http://collections.si.edu/search/results.htm?q=%23E233091. For the full analysis of this map, see Meadows, William C. "'We want our land as it is': Black Goose's Map as an Example of Kiowa Political Cartography." In D. G. Cole and I. Sutton (eds.). *Mapping Native America: Cartographic Interactions between Indigenous Peoples, Government, and Academia; Volume III, Cartography and Indigenous Autonomy*. Charleston, SC: CreateSpace, 2014, 35–67. Amos Bad Heart Bull's work is comprehensively covered in Blish, Helen H. *A Pictographic History of the Oglala Sioux, 50th Anniversary ed.* Lincoln: University of Nebraska Press, 2017. Maps of the battle of the Little Big Horn are seen on pp. 215 and 256; battle of the Rosebud on p. 188; fights with the Crow on pp. 185–186, and 396; of the Black Hills region on pp. 287–288; and a fight with the Cheyenne on p. 391.

36 Bolich's map of the proposed *State of Sequoyah* is from the LoC at https://www.loc.gov/resource/g4021g.ct011168/. The map by Sam Attahvich (*Indian Tribes, Reservations and Settlements in the United States)* is from the LoC at https://www.loc.gov/item/2009579474/.

37 Kroeber's foldout map is on p. 254 of his book, which is a PDF from the UC Berkeley Library at http://digitalassets.lib.berkeley.edu/anthpubs/ucb/text/ucp038-001-002.pdf. For more analysis of Kroeber's and other academics' work, see Sutton, Imre and Daniel Cole. "Introduction: Cartography and the Academy: Science and Innovative Efforts at Mapping Native America." In D. G. Cole and I. Sutton (eds.). *Mapping Native America: Cartographic Interactions between Indigenous Peoples, Government, and Academia; Volume II, Cartography and the Academy*. Charleston, SC: CreateSpace, 2014, 14.

38 Sturtevant's map is from the LoC at https://www.loc.gov/item/95682185/ and Goddard's map is from Reddit.com at https://external-preview.redd.it/oODTOK44SJ7kw_dP-Xq_e6vumV1wQ4dx5c4evWCftqE.jpg?auto=webp&s = 64a0292b202522e652b784f72d87fd04be0da65e.

39 The BIA's 2016 map is at https://www.bia.gov/sites/bia.gov/files/assets/public/webteam/pdf/idc1–028635.pdf. Their 2018 map is at https://biamaps.doi.gov/bogs/staticmaps.html and the current interactive online map is at https://biamaps.doi.gov/indianlands/. Additionally, an interactive map of the Tribal Leaders directory in the U.S. at https://www.bia.gov/sites/bia.gov/libraries/maps/tld_map.html. The legal actions referred to include: 1) See Osage Nation v. Irby, 597F.3d 1117 (10th Cir. 2010) for where the reservation and disestablished status is discussed. The Supreme Court denied certiorari in 2011; and 2) See DeCoteau v. Dist. County Court for Tenth Judicial District, 420 U.S. 425, for how the reservation was reduced and the unalloted land was restored to the public domain. Other current BIA interactive web maps include the BIA's Tribal Resilience Program Awards map, which provides access to BIA's Tribal Resilience Award details and Tribal Resilience Resources for every tribe or tribal organization that has received an award. The map is searchable by location or name. The map includes regional boundaries for related federal agencies, such as USGS' Climate Adaptation Science Centers, as reference at https://biamaps.doi.gov/tribalresilience/. The BIA's Operational Map provides users with a national perspective of the current weather conditions, watches, warnings, and advisories affecting BIA resources at https://biamaps.doi.gov/operational/.

40 The Census Bureau's map is at https://www2.census.gov/geo/maps/special/AIANWall2010/AIAN_US_2010.pdf; and the Indian Health Service's interactive map is at https://www.ihs.gov/locations/.

41 The U.S. Forest Service's interactive map is at https://usfs.maps.arcgis.com/apps/webappviewer/index.html?id=fe311f69cb1d43558227d73bc34f3a32.

42 Sappington, N., with D. Gadsden, A. Taylor and C. Thomas (eds.). *Tribal GIS: Suuporting Native American Policies with GIS*. Redlands: Esri Press, 2008; Taylor, A., D. Gadsden, J. Kerski and H. Warren (eds.). *Tribal GIS: Supporting Native American Decision Making.*

Redlands: Esri Press, 2012; Taylor, A., with H. Guglielmo and C. Thomas (eds.). *Tribal GIS: Supporting Native American Policies with* GIS. Redlands: Esri Press, 2013; Taylor, A., D. Gadsden, J. Kerski and H. Guglielmo (eds.). *Tribal GIS: Supporting Native American Decision Making*. Redlands: Esri Press, 2017.

43 Village Earth's website (http://www.lakotalands.net/) connects to the interactive map of Pine Ridge IR at https://villageearth.maps.arcgis.com/apps/MapTools/index.html?appid=bdfebb55407447f1b6d3705c3fe7e0bf.

## References


Bernstein, David, *How the West Was Drawn: Mapping, Indians, and the Construction of the Trans-Mississippi West*, Lincoln: University of Nebraska Press, 2018.

Blish, Helen H. *A Pictographic History of the Oglala Sioux, 50th Anniversary ed.*, Lincoln: University of Nebraska Press, 2017.

Cole, Daniel G., and Imre Sutton. "A Cartographic History of Indian-White Government Relations during the Past 400 Years," *American Indian Culture and Research Journal*, *37*, 1 (2013): 5–77.

Fredlund, Glen, Linea Sundstrom, and Rebecca Armstrong. Crazy Mule's Maps of the Upper Missouri, 1877–1880, *Plains Anthropologist*, *41*, 155 (1996): 5–27.

Kroeber, A.L. *Cultural and Natural Areas of Native North America*. Berkeley: University of California Press, 1939.

Lewis, G. Malcolm. *"Indian Maps: Their Place in the History of Cartography," Great Plains Quarterly*, Spring 1984: 91–108.

Lewis, G. Malcolm. "Maps, Mapmaking, and Map Use by Native North Americans," in Woodward, D. and G. Malcolm Lewis, eds., *History of Cartography, Volume 2, Book 3, Cartography in the Traditional African, American, Arctic, Australian, and Pacific Societies*, Chicago: University of Chicago Press, 1998, pp. 51–182.

Lucchesi, Annita Hetoevehotohke. "Indians Don't Make Maps": Indigenous Cartographic Traditions and Innovations, *American Indian Culture and Research Journal*, *42*, 3 (2018): 11–26.

Meadows, William C. "'We want our land as it is': Black Goose's Map as an Example of Kiowa Political Cartography" in Cole, D.G. and I. Sutton, eds., *Mapping Native America: Cartographic Interactions between Indigenous Peoples, Government, and Academia; Volume III, Cartography and Indigenous Autonomy*, Charleston, SC: CreateSpace. pp. 35–67, 2014.

Rice-Rollins, Julie A. "The Cartographic Heritage of the Lakota Sioux," *Cartographic Perspectives*, *48* (2004): 39–56.

Ronda, James P. "A Chart of His Way: Indian Cartography and the Lewis and Clark Expedition," *Great Plains Quarterly*, Winter 1984, *4*: 43–53.

Ronda, James P. *Lewis and Clark among the Indians*, Lincoln: University of Nebraska Press, 2002

Sappington, N., with D. Gadsden, A. Taylor, and C. Thomas, eds. *Tribal GIS: Supporting Native American Policies with GIS*, Redlands: Esri Press, 2008.

Steinke, Christopher. "'Here is my country': Too Né's Map of Lewis and Clark in the Great Plains", *William and Mary Quarterly*, *71*, 4 (2014): 589–610.

Sutton, Imre, and Daniel Cole. "Introduction: Cartography and the Academy: Science and Innovative Efforts at Mapping Native America" in Cole, D.G. and I. Sutton, eds., *Mapping Native America: Cartographic Interactions between Indigenous Peoples, Government, and Academia; Volume II, Cartography and the Academy*, Charleston, SC: CreateSpace, 2014, pp. 9–20.

Taylor, A., D. Gadsden, J. Kerski, and H. Warren, eds. *Tribal GIS: Supporting Native American Decision Making*, Redlands: Esri Press, 2012.

Taylor, A., with H. Guglielmo, and C. Thomas, eds. *Tribal GIS: Supporting Native American Policies with GIS*, Redlands: Esri Press, 2013.
Taylor, A., D. Gadsden, J. Kerski, and H. Guglielmo, eds. *Tribal GIS: Supporting Native American Decision Making*, Redlands: Esri Press, 2017.
Viola, Herman J. "Smithsonian Expert Discusses Comanche Map," *Friends of the Geography and Map Division*, 12, 4 (2014): 3.
Warhus, Mark. *Native American Maps and the History of Our Land*, New York: St. Martin's Press, 1997.

# 7 Mapping with Indigenous Peoples in Canada

*D. R. Fraser Taylor*

## Introduction

The Geomatics and Cartographic Research Centre (GCRC) at Carleton University in Ottawa has been creating cybercartographic atlases with Indigenous communities in Canada for almost two decades. The initial approach, as outlined in *Cybercartography: Theory and Practice*,[1] described the theory and practice of cybercartography and how the concept originated, but at that time there were few examples of cybercartography being used in relationship to Indigenous mapping. Since that initial development phase, the GCRC has steadily increased the Indigenous mapping element of its work, and in 2014 published *Developments in the Theory and Practice of Cybercartography: Applications and Indigenous Mapping*.[2] In that volume, detailed descriptions of a number of Indigenous atlases were given. In 2019, a study entitled *Further Developments in the Theory and Practice of Cybercartography: International Dimensions and Language Mapping*[3] was released and again Indigenous mapping was a central element of the book. In addition to examples from Canada's North, examples of Indigenous mapping in Mexico, Brazil and Kyrgyzstan were described. The year 2019 also saw the publication of a description of some of GCRC's work on the topic of reconciliation with Indigenous Peoples—*Cybercartography in a Reconciliation Community: Engaging Intersecting Perspectives*.[4]

This chapter will outline some of the lessons learned from this ongoing process as well as describe the development of the innovative mapping and data management framework, Nunaliit, used to create the cybercartographic atlases. A cybercartographic atlas is quite different from a conventional atlas, and is a metaphor for all kinds of qualitative and quantitative information linked by location. These are interactive, multimedia, and multisensory online products. Over the years, the atlases have become increasingly community controlled and community generated, and telling stories from a community perspective is one of their central features. Storytelling is a key aspect of all Indigenous societies, and cybercartography's ability to tell these stories using sound and videos is one of the reasons it is so useful in Indigenous mapping, and why it has been so enthusiastically adopted by Indigenous peoples. Nunaliit means community in Inuktitut, and has been designed so that communities can produce their own atlases. It is an open source framework freely available to all users and can be learned by individuals with limited computer expertise in a matter of hours. Community ownership and control is of great importance, and creating the atlases often requires a decentralized and distributed data management approach. The process by which atlases are produced by Indigenous communities is equally if not more important than the atlases themselves, especially in relation to inter-generational interactions created in the communities.

## Cybercartography and Cybercartographic Atlases

Cybercartography was first introduced as a concept in 1997.[5] Since that time, both the theory and practice has evolved substantially, and much of this evolution has been the result of interactions with Indigenous communities.[6] In 2019, Cybercartography was redefined as

> … a complex, holistic, user centred process which applies location-based technologies to the analysis of all types of topics of interest to society and the presentation of the results in innovative ways through cybercartographic atlases. A cybercartographic atlas is a metaphor for all kinds of qualitative and quantitative information linked by location and displayed in innovative, interactive multimodal and multisensory formats.[7]

Atlases empower Indigenous communities to tell their own stories. Both mapping and storytelling are basic human instincts and are a central part of the holistic nature of Cybercartography.

Although the formal definition of cybercartography has changed, the six central ideas which underpin it are still relevant. These are:

- Individuals use all of their senses when observing what is around them. Cybercartography explores the use of all five senses, and is now moving into affective issues to include emotion. In Indigenous mapping where storytelling is of central importance, sound, especially narration, is being widely used, as well as is music.
- Individuals have different learning preferences and prefer teaching and learning materials in different formats. Cybercartographic atlases provide the same information in multiple formats. For Indigenous mapping, narration has proved to be the format best suited to the informal learning process of transmission of information from the elders, which is of central importance. For formal education in a school setting, vision and text are more popular, although selective use of narration is still important.
- Educational theory suggests that individuals learn best when they are actively rather than passively involved. Cybercartography engages and facilitates interaction. Cybercartographic atlases include a wide variety of representations of Indigenous community life, including art, music, place names, photographs, videos, ceremony, and socioeconomic and cultural activities, both past and present. Special attention is given to the design of user interfaces to facilitate interaction.
- The social media revolution has given people the power to create their own maps and narratives. The Nunaliit cybercartographic atlas framework is a data management framework that allows Indigenous communities to enter the information they consider important from a community perspective, which is often quite different from that which outsiders would consider important. The framework is open source, provides a built-in meta data structure for the

information, and does not require special knowledge to enter the information. After a few hours training, community members can learn. Mastering Nunaliit is no more difficult than using the ubiquitous smart phones now common in Indigenous communities, especially with young people.

- Many topics of interest to society are complex, and the same set of "facts" on issues of interest to Indigenous communities, such as environmental change and the health of species such as fish or caribou, are open to a number of interpretations. Cybercartography allows the presentation of different ontologies and narratives on the same topics without privileging one over the other. Indigenous communities want their knowledge and experiences to be treated as equally important to that of Western science, and cybercartography allows this to happen. Traditionally, the map was an authoritative source of information, and what was mapped and how that was represented lay in the hands of those producing the maps who were almost without exception from outside the community.
- As the earlier comment suggests, traditional cartography was supply driven. National mapping agencies supplied definitive and authoritative maps which decision makers and others used. Technological change has more recently allowed a demand approach. Cybercartography takes this one step further and empowers individuals and communities to create their own maps, including the choice of what to map or not map. Cybercartography democratizes mapping in new ways. Indigenous peoples, until recently, have often been largely "invisible" on maps or have been represented by others. Cybercartography gives voice to Indigenous peoples and other community groups both literally and metaphorically.

## The Nunaliit Cybercartographic Atlas Framework

Nunaliit means community in Inuktitut, and the name of this open source framework reflects the underlying principles of cybercartography as outlined in earlier Section. The Nunaliit Cybercartographic Atlas Framework (https://nunaliit.org) is an open-source, web-based development framework[8] and is the key software framework for the creation of cybercartographic atlases. The framework has been specifically designed to meet the needs of Indigenous communities although it does, of course, have many other applications (Figure 7.1).

Nunaliit uses location, especially maps, as a unifying framework to link all kinds of qualitative and quantitative information. It is a means of telling stories in a compelling fashion, and it can provide interactions with information in multiple forms. A curated collection of information to tell a story on a single web page is called a module, and a linked collection of such modules is a cybercartographic atlas. Nunaliit development has been driven by the needs of its users in a collaborative and iterative process that combines the way the framework has been developed with capacity building for the user community. This process is particularly important for mapping with Indigenous communities. Existing tools and off the shelf software rarely adequately meet user needs, so our approach has been to develop a software framework collaboratively with Indigenous communities which is flexible, extremely easy to use, user controlled and inexpensive, and which does not require programming skills to operate. Nunaliit has also been developed to have the interactive elements driven by data without the need to refactor these data elements for each different use. It uses

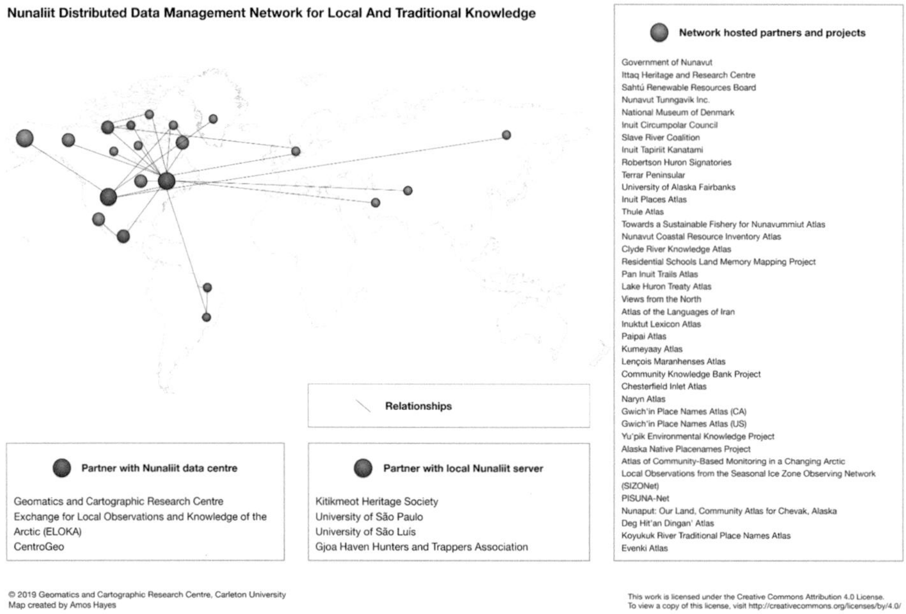


*Figure 7.1* Map of the Distributed Data Management Network for local and traditional knowledge.

open standards and is designed to work with data residing in different locations while facilitating new connections and the telling of new stories. Its development also reflects inter-project iterative development, where innovations in one atlas can lead to evolution in another. This, for example, is the case where the timeline development work on the Lake Huron Treaties Atlas[9] contributed to the geonarrative timeline technological development of the Thule Atlas.[10]

Nunaliit is fundamentally different from other interactive story driven systems now on the market. These systems, such as ESRI's Story Map, have embedded data and structures in the application logic which require coding skills to update with new information. It is rare to find such skills within many Indigenous communities, so the use of such systems including Google Mash Ups often leads to a loss of control on the part of the Indigenous communities concerned, as well as to a loss of control over the data used to create the map. Data received by Nunaliit are stored in a platform neutral way, allowing users to employ other tools without having to go through Nunaliit or have any knowledge of Nunaliit data structures. This is a significant difference from other open source data management systems like Mediawiki and Drupal, which require the brokering of all requests for information through their own programming interfaces. Nunaliit has been designed to avoid dependence on rigid proprietary data structures and to encourage maximum interaction with other systems. Nunaliit is very flexible and components can operate independently and can easily be integrated into particular applications. It uses the new BSD (Berkeley Software Distribution) open source license and emphasizes open standards for both data use and data sharing. Among its many attributes is its ability to ingest almost any kind of information. It can accommodate and use Indigenous languages, and this feature is of special value when mapping with Indigenous communities. Because it uses a flexible document-oriented data base, it has none of the rigidity of a system with a fixed schema approach. Often when mapping within an Indigenous community, the particular content requirements of that community emerge over time, and often include data types which may not have been expected in advance. For example, the (now archived) Atlas of Arctic Bay includes a rap video by local youth entitled "Don't call me Eskimo" whose lyrics are a compendium of the many issues facing young people in a small northern community. As Nunaliit continues to develop, new features are being added on a continuing basis. One of the more interesting recent developments has been the integration of the Garmin In Reach communication device with Nunaliit.[11]

As part of a project with the community of Gjoa Haven, a team of academic researchers and an Inuit community organization were working on research on sustainable fisheries in the region. They were using Nunaliit to digitize, preserve, and make available the mapped results of inventories on observed fish stock and related harvest information. As part of this research, a harvest study was carried out where hunters had Garmin In Reach GPS communication devices on their snow mobiles. In addition to locations, these devices have the ability to send positional information and messages via satellite to the vendors' website where they could they would be extracted manually from the onscreen information. A new function in Nunaliit was created to allow information to be directly fed into the atlas. In addition to the research values, the community could see where hunters were real time on the map and to respond to their messages by email.

Nunaliit continues to be built incrementally in cooperation with a number of Indigenous communities. Joint development of the software is accompanied by training and educational sessions. We have already had one session bringing twelve user communities and organizations together to learn from each other and to share experiences. Nunaliit documentation has been translated into Spanish and is in use by communities in Mexico[12] and plans are underway for its use by Indigenous communities in Brazil.[13]

Development of this innovative framework will continue in response to the needs of Indigenous communities and new ways will be found to tell their important stories. Unlike many information systems, Nunaliit uses a distributed data management system which allows communities to host and control their own information. It does not disappear into an anonymous cloud, controlled from outside. The centralized cloud-based systems are technologically very efficient, but a distributed data system is reflective of the strongly expressed desire by Indigenous communities to control their own information and to help avoid the misuse of that information, which has unfortunately been so common in the past. The important ethical and legal issues involved will be discussed more fully later in this chapter.

## Mapping with Indigenous Peoples—Some Examples

Over the last 15 years, the Geomatics and Cartographic Research Centre (GCRC) has been involved in the production of cybercartographic atlases in cooperation and partnership with Indigenous communities and organizations. These can be viewed in the GCRC website at https://gcrc.carleton.ca/index.html.

The communities and organizations involved include Arctic Bay, Sahtu, The Kitikmeot Heritage Society, Gwich'in, Gjoa Haven, Clyde River, Chesterfield inlet, Inuit Tapiriit Kanatami, the Nunavut Coastal Research Inventory of the Government of Nunavut, the National Centre for Truth and Reconciliation of Canada, Shingwauk Residential Schools Centre, and the Children of Shingwauk Alumnus Association, amongst others. In many instances, the concept of the atlases has been generated within the communities themselves, who have approached GCRC for help in creating an atlas to respond to their own perceived needs. The community often provides the funds for atlas creation and in most cases the resulting atlas resides on a community-controlled server with a mirror site at GCRC. When the community uses its own resources, it is an indication that the atlas is being produced as a result of community needs, rather than at the initiative of outsiders. This "bottom up" approach is an important element in the theory and practice of cybercartography as outlined in the previous Section and further elaborated in Taylor and Pyne[14] and Taylor, Anonby and Murasugi.[15] In addition to the community and organization websites, and the GCRC website indicated above, descriptions of many of these atlases can be found in Taylor and Lauriault.[16]

For the purposes of this chapter, a detailed description of the innovative approaches of the Kitikmeot Heritage Society (KHS) has been chosen, as it best illustrates how a community driven mapping process has developed over the last 15 years. This summary is based on the chapters by Keith, Crockatt, and Hayes.[17]

The KHS was incorporated in 1995 as a volunteer heritage organization with an elected board of directors primarily made up of Elders from Cambridge Bay in Nunavut, Canada. Cambridge Bay is the center of the Kitikmeot region with a mixed population, primarily Inuit beneficiaries of the Nunavut land claims agreement, which was the basis on which the Territory of Nunavut was created. Nunavut is one of three territories of the Canadian federation. Initially, KHS served the community of Cambridge Bay, but over time has expanded to include the whole region. In 2001, KHS helped build a community library and cultural center which acts as the headquarters of the Society and houses museum quality cultural and archaeological exhibits. "The KHS preserves, protects and celebrates the history, culture, language and diversity of the people of the Kitikmeot region."[18] The Society does this through a variety of means, including archiving the oral traditions and history of the Elders and a variety of other activities.

An important element in these activities is mapping. In 2000, the Society began the systematic collection of Inuit place names and land use features of the Cambridge Bay region by consulting with Elders and recording these on 1:250,000 topographic maps as well as video recording the stories related to each place. The KHS designed and printed its own 1:250,000 maps and distributed these to the various communities of the region. Keith, Crockatt, and Hayes comment,

> While these maps were very popular and useful ... they were not the right vehicle to ensure that all of the oral traditions collected during the project were made available to the public. Another form of cybercartographic product was needed to deliver on the KHS mandate to actively make available recorded oral traditions to promote their continued use by younger generations.[19]

In 2005, the KHS discovered Nunaliit and provided financial support to GCRC to build the Kitikmeot Place Names Atlas to respond to the goals and context set by KHS.

All of the goals of the atlas were met and include a continuous zoomable map, a talking map, a multimedia map, and a virtual visit. All of these functionalities were developed by GCRC in consultation with KHS. A Wiki map content was added to allow direct input from the community, and a dedicated atlas kiosk was established in the Community Centre to facilitate this. The atlas kiosk does not require knowledge of how to use a specific operating system or graphic user interface and is very easy to use. As a result of this function, the Kitikmeot Place Names Atlas is a "living atlas" which is constantly being updated by the community, who add new information and content. The decision on whether to finally incorporate the information provided is made by the KHS editorial board, which includes the Elders of the communities involved.

The atlas resides on a server based in the cultural center and can be viewed at the KHS website https://Kitikmeotheritage.ca. There is a mirror site at Carleton University and the content of both sites is coordinated so updates take place at both sites in almost real time. It is important to KHS that the atlas resides in the community where it can be easily accessed. Band width and connectivity in the Canadian North, although improving, are still problematical, and it is important that the atlas be readily available to the Inuit communities it serves. A distributed data base approach is more effective in this respect than the technically more elegant cloud-based approach. Keith, Crockatt, and Hayes comment,

> The Kitikmeot Place Names Atlas is an important example of how community-based heritage groups and academic institutions can collaborate to realize advancement in how technologies can be used by Inuit and First Nations to preserve and promote their language and culture. Through its collaboration with the GCRC the KHS has been able to accomplish one of its fundamental goals, which is to mobilize the products of cultural preservation projects so that they can contribute to the ongoing transmission of Inuit knowledge.[20]
>
> KHS has been able to develop an interactive, multimedia cartographic product that allows for the transmission of Inuit traditional knowledge through the voices of Inuit knowledge holders themselves. In the Kitikmeot Place Names Atlas, we see the beginnings of an Inuit adaptation of the technology of (cyber) cartography as an authentic vehicle for the transmission of place related traditional knowledge.[21]

Since 2014, the KHS has continued to add content to the atlas, and has now expanded it and renamed the atlas as the Atlas of Inuit Place Names to reflect the fact that the atlas is being expanded to include place names from across the Arctic, not just the Kitikmeot region.

In 2014, KHS began a new and innovative mapping venture in cooperation with the GCRC and the National Museum of Denmark to produce the Thule Atlas.[22] Between 1921 and 1924, a Danish/Greenlandic expedition led by Knut Rasmussen completed the first comprehensive recording of Inuit societies in Canada. The expedition collected a vast amount of information in a variety of forms, and is a unique and comprehensive record of Inuit life at a time when Inuit were still living a pre-Christian world view and material lifestyle. The year 2021 will mark the 100th anniversary of Rasmussen's Fifth Thule Expedition, and KHS was looking for a way to commemorate this event. They chose to develop an interactive multi-media atlas called the Thule Atlas to digitally return the wealth of cultural knowledge that resides predominantly in institutions in Denmark to contemporary Canadian Inuit. To do this, KHS formed a partnership with the National Museum of Denmark and the GCRC. The National Museum of Denmark agreed to provided information from their large collection of documents and artifacts on the Fifth Thule Expedition, and discussions and interaction on the digital return issue continue. There have been two visits of KHS researchers to Denmark, one of which included Inuit Elders. GCRC researchers have also visited Denmark to discuss technical transfer issues.

KHS outlined the four central requirements they needed to meet the project's goals. These were:

1. Provide digital access to Inuit knowledge gathered on the expedition;
2. Provide opportunities for Inuit to verify and enhance knowledge collected by the expedition;
3. Link the results of contemporary research and Inuit experiences to expedition findings;
4. Create opportunities for Nunavummiut to interact with expedition objectives and environments in augmented reality environments.[23]

The Nunaliit Cybercartographic Atlas Framework described earlier in this chapter was the foundation of the approach developed to meet the four objectives. It tells stories and highlights relationships between multiple types of information and sources[24] using location to connect and interact with information sources. The fact that Nunaliit does not have a fixed schema was particularly important to this project. This allows the flexible organization of data and results in a large non-linear web of knowledge which can be presented in many ways. "This unique design becomes particularly important in terms of a desire to create a database of Inuit knowledge that is actually amenable to Inuit structures for storing, teaching and using knowledge."[25]

The interface design of the atlas was created to meet the needs of different audiences including Inuit Elders, community members, and academic researchers. This resulted in three different but interconnected interface forms as shown in Figure 7.2.

The first interface allows users to access all information cartographically; the second provides access to interactive PDF versions of all published Fifth Thule Expedition reports. A parallel version of these reports has been designed to allow community-based input whether this be a photograph, a book page, or a transcribed song. This allows the amendment of the expedition information and placing it in a contemporary Inuit context by uploading videos, songs, and other material. The third kind of interface allows users to access information by subject categories such as oral traditions, place names, photographs etc. This interface consists of a series of visual tools and is useful for research users and others interested in a specific topic. This interface exists as a series of visual 'tiles,' which can be navigated according to the user's specific interests.

The Thule Atlas (www.Thuleatlas.org) is a work in progress. It is close to meeting one of its original goals of digitally returning the information on the Copper Inuit held in the collections of the National Museum of Denmark to the people of Nunavut. In this respect the partnership with the National Museum of Denmark is of great importance. The Museum, KHS ,and GCRC are breaking new ground with innovative new partnership approaches. In December 2017, a delegation from KHS, including two Elders, visited the Museum to study the collections. The Elders were able to provide information on many objects which they had seen used in their

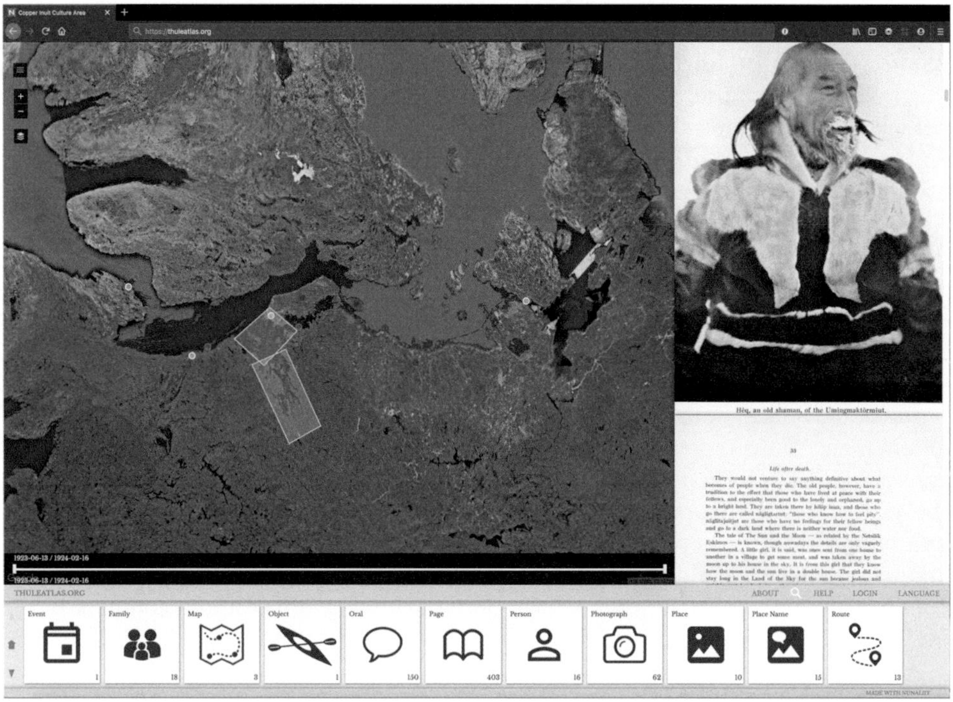


*Figure 7.2* Select categories of knowledge tiles available to be explored by Atlas users.[26]

childhood by their parents and grandparents. The Elders are "living archives," and their involvement is of great value to the project. In June 2018, the GCRC hosted a joint workshop with KHS and the Museum during which a new ethnographic collections module was designed.

Digital return is becoming increasingly popular in the Museum sector, but the Fifth Thule Atlas is a unique example. The atlas is designed to guide Inuit with access to the knowledge collected in the expedition, but also goes well beyond this function by enabling Inuit to verify and enhance this knowledge through the contributions and comment functions in the atlas design. The atlas also enables Inuit to interact with the exhibition objects, albeit in digital form, and to reacquire skills lost as a result of the colonial experience the Inuit have faced since the Thule Expedition took place. Keith, Griebel, Gross, and Jorgenson comment,

> In the Fifth Thule Expedition Atlas, the KHS and its partners have developed an innovative test of digital return that both provides access to Inuit knowledge recorded among the Knowledge holders of the 1920s, and creates a space for commentary and discussion of contemporary Inuit concerning the correction or reinterpretation of expedition findings. If Inuit choose to adapt the online application as a primary resource, it holds potential to significantly expand the body of traditional knowledge available to contemporary Inuit to be selectively integrated into their emerging expressions of their language and culture.[27]

## Legal and Ethical Issues of Mapping With Indigenous Peoples

Legal and ethical issues related to mapping are of central importance to all kinds of relationships with Indigenous peoples, and these issues are receiving increasing and much needed attention. In Canada these issues have been addressed by strict ethical guidelines created by academic research funding agencies and others, including several Indigenous organizations and Indigenous communities who have developed their own consent guidelines. As a result, the situation where traditional knowledge of Indigenous peoples was appropriated and often misused by outside researchers has improved. There are numerous examples of such unethical behavior by researchers, and although this risk has been reduced by heightened awareness and tightened control and regulations in recent years, it has not been completely removed. One of the major problems is that traditional knowledge is rarely protected by national legal systems. Despite U.N. recognition current intellectual property, copyright and traditional law are totally incapable of protecting traditional knowledge. As Ogumanan[28] has pointed out, such laws are based on Western conceptions of what constitutes innovation and creativity, and are based on individual ownership. Given the legal gap in this respect, the solution must lie in "soft law" and ethical guidelines and permissions, and even here there are many problems. Many of these guidelines and requirements were developed before the internet era, and issues such as informed consent and withdrawal of that consent are much more difficult and complex in the on-line mapping environment.

A comprehensive consideration of all of the issues involved is beyond the scope of this chapter, but they include consent, data storage, data access, data ownership, and appropriate dissemination tools. The GCRC has given all of the issues involved very careful consideration. Details of the issues and approaches used are given in Browne and Ljubicic.[29] A discussion of the approaches used by GCRC to address these challenges is given here.

Browne and Ljubicic point out that the root of the problem is a power imbalance between Indigenous peoples and the outside researchers involved in the mapping process. Building a working relationship of mutual trust with the Indigenous communities concerned is an important starting point, and to do this well takes time. It is an important and integral part of the process by which GCRC atlases are created. Indigenous communities have too often been the objects of research, and GCRC is attempting to address this imbalance by finding ways to make them the subjects of such research endeavors. The most effective approach, as illustrated by the description of our relationship with KHS given earlier, is to redress the power imbalance and give the communities control of the mapping process. Capacity building required to achieve this is an integral part of the mapping process. GCRC has carried out several training workshops with KHS in both Cambridge Bay and Ottawa. The software and equipment required to both produce and display the map including a local server has been installed in the cultural center in Cambridge Bay. The training workshops also included another important aspect of the atlas creation process, which is the intergenerational transfer and sharing between the elders, who are the main knowledge holders, and young people in the community. The elders are rarely interested in using the computer to create maps, but they are eager to pass on their knowledge to the younger members of the community. Young people are more interested in the technology, and as a result the mapping process is mutually rewarding.

Consent, data storage, data ownership, and appropriate dissemination are all in the hands of KHS, as is the choice of what to map. KHS has a clear idea of what it wants to achieve in the mapping process and it controls the agenda. GCRC's role is to respond to this agenda and to make suggestions on how it can be achieved. Decision making in this respect is joint. KHS also provides the funding to GCRC to carry out specific tasks, and this is a feature in almost all of the Indigenous ceremony atlases produced by GCRC. Scassa and Taylor observe that "These atlases in some instances are the first official recording of the aurally transmitted knowledge and elders and communities have authoritatively endorsed each record. The communities who have contributed to and authorized them regard the atlases as living archives."[30] This is certainly the case with KHS, which sees preservation of Inuit knowledge as a key aspect of its mandate.

In many ways having atlases produced by a community organization like KHS has advantages over others produced by individual communities. The Heritage Society provides an institutional continuity which is important for the ongoing creation of the atlases, and community processes which are used by KHS to create content can be continued over time, helping resolve the issue of the need for consent by future contributors to the atlases.

Preservation and archiving of the interactive atlases over time remains a challenge. The GCRC approach to the challenge is outlined by Taylor and Lauriault,[31] where archiving is seen as an integral part of the life cycle of atlas creation. This proactive archiving approach holds great promise to ensure that the atlases will continue despite changes in technology which result in many websites becoming unusable over time. With effective archiving being considered as an integral part of the atlas creation and updating process, the likelihood of preservation is much improved, but still faces challenges.

Many Indigenous communities assume that they own the data included in the maps produced. This is rarely the case. A digital multimedia atlas is a "creation" and is automatically protected by copyright, so what is protected is the expression of that information, not the source. This distinction is not always fully appreciated. If an outside organization produces a map or atlas, then the copyright is owned by that organization, and this can include the content. The community has de facto given up ownership of their own data. All GCRC atlases include a statement that the data in the atlases are owned by those contributing it. In the case of KHS, the copyright for the atlases lies with the society, and the atlases have their own URL on the KHS server. Mirror images are on the GCRC site, but the copyright and data ownership situations are clear.

Licensing is also a potential source of protection for Indigenous communities, and several communities and organization are defining their own licensing procedures. In Nunavut, for example, the Nunavut Research Institute is responsible for licensing all research projects. The GCRC has consulted widely with organizations and agencies responsible for the process, and has collaborated with the Canadian Internet and Public Policy Interest Clinic (CIPPC) in a report on licenses for local and Traditional Knowledge (LTK) in the context of digital cartography.[32] These important issues are now receiving international attention. In May 2019, the World Intellectual Property Organization (WIPO) met in Iqualit at a meeting organized by the federal government and Nunavut Tunngavik Inc. to identify and address gaps where the law inadequately deals with traditional knowledge. This builds on the substantial work on this topic in

Australia by Mukurtu content management system (CMS) (http://Mukurtu.org) and also includes consideration of an interactive on-line licensing system.

## Some Lessons Learned

Mapping *with* Indigenous people is very different from mapping *for* Indigenous people. The KHS example given in the previous section illustrates the importance of the process by which the maps were created, and is a feature of almost all the atlases which GCRC has helped create over the years. This is much more than simply the use of technology to produce maps. Partnership and trust take time to develop, and to be successful, this time must be productively spent and requires a high degree of human interaction.

Mapping must respect and be driven by community needs. What is to be mapped and how it is to be mapped should be decided at the Indigenous community level. Often their desires can be a surprise to outsiders. One example of interest in this respect is the now archived cybercartographic Atlas of Arctic Bay. Here we learned that what was important to the community was quite different than what GCRC researchers expected: the small community of Arctic Bay decided that they wanted to include topics such as the important sled dog races in addition to the rap video. Three young children wanted to be sure that "Dead Dog Lake," which was their fishing lake, was included in the atlas. Frank Street was added as the first street name in Arctic Bay by a Frank who named the street on which he lived. Sometimes topics are controversial. Mapping systems must be flexible enough to ingest a variety of different forms of information and systems with fixed schema. Commercial mapping programs are rarely able to do this. The ability to listen is of great importance in community driven mapping, and GCRC is responsive to local needs. Note that many components of the Arctic Bay Atlas are now incorporated in a larger atlas, the Inuit Places atlas, that is bringing together the traditional place names of Inuit Peoples: https://inuitplaces.org.

The ideal situation is where the community can produce its own atlases. Nunaliit is designed to be used by individuals who have no special knowledge of geographic information processing systems or techniques. Although Nunaliit is very easy to use, capacity building is still required, and workshops to achieve this need to be included in the atlas creation process. Community dynamics are important. The main knowledge holders in many communities are the Elders. Few Elders are familiar with computer technology, but they are anxious to share their knowledge with younger generations. Young people are keen to learn from the wisdom of the Elders and at the same time are comfortable with computer technologies. This has proven to be an ideal situation for the creation of atlases, as is illustrated by the examples given in this chapter. The ontologies of Indigenous peoples are quite different from those of Western science. These ontologies and the knowledge they contain and reflect are a parallel system of knowledge which must be accepted as such. It is not enough only to include traditional knowledge when it helps illustrate or expand on Western "scientific" research. In working with the Inuit, for example, it is critical to fully appreciate Qaujimajatuqangit, the term used to describe Inuit epistemology or the Indigenous knowledge of the Inuit. This requires an ability by Western researchers to listen, with some authorities emphasizing the need for "deep listening" in this respect. Semantics are challenging here as English translations of Indigenous concepts often do note capture the full meaning. Humans are only one actor in the complexity of

Indigenous concepts of environment and not the center of that concept as they are in Western science.

## Conclusion

Oral traditions and storytelling are both central to Indigenous societies. Mapping, which is only visual in nature, rarely captures this effectively. At a minimum, effective "mapping" must include an ability to capture and tell these stories. The multimedia and multisensory nature of cybercartography is valuable in this respect. The ethical and legal issues surrounding the mapping of traditional knowledge are implicit. As argued in this chapter, an effective response requires a change in the current unequal power relationships between Indigenous communities and Western mapping "experts." At a minimum, this requires the development of an equal power sharing relationship. A better situation is one where decision making power resides with a community organization such as the example of KHS given in this chapter. Such an approach has the advantage of helping resolve many legal and ethical issues, such as the important element of informed consent. Mapping is a means of preserving traditional knowledge. Many of the knowledge holders are Elders who are the "living metadata" of Indigenous mapping exercises. The traditional knowledge of individual Elders is often lost at their death. Preservation and archiving takes on new urgency in such circumstances. Technology is not an objective, value-free concept. In mapping with Indigenous communities, the most technologically elegant choice is not always the best choice. Many Indigenous communities like to see and control their own data, making concepts such as central computing capacity not acceptable for social rather than the technological reasons. Decentralized community centered approaches seem to work best. One shot Indigenous mappings are of limited value. A "Living Atlas" approach in which the community takes responsibility for the continuous updating of atlas content has merit. Ease of use is key. Systems which depend upon content expertise with an "expert" between the community and the map are less sustainable. Systems which accommodate Indigenous languages are especially useful.

The GCRC has been working in partnership with Indigenous peoples for over 15 years. The results of this partnership have been mutually beneficial, especially in relation to the development of theory and practice of cybercartography and the development of the Nunaliit framework used to create the atlases, and the demonstrated innovative benefits to communities that application of these to community needs have brought. This is an ongoing process and a mutual learning experience. A recent example of this process is the Clyde River Knowledge Atlas https://clyderiveratlas.ca. Jaypoody, Kautjk and Fox[33] are using Nunaliit in an outstanding example of a community driven and controlled atlas.

In June 2019, the Senate of Canada released a groundbreaking report entitled "Northern Lights: A Wake Up Call for the Future of Canada."[34] The report does not consider mapping in particular, but does look at research, traditional knowledge and the importance of community involvement and control. The report contains a number of recommendations for action by the Government of Canada. It proposes the wider adaptation of many of the approaches described in this chapter, already in use in mapping with Indigenous Peoples. Perhaps new approaches to mapping can lead to the more fundamental changes required in relationships with Indigenous Peoples worldwide.

## Notes

1 Taylor, D. R. F. (ed.). *Cybercartography: Theory and Practice.* Amsterdam: Elsevier, 2005, 574.
2 Fraser, D. R. F. and T. P. Lauriault (eds.). *Developments in the Theory and Practice of Cybercartography: Applications and Indigenous Mapping.* Amsterdam: Elsevier, 2014.
3 Taylor, D. R. F., E. Anonby and K. Murasugi (eds.). *Further Developments in the Theory and Practice of Cybercartography.* San Diego: Elsevier, 2019.
4 Pyne, S. and Taylor, D. R. F. (eds.). *Cartography in a Reconciliation Community: Engaging Intersecting Perspectives, Volume 8.* Amsterdam: Elsevier, 2019.
5 Taylor, D. R. F. "Maps and Mapping in the Information Era." In L. Ottoson (ed.). Proceedings of the 18th ICA International Cartographic Conference (ICC 97), Stockholm, Sweden, Vol. 1. Gavle: Swedish Cartographic Society, 1997, 3–10.
6 Taylor, D. R. F. and T. P. Lauriault, *Developments in the Theory and Practice of Cybercartography*, 2014.
7 Taylor, D. R. F. In D. R. F. Taylor, E. Anonby and K. Murasugi (eds.). *Further Developments in the Theory and Practice of Cybercartography*, San Diego: Elsevier, 2019.
8 Hayes, A., J. P. Fiset and P. L. Pulsifer. "The Nunalaliit Cybercartographic Atlas Framework." In D. R. Fraser and T. P. Lauriault (eds.). *Developments in the Theory and Practice of Cybercartographic Applications and Indigenous Mapping.* Amsterdam: Elsevier, 2014; Hayes, A. and D. R. F. Taylor. "Developments in the Nunaliit Cybercartographic Framework." In D. R. F. Taylor, E. Anonby and K. Murasugi (eds.). *Further Developments in the Theory and Practice of Cybercartography.* San Diego: Elsevier, 2019.
9 Pyne, S. "Lake Huron Treaty Atlas." https://gcrc.carlteton.ca.
10 Keith, D., B. Griebel, P. Gross and A. M. Jorgenson. "Digital Return of Inuit Ethnographic Collections using *Nunaliit*." In Taylor, D. R. F., E. Anonby and K. Murasugi (eds.). *Further Developments in the Theory and Practice of Cybercartography.* San Diego: Elsevier, 2019.
11 Hayes and Taylor, "Developments in the Nunaliit Cybercartographic Framework," 2019.
12 Andrade-Sanchez, J. A. and R. Eaton-Gonzales. "Cybercartography as a Transdisciplinary Approach to Solve Complex Environmental Problems: A Case Study of the Kumeyaay Peoples of Baja California and the Conservation of Oak Trees." In D. R. F. Taylor, E. Anonby and K. Murasugi (eds.). *Further Developments in the Theory and Practice of Cybercartography.* San Diego: Elsevier, 2019.
13 Machado, R. P. P. and U. D. Vieira Souza. "The Potential of Cybercartographic Atlas for Lençóis Maranhenses National Park, State of Maranhão, Brazil." In D. R. F. Taylor, E. Anonby and K. Murasugi (eds.). *Further Developments in the Theory and Practice of Cybercartography.* San Diego: Elsevier, 2019.
14 Taylor, D. R. F. and S. Pyne. "The History and Development of the Theory and Practice of Cybercartography." *International Journal of Digital Earth* 3, no. 1, 2010.
15 Taylor, Anonby and Murasugi, *Further Developments in the Theory*, 2019.
16 Taylor and Lauriault, *Developments in the Theory and Practice of Cybercartography*, 2014.
17 Keith, D., K. Crockatt and A. Hayes. "The Kitikmeot Place Names Atlas." In D. R. F. Taylor and T. P. Lauriault (eds.). *Developments in the Theory and Practice of Cybercartography: Applications and Indigenous Mapping.* Amsterdam: Elsevier, 2014, 227.
18 Keith, Crockatt, and Hayes, "The Kitikmeot Place Names Atlas," 2014, 220.
19 Keith, Crockatt and Hayes, "The Kitikmeot Place Names Atlas," 2014, 221.
20 Keith, Crockatt and Hayes, "The Kitikmeot Place Names Atlas," 2014, 226.
21 Keith, Crockatt and Hayes, "The Kitikmeot Place Names Atlas," 2014, 227.
22 Keith, Griebel, Gross and Jorgenson. "Digital Return of Inuit Ethnographic Collections using *Nunaliit*," 2019.
23 Keith, Griebel, Gross and Jorgenson, "Digital Return of Inuit Ethnographic Collections using *Nunaliit*," 2019, 308.
24 Hayes and Taylor, "Developments in the Nunaliit Cybercartographic Framework," 2019.
25 Keith, Griebel, Gross and Jorgenson, "Digital Return of Inuit Ethnographic Collections using *Nunaliit*," 2019), 308.
26 https://gcrc.carlteton.ca.

27 Keith, Griebel, Gross and Jorgenson, "Digital Return of Inuit Ethnographic Collections using *Nunaliit*," 2019, 313.
28 Oguamanan, C. *Intellectual Property in Global Governance*. New York: Routledge, 2011.
29 Browne, T. Di L. and G. J. Ljubicic. "Considerations for Informed Consent in the Context of Online, Interactive Atlas Creation." In D. R. F. Taylor and T. P. Lauriault (eds.). *Developments in the Theory and Practice of Cybercartographic Applications and Indigenous Mapping*. Amsterdam: Elsevier, 2014.
30 Scassa, T., T. P. Lauriault and D. R. F. Taylor. "TK in International Law." In D. R. F. Taylor and T. P. Lauriault (eds.). *Developments in the Theory and Practice of Cybercartography: Applications and Indigenous Mapping*. Amsterdam: Elsevier, 2014, 283.
31 Taylor and Lauriault, *Developments in the Theory and Practice of Cybercartography*, 2014.
32 CIPPC and GCRC (2016). https://cippic.ca/sites/default/files/file/CIPPIC_GCRC-TK_License_Proposal-July_2016.pdf.
33 Jaypoody M., R. Kautjk, S. Fox, P. Pulsifer and A. Hayes. "Clyde River Atlas: Linking Inuit Knowledge and Technology to Mobilize Knowledge at the Community Level" (Presentation to Artic Net Conference, Ottawa), December 2018.
34 https://www.sencanada.ca/en/info-page/parl-42-1/arct-northern-lights/.

## References

Browne, Timothy Di L. and Gita J. Ljubicic. "Considerations for Informed Consent in the Context of Online, Interactive Atlas Creation." In *Developments in the Theory and Practice of Cybercartography: Applications and Indigenous Mapping*. Edited by D. R. Fraser Taylor and T. P. Lauriault. Amsterdam: Elsevier, 2014.

Canadian Internet and Public Policy Interest Clinic (CIPPIC). *Licenses for Local and Traditional Knowledge in the Context of Digital Cartography*. Ottawa: CIPPIC and GCRC, 2016.

Engler, Nate J., Scassa, Teresa and D. R. F. Taylor. "Cybercartography and Volunteered Geographic Information." In *Developments in the Theory and Practice of Cybercartography: Applications and Indigenous Mapping*. Edited byD. R. Fraser Taylor and T. P. Lauriault. Amsterdam: Elsevier, 2014.

"Geomatics and Cartographic Research Centre" (GCRC). Accessed May 2019. https://gcrc.carleton.ca/index.html.

Hayes, Amos, Fiset, J. P. and Peter L. Pulsifer. "The Nunalaliit Cybercartographic Atlas Framework." In *Developments in the Theory and Practice of Cybercartography Applications and Indigenous Mapping*. Edited byD. R. Fraser Taylor and T. P. Lauriault. Amsterdam: Elsevier, 2014.

Hayes, Amos and D. R. F. Taylor. "Developments in the Nunaliit Cybercartographic Framework." In *Further Developments in the Theory and Practice of Cybercartography*. Edited byD. R. F. Taylor, Erik Anonby and Kumiko Murasugi. San Diego: Elsevier, 2014.

Jaypoody, Mike, Kautuk, Robert, Fox Shari, Pulsifer Peter and Amos Hayes. "*Clyde River Atlas: Linking Inuit Knowledge and Technology to Mobilize Knowledge at the Community Level*" (Paper Presented at Artic Net Conference, Ottawa, ON), December 2018.

Keith, Darren, Crockatt, Kim and Amos Hayes. "The Kitikmeot Place Names Atlas." In *Developments in the Theory and Practice of Cybercartography: Applications and Indigenous Mapping*. Edited by D. R. Fraser Taylor and T. P. Lauriault. Amsterdam: Elsevier, 2014.

Keith, Darren, Griebel, Brendan, Gross, Pamela and Anne M. Jorgenson. "Digital Return of Inuit Ethnographic Collections using *Nunaliit*." In *Further Developments in the Theory and Practice of Cybercartography*. Edited byD. R. F. Taylor, Erik Anonby and Kumiko Murasugi. San Diego: Elsevier, 2019.

Lauriault, T. P. and D. R. F. Taylor. "The Preservation and Archiving of Geospatial Data and Cybercartography as a Proactive Process." In *Developments in the Theory and Practice of*

*Cybercartography: Applications and Indigenous Mapping*. Edited byD. R. Fraser Taylor and T. P. Lauriault. Amsterdam: Elsevier, 2014.

Machado, R., P. Pérez and Vieira Souza, U. D. "The Potential of Cybercartography in Brazil: A Cybercartographic Atlas for *Lencóis Maranhenses* National Park, *Maranhao* State, Brazil." In *Further Developments in the Theory and Practice of Cybercartography*. Edited by D. R. F. Taylor, Erik Anonby and Kumiko Murasugi. San Diego: Elsevier, 2019.

2019 "Mukurtu – Home." *Mukurtu*. Accessed June 2019. www.mukurtu.org.

2019 "Northern Lights: A wake-up call for the future of Canada." *Senate of Canada*. Accessed June 2019. https://www.sencanada.ca/en/info-page/parl-42-1/arct-northern-lights/.

Oguamanan, C. *Intellectual Property in Global Governance*, Routledge, New York, 2011.

Pyne, Stephanie and D. R. Fraser Taylor. *Cartography in a Reconciliation Community: Engaging Intersecting Perspectives, Volume 8*. Amsterdam: Elsevier, 2019.

Scassa, Teresa and D. R. Fraser Taylor. "Legal and Ethical Issues around Incorporating Traditional Knowledge in Polar Data Infrastructures." *Digital Science Journal 16*, no. 1 (2017).

Scassa, Teresa, Lauriault, T. P. and D. R. F. Taylor. "Cybercartography and Traditional Knowledge: Responding to Legal and Ethical Challenges." In*Developments in the Theory and Practice of Cybercartographiy: Applications and Indigenous Mapping*. Edited by D. R. Fraser Taylor and T. P. Lauriault. Amsterdam: Elsevier, 2014.

Scassa, Teresa and D. R. F. Taylor. "Intellectual Property Law and Geospatial Information: Some Challenges." *WIPO (World Intellectual Property Organization) Journal Special Issue on Analysis of Intellectual Property Issues 6*, no. 1 (2014).

Taylor, D. R. F. "Maps and Mapping in the Information Era." In *Proceedings of the 18th ICA International Cartographic Conference (ICC 97), Stockholm, Sweden, Volume 1*. Edited by Lars Ottoson. Gavle: Swedish Cartographic Society, 1997.

Taylor, D. R. F. (Ed.). *Cybercartography: Theory and Practice*. Amsterdam: Elsevier, 2005.

Taylor, D. R. F. and Stephanie Pyne. "The History and Development of the Theory and Practice of Cybercartography." *International Journal of Digital Earth 3*, no. 1 (2010).

Taylor, D. R. F., and T. P. Lauriault (Eds.). *Further Developments in the Theory and Practice of Cybercartography: Applications and Indigenous Mapping*. Amsterdam: Elsevier, 2014.

Taylor, D. R. F., Anonby, Erik and Kumiko Murasugi (Eds.). *Further Developments in the Theory and Practice of Cybercartography: International Dimensions and Language Mapping*. San Diego: Elsevier, 2019.

Taylor, D. R. F. "Cybercartography Revisited." In *Further Developments in the Theory and Practice of Cybercartography: International Dimensions and Language Mapping*. Edited by D. R. F. Taylor, Anonby, Erik and Kumiko Murasugi. San Diego: Elsevier, 2019.

# 8 Early California Cultural Atlas

## Visualizing Uncertainties Within Indigenous History

*Steven W. Hackel, Jeanette Zerneke, and Natale Zappia*

### California Mission Records

The ECCA is a spatial history project initiated in 2004 by Steven W. Hackel and Jeanette Zerneke (Natale Zappia joined the project in 2009) to create innovative digital mapping techniques in the humanities and provide vital tools for the study and teaching of the history of colonial California in America and beyond.[1–5] During the early modern period, the establishment of missions amongst Indigenous peoples initiated dramatic demographic, environmental, religious, and social change. Beginning in the 15th century, and continuing through the present, missionaries traveled throughout the Americas to convert Native peoples to Christianity and encourage them to congregate in mission communities. Soon after they left their ancestral villages, Native peoples often suffered dramatic population decline. Compounding this catastrophe, colonists arrived with different settlement strategies and patterns of land use. This was especially true in Alta California, where Franciscans established 21 missions and baptized more than 80,000 Natives, paving the way for thousands of Spanish settlers to establish hundreds of ranchos between San Diego and San Francisco.

The ECCA, then, has its origins in historical documents and complex interactions initiated more than 250 years ago with the establishment of Mission San Diego on July 16, 1769. On that fateful day, the Franciscans inaugurated a chain of missions that would eventually stretch from San Diego to just north of San Francisco. More than a year and a half after the establishment of Mission San Diego, in the first months of 1771, the Kumeyaay allowed the missionaries to baptize a handful of their children.[6] The first Native baptized at San Diego was a three-year old boy who the padres named "Francisco Antonio"; Francisco Antonio was the son of a village leader who himself would be baptized in 1771 and named "Carlos" in honor of Spain's King Charles III.

In keeping with Catholic practice after the Council of Trent, the Franciscans recorded the names of these Kumeyaay and the sacraments they granted them in the mission's baptism register. The full details of these first baptisms at San Diego and those of some 500 other Kumeyaay are not known to us today in full, because the register in which the padres recorded them was destroyed in 1775 when Carlos and his brother led an attack that killed one missionary and burned Mission San Diego to the ground.[7] However, good recordkeepers that they were, the Franciscans, soon after the attack, painstakingly reconstructed the baptism register from memory in a new volume that survives to this very day. By the time missionaries began anew with the baptism register of Mission San Diego, Franciscans had established four more missions in Alta California: San Carlos Borromeo (1770), San Antonio de Padua (1771), San Gabriel Archangel (1771), and San Luis Obispo (1772). At these missions in these early years, missionaries baptized more than 1,000 Natives, believing that they were cultivating in them the rudiments of a Spanish Catholic identity.

In the coming decades, the padres would establish 16 more missions as well as a chapel at the presidio of Santa Barbara and a church in the pueblo of Los Angeles. At all of these sites, missionaries dutifully administered and meticulously recorded tens of thousands of sacraments. In so doing, they built a set of records that would encompass nearly the entirety of the region's colonial population. While there is no countervailing set of records from Natives describing colonial California, these baptism, marriage, and burial registers constitute an amazingly thorough record of the lives of the Natives, soldiers, settlers, and missionaries of colonial California. And although there has been heated debate during the last 250 years over the goals, the morality, and the legacy of the mission enterprise and its effects on California Natives, scholars have never doubted the importance of the sacramental registers to an understanding of the people of Alta California.[8]

Collectively, these mission registers contain not only much of the information necessary to reconstruct the lives of tens of thousands of Natives and settlers but also the raw materials for the discovery and writing of larger and more intimate histories of colonial California. Moreover, these records are of added importance given that the overwhelming majority of the documents generated in Spanish and Mexican California were destroyed in the earthquake and fire of 1906. In many ways, these projects are an attempt to bring new sources to bear on a field whose history went up in smoke before it was ever written.

From the spatial historians' perspective, the surviving mission records are incredibly rich. When California missionaries baptized an individual—be it a Native adult or a newborn Spanish child—they recorded that person's given Spanish name, godparents, place of origin, age, parents, siblings, and, when applicable, the individual's children, marital status, as well as any other family relations or facts about that person—such as political or occupational status—that they considered relevant to his or her identity. More importantly, though, for each sacrament they performed, the Franciscans recorded the date and location where it was administered thereby providing historians of the future with crucial patio-temporal data. Furthermore, in California, as an aid to recordkeeping, missionaries assigned each baptism a unique number, beginning with the number one for the first baptism at each mission. Similarly, when they married or buried someone and performed the associated sacraments, missionaries recorded that individual's Spanish name, age, marital status, place of baptism, family relations, and, when known, their baptism number and mission of baptism. Franciscans also assigned marriage and burial records unique numbers at each mission. Thus, at any given time, the missionaries could determine how many baptisms, marriages, and burials they had performed at each mission. And each year they used their tabulations of sacraments to justify their work here to a skeptical political leadership in central Mexico. The padres also believed that these records gave them a clear way of identifying individual Natives within the growing ranks of neophytes at each mission. Remarkably—given the political instability of California as it shifted from Spanish to Mexican rule, and then to U.S. statehood—the baptism, marriage, and burial records for California's missions survive with but a few exceptions. Altogether, the sacramental registers from these sites contain records on more than 100,000 baptisms, 70,000 burials, and 28,000 marriages that the Franciscans administered between 1769 and 1850. These records are only as useful as they are accessible, and until recently they were pretty inaccessible. Available only on microfilm at selected archives throughout the state, and unindexed and often barely legible, they were used only selectively by a small group of scholars and clergymen.

A prerequisite to the full exploitation of these registers was a workable system that would allow scholars to access the information in the registers easily and systematically. Without this comprehensive and integrated database, it was impossible, many of us reasoned, to see the movements of Natives and settlers from one part of the province to another or to grasp various social processes and patterns of historical change that unfolded across all of the missions, presidios, and pueblos of colonial California. Out of this need was thus born the seminal Early California Population Project (ECPP), which attempted to make this sacramental data more readily available. Initiated by Steven W. Hackel in 1995 and supported by the Huntington Library, the database went online in 2006 and since then, it has become an indispensable resource reshaping how we understand population change, land use, Native culture, naming practices, and god parentage in Alta California. The ECPP includes records from all 21 missions and has more than 200 fields that detail many facets of the lives of California Natives. As suggested above, ECPP data naturally lends itself to spatial and temporal analysis. Every field in every record describes a person or event that can be situated in a Native village, a Franciscan mission, a civilian settlement, or a military fort, and just as important, everything the padres recorded in the records has a date, or a precise temporal component. Because the separate baptism, marriage, and burial registers for all of California's 21 missions are largely complete, consistently thorough, and in many ways cross-referenced, records from different missions and registers can be linked and sorted by individual. The California mission registers as captured in the ECPP, therefore, contain the information necessary to reconstruct not only the individual lives of tens of thousands of early Californians, but also temporal and spatial dimensions of early California that may not have been seen before.

## ECCA—Project Scope

Based upon a visualization and enhancement of the data in the ECPP, through the ECCA we have created dynamic, diachronic, and interactive maps showing the movement of California Natives to the missions, the resettlement of their lands by colonists, and the new ways that Spaniards and Mexicans bounded and used the land. Thus, while our study has focused on colonial California, the ECCA illustrates in stark detail colonial processes that have unfolded across the globe and involved millions of people over five centuries. In a broader context, the spatio-temporal mapping of multiple dimensions of demographic and geographic change over time developed by ECCA provides a technical model for the study of dramatic changes that have occurred across the globe. Yet the project goes beyond the creation of a powerful new tool for understanding change in the early modern period. It challenges the false certainty of traditional mapping techniques, embraces spatial and temporal ambiguities, and pioneers new methods of integrating primary source materials into diachronic web-based visualizations.

In its conceptualization and implementation, the ECCA pioneers techniques that visualize change over time and deploy spatio-temporal technologies. It links historical data to dynamic maps, builds collaborations between California Natives, historians, anthropologists, educators, and experts in data management and interface design, and facilitates innovative understandings of the impacts of Spanish settlement upon California's Native peoples. These partnerships and visualizations can serve as models for the study of colonial regions and processes elsewhere in the world. The ECCA, by integrating a range of sources and technologies, enables students and scholars to better understand how European colonization affected Native peoples. It illustrates and examines relationships between time and space. In doing so it shows how adding the spatial component to temporal analysis leads to a deepening of humanistic inquiry and a reformulation of the inquiry itself. Finally, the ECCA offers new ways of mapping spatial and temporal uncertainty and contested data.

## An Interdisciplinary and Collaborative Approach

The ECCA's interactive website integrates and manages historical resources, visually illustrates historical data related to California between 1769 and 1850, displays research results in the form of maps and other visualizations, and, through various educational tools, educates students from elementary school to the university classroom. This project is interdisciplinary and collaborative; it draws upon the expertise of research scholars, librarians, archivists, software engineers, technical experts, California Natives, and primary school teachers. Fundamentally, the project represents a new partnership between existing programs. It combines the extensive and unique database of the Early California Population Project (ECPP) with the technical expertise of the Electronic Cultural Atlas Initiative (ECAI) to enable new understandings of California and its peoples between the founding of California's first mission in 1769 and the admission of California into the Union in 1850. Since beginning the project in 2008, our intent has been to create an innovative website that improves how we understand a pivotal epoch in California history, visualizes and explores Indigenous perspectives on these events, and demonstrates to humanists how emergent technologies can deepen and foster humanistic inquiry.

More specifically, the ECCA visualizes historical transformations in California before 1850 by mapping the migration of Natives from their native villages to Franciscan missions, the immigration of soldiers and settlers to California from northern Mexico, the initiation and growth of domestic agriculture and animal husbandry at the missions, and the transfer of huge parcels of land from California Natives to Spanish and Mexican landholders. The ECCA does not merely deliver material and ideas that are already available in print. Rather, it is creating new knowledge about where Native peoples lived in California, the extent to which colonization disrupted their lives, and how the human and social geography of much of California was radically altered in the century after 1769.

## Funding and Building the ECCA

The ECCA has been supported by two National Endowment for the Humanities (NEH) Digital History Start-Up grants and a UCR Chancellor's Initiative grant. This support allowed the ECCA to assemble our team of scholars and consultants, and it allowed us to begin our work. With Level I funding, the ECCA team created a Phase I website with Google Earth visualizations. The team focused on two central California missions, San Carlos Borromeo and San Juan Bautista. For San Carlos, the team relied upon single locations for Native villages. In adding San Juan Bautista to the study, the team plotted multiple locations for villages and devised a register that explains the sources of information for this mapping as well as the relative certainty of these locations. In this work, the team was not simply visualizing existing information but generating new knowledge about village locations and their movements over time. Furthermore, ECCA staff created in this new prototype a means by which simply clicking the cursor on a Native village allows access to basic information on Natives who lived in these villages. An innovative implementation of the Google Earth time bar showed the movement of Natives to missions during every year until the village was in all likelihood depopulated. ECCA staff added to the website geo-registrations of historic maps and boundaries of ranchos created in California during the Spanish and Mexican periods. Thus, with Level I funding the ECCA devised new ways to visualize and understand historical transformations that heretofore had only been represented on paper and in ways that did not adequately consider spatial and temporal relationships.

With Level II funding, ECCA staff enhanced the project's IT infrastructure by expanding the project database, setting up the initial website, and initiating use of the ECAI Data Portal for sustainably hosting datasets. Staff created more complex visualizations, including the integrated visualization for Los Angeles, which includes four missions and multiple Native group affiliations. With the help of Stephen O'Neil and local Native consultants, the team mapped locations for villages in the Los Angeles.[9] Staff increased the project's complexity by addressing the uncertainty and ambiguity issues surrounding native village locations. The ECCA fostered partnerships with UCLA's National Center for History in the Schools (NCHS) to develop teaching applications for the ECCA and with the Stanford Spatial History Project to provide peer review and a visualization of Native migration to the missions of the Los Angeles region. Most important, in Phase II we constructed a Google Earth dynamic map for the LA region that is now our prototype as we go forward. The prototype incorporates five distinct datasets with customized cartographic representation, direct links to data source information, related resources and contextual information, and interactive control of the time range displayed. This prototype can be displayed as a video sequence. At the time we developed our project, Google Earth provided the most flexible platform for developing visualizations, which incorporate a timeline and diverse data layers while maintaining the data in an open source sustainable format and enabling interactive display on the web.

In this new map we tested and trouble-shot the implementation of the following data layers: California Native tribal regions and village locations; locations for four missions (including place of establishment and relocation); active and accessible baptism records for Natives from villages and those born at the missions; founding dates; and boundaries and online maps of Spanish and Mexican ranchos. We discovered that the incorporation of four missions simultaneously added significant complexity and interpretive power to the project but did not pose insuperable technological challenges. We deployed new spatial visualization techniques including Google Fusion maps to create customized visualizations for subsets of the data. As we learned in this second phase of our work, the single wave of baptisms originating at villages closest to missions and spreading to those farther away over time as seen in the Phase I work for Central California was not apparent in the Los Angeles region. Instead we discovered that Natives from the same village often went to different missions depending on their date of migration.

To illustrate this complexity, we developed new symbols to indicate which mission registered baptisms from a particular village and we implemented a special interface to explore this facet of Native migration. Documenting mission-born baptisms that were orders of magnitude larger than those of the individual villages posed a new challenge, which we have currently solved by adding to our visualization vertical polygons representing the population of each mission. This feature allowed us to represent what had not been seen before: the growth of the mission-born population as it outnumbered new recruits from villages. We intend to explore additional ways of displaying population and other mission information, such as production of commodities that change in parallel with the spatial changes being represented on the map.

In Phase II we also learned yet again how adding a spatial component to temporal analysis leads to new levels of inquiry. In the 1820s and 1830s, Natives came to the missions from the interior of California, an area far less affected than the coastal region by the growth of mission agriculture and livestock and the creation of Spanish and Mexican ranchos. Thus, we are now asking: If mission encroachment on Native subsistence drove Native movement to the coastal missions before 1820, what led Natives from the interior of California to the missions after 1820? Furthermore, now that we can see the spatial and temporal patterns of mission recruitment for the regions of Monterey and Los Angeles, we have begun to ask how these patterns might differ from those of the recruitment areas of missions in San Diego, Santa Barbara, and San Francisco, all regions with different tribal organizations, natural environments, Spanish settlement, and cultural diversity. During Phase II we also held community meetings with Southern California Native groups to ensure our work is respectful and supportive of their interests. We will continue this practice in the expansion of the project to additional regions. Finally, during our Level II work we developed a project website. This portal provides access to data created and collected by the project as well as links to project resources and documentation of project methodology. It displays the various prototypes developed as well as various experiments with map design. Since the technology of visualizations is changing rapidly, we created multiple trial visualizations to provide users many possible views on the past. In addition, each primary dataset used in the integrated maps is displayed independently to highlight the flexibility of the system and potential for the development of customized web interfaces for research and curriculum development.

## Mapping Villages and Spatial Ambiguity

Our sense is that the ECCA not only deepens our understanding of colonial California but that it does something more: It deepens humanistic inquiry by calling attention to the way databases tend to erase or render meaningless geographical information. Databases are increasingly common in the humanities and social sciences. They are often full of geographic information, but coding and database structures can reduce this information to mere place names and administrative units. This leaves records with spatial information that can only be deduced or understood by experts who have local knowledge and special maps. The ECCA, by linking to a dynamic map the spatial and temporal attributes in the vital records of individuals found in the Early California Population Project (ECPP), reintroduces the geographical component into the study of early California and its peoples and creates a foundation for historical and geographical inquiries while resisting the false certainty reflected in nearly all maps.

While our project focuses on early California, one of our main objectives is to explore and resolve important issues that have often made it difficult for humanities scholars to adopt emergent technologies. Thus, the ECCA has at its core a willingness to acknowledge and map ambiguity, an issue that has bedeviled humanists' attempts to use new mapping technologies. Typically, computer programs that historians use to map events spatially and temporally do not have a means to handle ambiguity. By addressing this issue, the ECCA contributes to the movement toward deep mapping (or thick mapping), defined as an effort to bend geo-spatial technologies to the needs of humanists to create rich, multi-scalar, and dynamic maps embracing uncertain qualitative data.

Over the past decade, scholars have begun to utilize spatially aware technologies for a broad range of applications. Humanists (and especially historians), however, have only recently adopted technologies to account for spatial ambiguity. Most people still tend to like maps that situate people, cities, and villages in one place even though this creates a false sense of certainly and can mislead. We believe that new spatio-temporal technologies allow us to move beyond the simplistic, one-dimensional maps of the past. The ECCA has developed methodologies to handle these issues and to create visualizations that support more complex narratives. Each of the sources of data used by the ECCA has its own characteristics of ambiguity. Acknowledging this can assist in choosing appropriate scale and attributes for the datasets.[10]

The ECCA contributes to advances in deep mapping, which fuses the humanities' traditional focus on nuance, voice, experience, text, and image with the systematic approaches of geographic information science and computer modeling to link time, space, and culture dynamically. To ensure that we will draw effectively from and contribute to this multi-disciplinary effort, ECCA enlisted David Bodenhamer, a pioneer of deep mapping who has published extensively on the subject. We also employed several multi-layered methods during our Level I and Level II phases. First, we tested several map designs to determine appropriate visualization methods to support users' understanding of the content. Then we created and collected the data for curation and long-term use. Visualizations based on this data are flexible and can adapt to users' requirements and technical changes. Our Level I and Level II websites used demographic data from the ECPP, historical maps from online archives, shape files for historical ranchos, and a reference dataset of the locations of the California missions. We then linked the demographic data from the ECPP (the age, sex, origin, and marital status of an individual at baptism, marriage, and death) to the locations of Native villages that we determined through mission records, historical documents, and anthropological reports. This method allowed us to situate people in space at specific moments in their lives and to understand their relationships to other members of their families, villages, and larger communities.

Considerable effort went into the process of determining appropriate village locations mentioned in the historical records. The ECCA team consulted multiple sources and where appropriate developed an ontology of village types and locations. We overlaid maps on current satellite imagery and consulted with Native groups to discuss their understanding of village locations in the past (note that villages are intentionally represented imprecisely in order to deter would-be "pot hunters" pillaging ancient village sites). We then tested several ways to represent baptismal data. The integrated overview shows the probable locations of the villages both before and after active baptisms as static locations. The height of a polygon emanating from that village location corresponds with the same period when residents of the village were baptized. This allows a quick visualization of the number of people moving from the village to the missions each year. In addition, a pop-up window allows users to click on a village to see the number of baptisms recorded in a particular year. We also overlaid the expanding boundaries of Spanish and Mexican rancho lands to visualize changes in land use (Figure 8.1). This visualization, in particular, has received widespread positive interest at both informal meetings and academic conferences as well as from K–12 teachers and California Native groups.

We developed our technical infrastructure as a lab for testing multiple technologies and methodologies. The lab approach proved important. With continuing changes in available technologies to implement project goals, it allowed us to be creative and migrate to new technologies. Initially, we imported ECPP demographic data to a

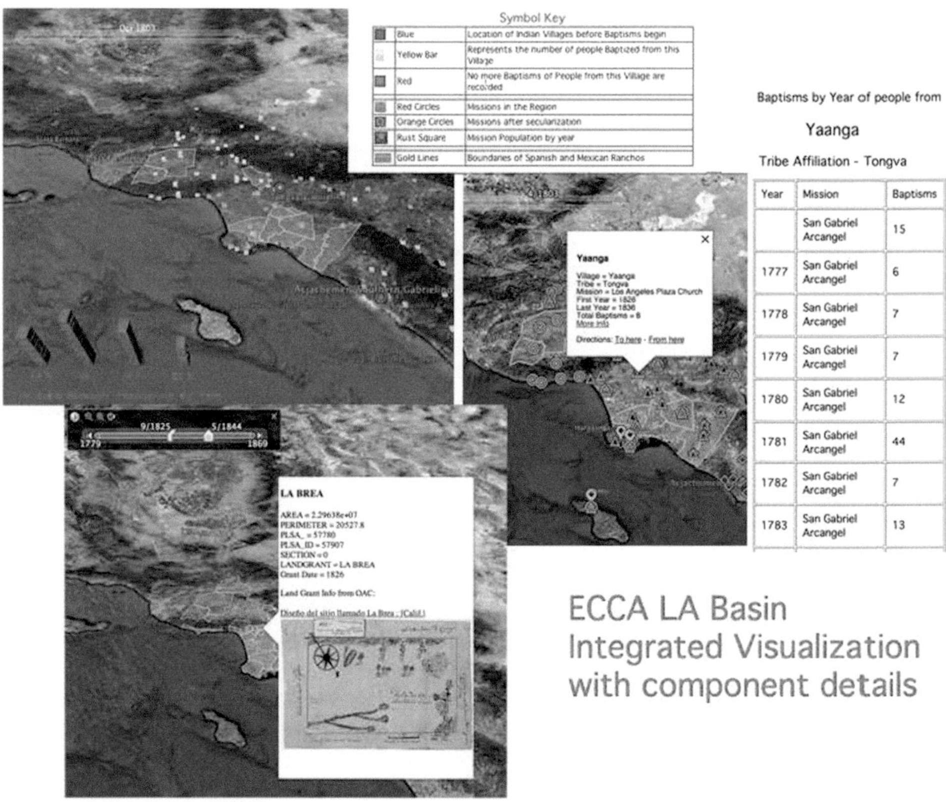


*Figure 8.1* ECCA LA Basin Integrated Visualization, ECCA, 2013.

Microsoft Access database for the project. Using this platform, staff generated multiple queries, some of which were exported to create GIS data for map layers. Exported datasets then could be used in multiple systems. Baptism IDs in the ECPP data enabled linkage to the Huntington Library's site to access full profiles of individuals. We then added map layers for data subsets to Google Earth. Customized Google Earth views were hosted on the website using an embedded Google Earth online app. For the Los Angeles Basin, Google Fusion was used to upload data to spreadsheet formats with the ability to output KML files for Google Earth as well as embed maps in webpages. The Google Fusion online maps allow users to interact with the data and link to related online data and data sources.

We chose to use Google Earth as our map development platform to generate a range of interactive visualizations. Google Earth, at the time that we developed ECCA, was freely available, easy to learn, had a large user base, and what we assumed to be a sustainable data format. It allowed us to create complex visualizations with a data-driven timeline, flexible layering and cartography, links to related resources, and three-dimensional representation of data. With our platform in place, we began customizing sets of data layers to create dynamic maps that could be used for teaching and research. One example is a video of a dynamic map created by ECCA staff expressly for the 2013 Huntington Library exhibition, "Junípero Serra and the Legacy of the California Missions."[11] As we expand the project to cover the entire California mission system, we intend to deploy a web interface allowing users—including those exploring general network analysis and trends (such as migration between missions) and descendants tracing their ancestry—to select data layers and locations and create their own visualizations. For collaborations with user communities we have the capability to create custom visualizations of one particular mission or sets of missions, or approach the material from different perspectives.

## Layers of Native Ethnogeography

One of the most interesting aspects of the ECCA project is our attempt to provide and visualize multiple layers of Native ethnogeography. Native villages, political affiliations, seasonal locations, migration patterns—all of these variants played a role in our methodological approach. The ECCA has consistently taken an inclusive approach with our data. The more information gathered, the more complex and dynamic our maps. Thus, almost all of our data have been incorporated in some way. The most interesting moments in the project came when ambiguous or contradictory information collided and new questions emerged.

An excellent example of our approach to Native ethnogeography can be seen in our treatment of the area around Mission San Juan Bautista, because it represented the most complex cultural interface between Franciscans and Natives and also among Indigenous communities. Wedged between the coastal plains and river valleys near the Monterey Peninsula to the west and the San Joaquin Valley and Sierra foothills to the east, San Juan Bautista illustrates the complexities of Native territories and the impact of the mission on cultural geography. At the mission, Costanoan-speaking Mutsuns, Unijaimas, and Pagchins interacted with Penutian-speaking Yokuts. Trade and intermarriage had occurred for centuries before the missions, but San Juan Bautista introduced new regimes of labor, food, and culture that influenced the already dynamic patterns of interaction within the region.

Thus far, our master table for village locations and ethnographic information includes the original transcription of the village name, larger political entity, reference, location, accuracy of location, linguistic affiliation, second or third village name, source for the location, and coordinates (latitude/longitude) (Figure 8.2). Scholars continue to debate over the exact location of California Native villages inhabited before moving to the missions. ECCA's methodology embraced these unknowns.

As we discovered, the missionaries' transcriptions of village locations more than occasionally contained contradictory names, and this was coupled with geographic ambiguity. For example, the Calendaruc, who lived near the coast along the Pajaro River were sometimes listed as "Calendaruc—playa" (referring to villages along the coast) or "Calendaruc—Guachirron" (referring to inland villages along the river). Calendarucs lived in different geographic areas but still seemed to self-identify themselves (depending on the interpretation of the transcriber) as Calendarucs. We also found political ambiguity in the records. Again, the Calendarucs provide a good example. Several entries include "Calendaruc—Tamaros" or "Calendarruc—T i havun Thaum." The first entry represents two different political groups who were neighbors. Most likely, this individual represented intermarriage between villages. Many Costanoans identified themselves as part of more than one village. Villages sometimes represented specific political entities, while other border villages represented multiple groups. The second entry refers to a specific town—Tiuvta. This represents another sort of ambiguity. Certain group names referred to a specific

| Village | Tribe | Icon | Certainty |
|---|---|---|---|
| Aataveanga | Tongva | small_green | approx |
| Abalone Hill | Acjachemen | small_red | exact |
| Achooykomenga | Tongva | small_green | approx 3-5 miles |
| Ahunx | Acjachemen | small_red | approx-3 miles |
| Akuragna | Tongva | small_green | approx 3 miles |
| Akuuronga | Serrano | small_blue | general - approx. 3 miles |
| Alauna | Acjachemen | small_red | exact |
| Amauge | Acjachemen | small_red | guess - 5 miles |
| Amupubit | Tongva | small_green | approx 3-5 miles |
| Amuscupiabit | Serrano | small_blue | approx 2 miles |
| Angayaba | Serrano | small_blue | approx 3-5 miles |
| Anonga | Acjachemen | small_red | approx - 5 miles |
| Apachia | Tongva | small_green | general |
| Apiacobit | Serrano | small_blue | general - approx 5-10 miles |
| Apinjabit | Serrano | small_blue | general - approx 5-10 miles |
| Ashuukshanga | Serrano | small_blue | approx 3 miles |
| Atavsanga / Huam | Chumash | small_yellow | approx - 2 miles |
| Atongaibit | Serrano | small_blue | approx 5 miles |
| Atosemeie | Acjechemen | small_red | unknown - 10 miles |
| Axachme | Acjachemen | small_red | exact |

*Figure 8.2* Reference table of Native villages with tribe and location certainty annotation, ECCA, 2013.

village, while others referred to a regional entity. In the case of the Calendaruc, Tiuvta is the specific town and Calendaruc the larger entity.

For some of our village mapping, we consulted Randy Milliken's extensive research. His most recent work utilizes the Community Distribution Model (CDM) to assign little documented Native populations within hexagonal clusters that reasonably locate tribal groups within a geographic radius. This provided another layer of territory between and among villages and political entities. We also relied on other anthropologists like Chester King, who also pioneered studies on interregional trade between coastal and inland groups (Chumash and Tubatual), as well as the *Handbook of North American Indians*. Many villages were referenced in at least two or more of the sources and explained in different contexts. For example, what Milliken might specify as a "Costanoan/Penutian-speaking" village or political entity, another source might refer to as "Costanoan." These studies and our research have shown us that perhaps it is best to see villages and political entities intertwined but also separable depending on the season, economic activity, or political/cultural activity (i.e. inter-village marriage). The visual map representation of our data has helped us to revise our questions and develop a new understanding of our data.

## Looking Forward

As outlined above, the innovative features of the ECCA website provide ample opportunities for researchers inside and outside of the academy. The information and visualizations created also support learning for 4th grade students across California and teaching and learning at colleges and universities. Variations of the primary spatio-temporal interface can be constructed for targeted user groups, specific academic investigations, or new websites with specific functions. ECCA continues to work with its collaborators to create these specialized visualizations. While our efforts have initially focused on serving the educational needs of California's 300,000 fourth-grade students learning about California history each year, we believe that our site will ultimately have a national presence through our partnership with organizations like the National Center for History in the Schools. The ECCA team will continue to meet with Native groups, such as the Ohlone Tribe, the Kizh Gabrieleño, the Tongva Gabrielino, the Juaneño Band of Mission Indians (Acjachemen Nation), and the Santa Ynez Band of Chumash Indians, to disseminate the ECCA and help incorporate its visualizations into the Tribes' educational programs and museums. Ultimately, we hope that the ECCA—like all developing digital humanities projects engaged in historical mapping—will foster new research and understandings about Indigenous history in California and beyond.

## Notes

1 For California Native language recovery and the monumental efforts of John P. Harrington, see Hinton, Leanne. *Flutes of Fire: Essays on California Indian Languages*. Berkeley: Heyday Books, 1994.
2 See website: https://www.latimes.com/projects/la-me-tongva-map/.
3 Curwen, Thomas. "Tongva, Los Angeles' First Language, Opens the Door to a Forgotten Time and Place." *Los Angeles Times* (May 12, 2019).
4 For Tongva cultural geography, see McCawley, William. *The First Angelinos: The Gabrielino Indians of Los Angeles*. Malki Museum Press, 1996.

5 See Hackel, Steven W., Natale Zappia and Jeanette Zerkneke. *The Early California Cultural Atlas*. http://ecai.org/ecca.
6 See *The Early California Population Project: A Database Compiled and Developed at the Huntington Library*. Edition 1.0. General Editor, Steven W. Hackel, Lead Compiler, Anne M. Reid (The Henry E. Huntington Library, San Marino, California, 2006). https://www.huntington.org/ecpp for baptismal records for all of the California missions.
7 For an account of the attack, see Hackel, Steven W. *Children of Coyote: Missionaries of Saint Francis: Indian-Spanish Relations in Colonial California, 1769–1850*. Chapel Hill: University of North Carolina Press, 2005, 258–261.
8 See Johnson, John R., "Mission Registers as Anthropological Questionnaires: Understanding Limitations of the Data." *American Indian Culture and Research Journal* XII (1988): 9–30. See also, Hackel, Steven W., *Children of Coyote*, 2005, 449–455.
9 See ECCA website: Early California Cultural Atlas: Research into the Native American Village Locations and Attributes of the Greater Los Angeles Region During the Contact Period.
10 See website: http://ecai.org/ecca/ECCA-AddressingUncertaintyAmbiguity2013.html.
11 See website: http://ecai.org/ecca/Videos/GreatMigrationVideo.html.

## References

Bean, Lowell ed., *The Ohlone Past and Present*. Menlo Park, CA: Ballena Press, 1994.

Clark, Donald Thomas. *Monterey County Place Names: A Geographical Dictionary*. Carmel Valley, CA: Kestral Press, 1991.

Hackel, Steven W. *Children of Coyote: Missionaries of Saint Francis: Indian-Spanish Relations in Colonial California, 1769–1850*. Chapel Hill: University of North Carolina Press, 2005.

Heizer, Robert, (Ed.). "California Indian Linguistic Records: The Mission Indian Vocabularies of Alphonse Pinart," *University of California Anthropological Records* 15 (1952).

Hinton, Leanne. *Flutes of Fire: Essays on California Indian Languages*. Berkeley: Heyday Books, 1994.

Johnson, John R. "Mission Registers as Anthropological Questionnaires: Understanding Limitations of the Data," *American Indian Culture and Research Journal, XII* (1988).

McCawley, William. *The First Angelinos: The Gabrielino Indians of Los Angeles*. Banning, CA: Malki Museum Press, 1996.

Kroeber, Alfred L. "The Chumash and Costanoan Languages," *University of California Publications in American Archaeology and Ethnology* 9 (1910).

Kroeber, Alfred L. *Handbook of the Indians of California*. New York: Dover Press, 1976.

Milliken, Randall T. *A Time of Little Choice: The Disintegration of Tribal Culture in the San Francisco Bay Area, 1769–1810*. Menlo Park, CA: Ballena Press, 1995.

Shoup, Laurence H. and Randall T. Milliken, *Inigo of Rancho Posolmi: The Life and Times of a Mission Indian*. Menlo Park, CA: Ballena Press, 1999.

Wiener Stodder, Ann Lucy. "Mechanisms and Trends in the Decline of the Costanoan Indian Population of Central California," *Coyote Press Archives of California Prehistory 4* (1986).

Yamane Linda, ed., "A Gathering of Voices: The Native Peoples of the Central California Coast," *Santa Cruz County History Journal 5* (2002)Yamane, Linda.

# 9 Access to Government Information and Inclusive Stewardship of North America's Archaeological Heritage

*Eric C. Kansa, Sarah Whitcher Kansa, David G. Anderson, Joshua J. Wells, Kelsey Noack Myers, and Stephen Yerka*

## Introduction

This chapter explores some tensions and challenges in access to information about the heritage of Indigenous peoples. The notion that access to information is always good has seen rightful critique. Even well-intentioned, arbitrary exposure of information about the histories of colonized peoples can further reinforce colonialism. At the same time, information asymmetries also reinforce colonialism. In the United States, as well as many other nations, governments at the national, regional (state), and local level administer laws and regulations about archaeological and historical sites. Tribal Nations also have legal administrative jurisdiction over archaeological and historical places, but typically work with far less funding and staffing. Information flows critical to the protection of Indigenous heritage requires coordination among various federal, state, and Tribal Nation officials. Yet such coordination and information sharing is typically haphazard, leaving often under-resourced offices of tribal historic preservation with little information needed for decision making.

Digital data plays a key role in these administrative processes. Extrapolating from available government records, there are least two million recorded archaeological and historical sites across North America. In many cases, information about these sites is scattered across museum collections, published papers, and unpublished reports. Other government published documents describe regulatory decisions about these sites, especially decisions about repatriation, preservation, and legal custody. Efforts to collect and compile archaeological data have a long history, and information about archaeological sites and collections is maintained in every state and territory. However, this information is scattered and largely inaccessible, especially to descendent communities who may often lack access to government information systems or university libraries. Only rarely have these data been compiled and examined at large geographic scales, especially those crosscutting state lines.

In this context, the Digital Index of North American Archaeology (DINAA)[1] provides infrastructure for linking archaeological sites to other web-based resources that describe those places. Using transparency and access as a strategy to make stewardship of North American cultural heritage more inclusive, DINAA aggregates archaeological and historical data from state and tribal governmental authorities that manage United States cultural resources (Figures 9.1 and 9.2), providing the most comprehensive and detailed database documenting human settlement in North America currently available. The nation's investment in archaeology and historic preservation has produced a vast, widely dispersed, and variably curated literature. DINAA helps make the results of that effort,

| STATE | DATA COMPILED | NOTES** |
|---|---|---|
| *Alabama | 27,594 | |
| ^Alaska | 0 | |
| ^Arizona | 0 | |
| #Arkansas | 49,325 | |
| ^California | 4646 | Indexed Sites are from Other Sources |
| ^Colorado | 464 | Indexed Sites are from Other Sources |
| ^Connecticut | 0 | |
| ^Delaware | 0 | |
| *Florida | 31,041 | |
| *Georgia | 53,191 | |
| ^Hawaii | n/a | Not Part of Current Project |
| ^Idaho | 13 | Indexed Sites are from Other Sources |
| *Illinois | 59,610 | |
| *Indiana | 41,473 | |
| *Iowa | 26,850 | |
| #Kansas | 15,610 | |
| *Kentucky | 18,811 | |
| *Louisiana | 19,096 | |
| *Maine | 0 | Data Transfer in Progress |
| *Maryland | 13,167 | |
| ^Massachusetts | 1 | Indexed Site is from Other Sources |
| #Michigan | 24,985 | |
| #Minnesota | 13,037 | |
| #Mississippi | 33,539 | |
| *Missouri | 32,579 | |
| #Montana | 61,718 | |
| #Nebraska | 11,002 | |
| #Nevada | 126,491 | |
| ^New Hampshire | 0 | |
| *New Jersey | 0 | |
| ^New Mexico | 10 | Indexed Sites are from Other Sources |
| ^New York | 0 | |
| *#North Carolina | 52,479 | 11,155 have full site data indexed in DINAA |
| ^North Dakota | 41 | Indexed Sites are from Other Sources |
| *Ohio | 48,291 | |
| #Oklahoma | 25,317 | |
| ^Oregon | 117 | Indexed Sites are from Other Sources |
| *Pennsylvania | 21,815 | |
| *Rhode Island | 3695 | Data Transfer In Progress |
| *South Carolina | 28,261 | |
| ^South Dakota | 162 | Indexed Sites are from Other Sources |
| *Tennessee | 25,395 | |
| *Texas | 3421 | Partial Sample of Site Files |
| #Utah | 96466 | |
| *Vermont | 0 | Data Transfer In Progress |
| *Virginia | 42,480 | |
| #Washington | 32,991 | |
| ^West Virginia | 38 | Indexed Sites are from Other Sources |
| ^Wisconsin | 57 | Indexed Sites are from Other Sources |
| ^Wyoming | 40 | Indexed Sites are from Other Sources |
| TOTALS | 1,045,319 | |

* DINAA Partner. State information managers have provided full site data
# State information managers have provided a listing of site numbers by county to DINAA
^ State information managers have provided no information to DINAA
** Other Sources refers to repositories other than state site files

*Figure 9.1* Sites indexed or being incorporated into DINAA as of September 30, 2019 (n = 881,166 sites indexed, 1,045,319 sites compiled total).[8]

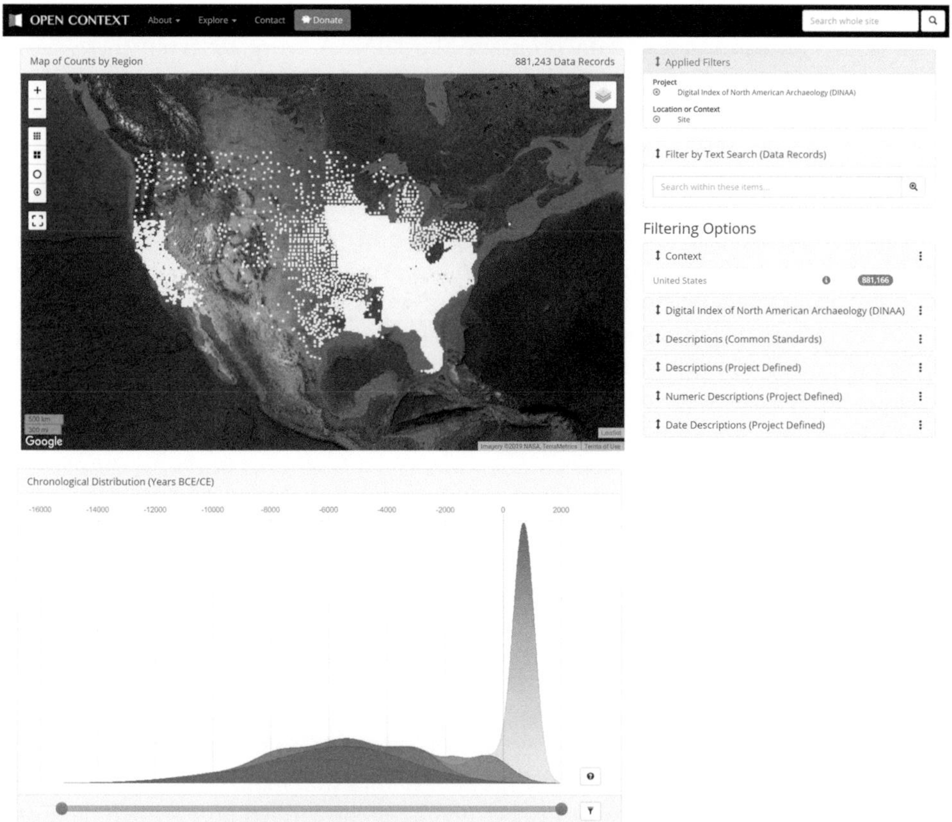


*Figure 9.2* DINAA has compiled data from 1,045,319 sites as of September 30, 2019. The total includes information provided by state site file managers and State Historic Preservation Officers (SHPOs) in the Eastern United States, as well as information obtained from other repositories, including museum collections, online research databases, and through text mining of journals like *American Antiquity* and the *Midcontinental Journal of Archaeology*, and gray literature such as the *Federal Register* and the *Index of Texas Archaeology*. For current data, visit: http://ux.opencontext.org/archaeology-site-data/dinaamap/.

often overseen by State Historic Preservation Officers (SHPOs), Tribal Historic Preservation Officers (THPOs), collections managers and curators, and a vast research community, more accessible to scholars, land managers, and the public alike.

DINAA publishes (highly redacted and generalized, see further) aspects of these data for anonymous open access without login or intellectual property barriers. Is such openness appropriate in this context? After all, data about people—especially people who have undergone a traumatic colonial history and continued oppression—is sensitive and problematic. Data can be used abusively, especially in the hands of powerful government officials or private companies. It is not only bureaucracies that can use data abusively. For example, location information about sites can be used by individual vandals, who may be motivated by hate and racism, to defile a place sacred to Indigenous peoples; furthermore, contested definitions of archaeological sites and their perceived importance, as instantiated in digital data, have the potential to spark tremendous political abuses and discord.[2]

How can an open access data program like DINAA work to avoid these risks? First, it is important to recognize that access barriers have their *own* risks that can compound the risks of managing sensitive data. Access barriers typically track personal logins and accounts of individual users as they access sensitive data. Appropriately authorizing the correct people, keeping their access credentials secure and secret, and responsibly managing data that tracks user behaviors all involve additional security risks and responsibilities. In other words, *user data is itself sensitive data.* Collecting user data in order to monitor access permissions or appropriate uses of data involves privacy risks. A program needs enough financial and technical resources to responsibly manage such risks. The funding constraints in archaeology make secure management of sensitive data doubly challenging.

These perspectives inform DINAA's open access strategy. A central goal of DINAA's open access strategy is to reduce risks of harm. *The most effective approach to protecting sensitive data is to avoid the collection and storage of such sensitive data in the first place.* For that reason, Open Context, the data management platform that hosts DINAA, collects and stores no user data. It avoids such common tracking mechanisms like logins, cookies, Google analytics, and the like. More importantly, as we describe next, DINAA only manages highly redacted and low-precision data. There is no password protected version of DINAA with more sensitive data that can be exposed by accident or hacking—the project only manages redacted, "low risk" information and it is all made public. Finally, DINAA serves as an "index," meaning it works as a finding aid that directs users to richer information resources stored elsewhere. Those other information resources can have additional protections and requirements, as judged necessary by the communities that manage them. Thus, in its capacity as an index, the DINAA project highlights how open access/open data can work collaboratively, in conjunction with systems and communities, especially Indigenous communities, that protect sensitive information.

This open data approach also contributes to recent collaborations among archaeologists, archivists, museum curators and American Indian Tribal Nations, tribal groups, traditional landowners, and other sovereign Indigenous groups, that attempt to better understand and address damage caused by colonialism. Executive Order 13175—Consultation and Coordination With Indian Tribal Governments (November 6, 2000), initiated federal government policies that further promote such partnerships, leading to collaborative land management, cultural heritage preservation, research, education, and community development programs.[3] DINAA builds upon and further enables these partnerships by making key data more accessible for descendent communities and Native American officials that manage the historical preservation efforts of sovereign Tribal Nations.

## Public Investments in Archaeology

Archaeological data constitute the direct evidence of past human behaviors and are essential for identifying and describing patterns of change in past human societies. Recent estimates demonstrate the magnitude of public investments in archaeology.[4] Conservatively, the public invests over $500 million per year to comply with historical and archaeological protection measures required by federal law. This level of investment nearly matches the total *combined* budgets in 2019 of the Institute of Museum and Library Services (roughly $240 million), the National Endowment for the Humanities (roughly $150 million), and the National Endowment for the Arts (roughly $150 million); the National Science Foundation budget for archaeology annually, in fact, is only a tiny fraction of the total directed to the nation's cultural resource management (CRM) programs. These surprising numbers demonstrate archaeology's relative importance in cultural heritage activity. Unfortunately, much of this work and investment goes largely unnoticed. Up to now, decades of effort directed to managing and protecting America's archaeological heritage has led to few publicly accessible impacts. Cultural resource management largely takes place within relatively opaque bureaucratic processes that regulate construction and development. CRM work has resulted in an estimated 350,000 reports nationwide as of 2004,[5] but because of limited access and cataloging, irreplaceable cultural heritage documentation in these "gray literature" reports goes ignored or at best underappreciated. Furthermore, because most CRM projects receive minimal attention, the vast majority of the reports produced see little external reuse in research or other publications that greater peer recognition and review would bring.

DINAA represents a required first step to encourage greater public knowledge and ideally accountability (vis-à-vis different public communities) for this tremendous public investment in cultural heritage. However, it would be naïve and unrealistic to impose a single data standard, expected to be broadly applicable for a continent full of archaeological sites collected by many different organizations for decades, and representing more than 13,500 years of differing cultures in widely varied environmental settings. Most of the state systems currently in place, in fact, encompass tens of thousands of sites and have been in place for more than half a century, leading to many separate database systems with unique constraints on data types and coding solutions. Often overworked and understaffed, those tasked with site file management lack the resources needed to completely restructure their datasets to meet external standards. In fact, the first of a series of DINAA workshops starting in 2014 gave many of our governmental data suppliers a rare chance for professional development alongside their counterparts from other jurisdictions.

## Site Security Measures

DINAA develops crosswalks between data sets from different sources to facilitate discovery across broad regions. However, we recognize that security of archaeological sites must be protected for ethical and legal reasons. In the United States, the locations of archaeological sites are highly sensitive data and their release could have grave repercussions. It is difficult to develop adequate information security measures for public-facing websites and prevent accidental data releases or data theft through hacking and other leaks. Even if we deployed appropriate security measures, our systems would need extensive auditing for compliance to Archaeological Resource Protection Act (ARPA) regulations and our project team would be legally liable for any release of sensitive data. For these reasons, managing sensitive site location data lies beyond the scope of the DINAA project, and no such information is released or even stored. To eliminate the risk of accidental or malicious disclosure of sensitive data, DINAA only stores and releases spatial coordinates at a reduced level of geographic precision. We negotiate the exact spatial resolution we use for public data with SHPOs, THPOs, and agency personnel; we expect it to be at the 20 km resolution used in the current

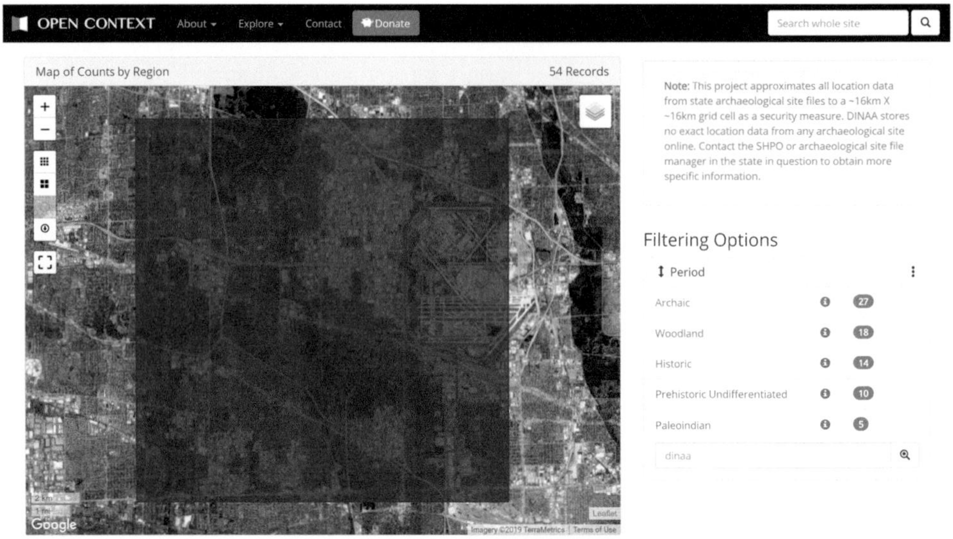


*Figure 9.3* DINAA map viewer showing the finest resolution of 20 × 20 km grid cell, in this example over O'Hare International Airport, Chicago. This particular grid cell represents 54 archaeological site records, including 5 Paleoindian, 27 Archaic, 18 Woodland, 14 historic, and 10 undifferentiated pre-contact components.[17]

iteration of DINAA (Figure 9.3), or no larger than county level, which was used in earlier efforts.[6] Though not useful for compliance reviews, DINAA's 20 km resolution facilitates important research programs and Linked Open Data applications (Figure 9.4). DINAA also associates appropriate SHPO contact information with each data record to enable qualified investigators to request higher resolution data from state officials for use in more specific geospatial research.

## DINAA's Approach to Data and Collaboration

Since 2012, the DINAA team has contacted SHPOs, THPOs, state archaeologists, and site file managers in continental North America, describing the project goals and seeking input and participation.[7] Our team has made these contacts on an annual or biannual basis, and the number of states participating by providing site information has been growing steadily as a result. This data expansion effort started in the eastern part of the continent and has grown to encompass the entire country. The DINAA team works with the archaeological site file databases held by SHPOs and allied federal and tribal agencies across North America, developing protocols for their linkage for research and management purposes. Site files contain data and metadata about the chronology, location, and function of sites, among other information used by government officials and the research community alike, and can include diagnostic artifact descriptions, radiocarbon data, and bibliographic citations.

DINAA currently documents 1,045,319 sites from 41 states (Figures 9.1 and 9.2), gathered either directly from agencies, through journal text-mining, or through links with museum collections and other online resources and repositories. We expect that

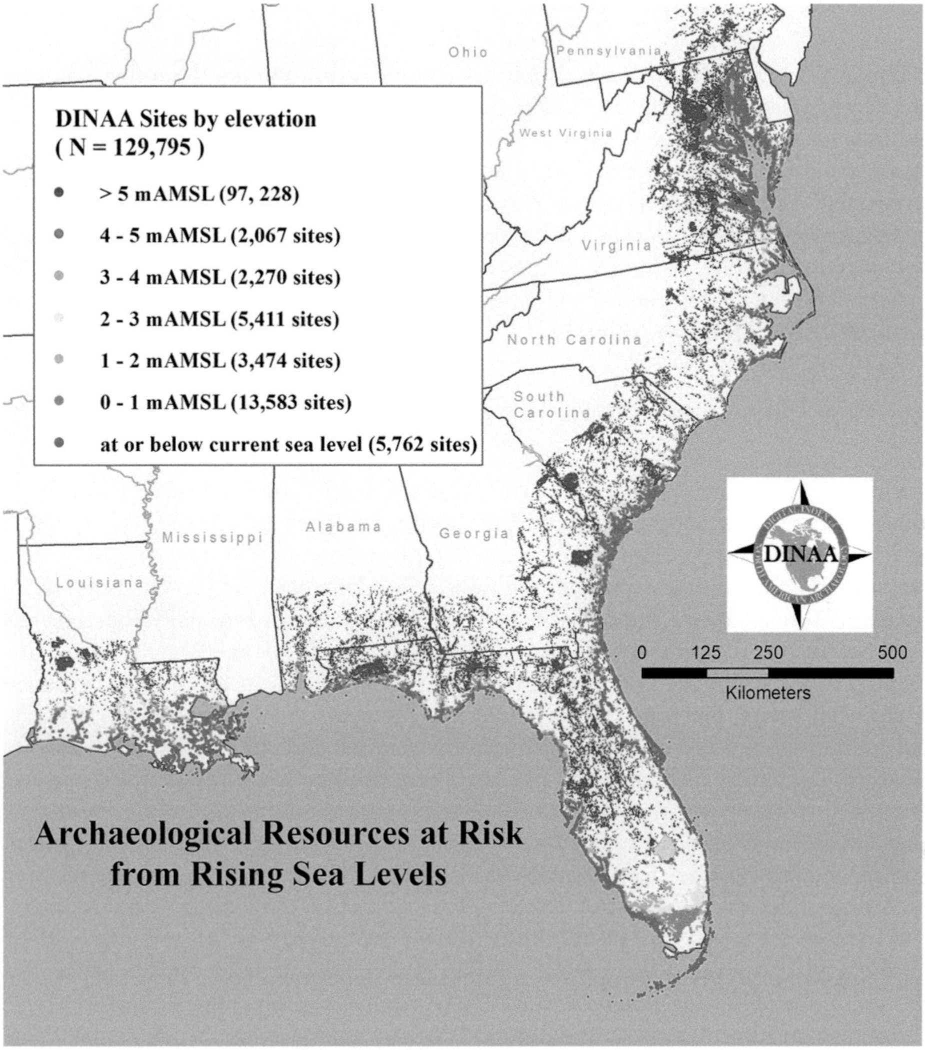


*Figure 9.4* Map showing site density as it relates to potential loss from sea-level rise and grouped by elevation in meters above present mean sea level, illustrating all sites within a buffer of 200 km from the present coastline in gray.[18]

total to rise dramatically, to approximately two million sites, when information managers in the remaining states join in the effort. As the utility and comprehensiveness of DINAA continues to grow, and as the results of efforts within the profession to make heritage information more generally available take hold, we expect DINAA to achieve its goal of encompassing most of the country. The public can download these records (with precise location and other sensitive data redacted, see following) free of charge, and free of intellectual property restrictions, via Open Context (opencontext.org), an open access data publishing service (Figure 9.2). In most cases, the unique identifier for each site is the Smithsonian Trinomial (although some states use their own identifier system). These records cross-reference reports, museum collections, bibliographic references, and other online datasets that reference the same trinomials.

While DINAA continues to add new sites from across the continent, a recent increase has come through linkages with a wide range of sources, in addition to what is held in state site file systems, such as in reports in tDAR (the Digital Archaeological Record),[9] radiocarbon dates (in the Canadian Archaeological Radiocarbon Database),[10] museum objects (in the Phoebe A. Hearst Museum of Anthropology), and research databases developed by individual scholars, such as compilations of attributes for projectile points or about prehistoric structures, like those in the Eastern Woodlands Household Archaeology Data Project,[11] and the Paleoindian Database of the Americas (PIDBA).[12]

## Community Input and Iterative Design

Since its inception, DINAA has turned toward user communities for guidance on how to improve search, navigation, and data export features so that people can use the dataset with greater ease and confidence. Such feedback is vital to the project's overlapping ethical guidelines in the domains of archaeology and open government data. As an archaeological project, we adhere to principles such as the Society for American Archaeology's Principles of Archaeological ethics, especially those principles regarding accountability and public outreach to tribal communities and affiliated peoples whose direct heritage is being addressed.[13] As an open government data project, we adhere to the principles of iterative communication with stakeholders in order to address positive benefits of scientific and cultural data sharing, but also to try to identify and minimize potential negative impacts.[14] The DINAA project designers also recognize that neither the archaeological community nor nation-state governments have historically been particularly responsive to the needs of Indigenous peoples, and we consider this iterative work to be a first step in important efforts to decolonize archaeological data.

As a result of a 2017 PLOS ONE article on the impact of sea level rise on US archaeological resources (Figure 9.4),[15] Tribal government heritage officials within the Seminole Tribe of Florida contacted the DINAA team expressing interest in the DINAA database as an information source, and DINAA technologies as infrastructure to inform design choices within the Tribal government. Thus, for the past three years DINAA has engaged in a long-running series of discussions with representative officials of the Seminole Tribe of Florida about potential uses of the project results; these discussions also include more formalized interviews about the usability of the project interface and data structures. A series of interviews have helped us to better understand user needs to identify and prioritize user interface improvements on Open Context in order to make DINAA a more effective tool.

The Seminole traditional ancestral territory is located in the Southeastern states with some of the densest coverage by the DINAA database. This area also includes states containing heritage resources under threat from rising sea levels, highlighted in the *PLOS ONE* article. Because DINAA is managed as an open information project, with no intellectual property restrictions hindering reuse, the Seminole have embarked on experimental exercises to test the capacity of DINAA to interoperate with their governmental GIS systems, to assist tribal heritage planning on massive scales;

they are also considering how DINAA technologies could be used to promote archaeological education and other heritage information to tribal members in ways that do not endanger protected archaeological site information and also respect cultural sensibilities regarding heritage resources. Seminole officials have chosen to maintain close contact with the DINAA project and have engaged in an ongoing series of conference calls as they investigate the potential of DINAA for their own purposes. These tests have also formed some of the basis for other use-testing interviews involving Seminole governmental heritage and geospatial experts, as the project assesses the overall usability and user-friendliness of web interfaces and data product organization.

Collaboration with the Chippewa Cree THPO (and additional consultation with the Eastern Band of Cherokee Indians THPO) has been essential in crafting and promoting ethical and responsive best practices in developing and using DINAA. The Chippewa Cree THPO works within the larger Chippewa Cree Cultural Resources Preservation Office on Rocky Boy's Indian Reservation, and has developed a tribal cultural monitoring program and consultation database that utilizes the traditional knowledge of tribal members in meeting the requirements of NHPA Section 106 CFR Part 800, to allow commentary by stakeholders on the effects of undertaking on identified historic and culturally significant properties. During the initiation of tribal consultation, THPO representatives receive detailed archaeological survey reports for each project generated by cultural resource management firms, but these often lack general background information on the archaeological resources previously documented in each area by SHPOs. Co-author Myers (at that time working with the Chippewa Cree THPO) identified the following needs:

- State agencies often do not allow access to their databases without archaeological credentials, in effect, gatekeeping information from tribal communities. DINAA must reduce these barriers.
- Tribal community members need multiple routes to find information. Straightforward user interfaces, direct links from THPO webpages, and other measures may be required.
- Tribal interests extend over multiple state boundaries. The Chippewa Cree monitor a ten-state area and the Eastern Shoshone monitor a 16-state area. By aggregating across state lines, DINAA can facilitate discovery of needed information with good search and mapping features.
- Technical jargon and complexity will often limit use. DINAA needs to develop clear and accessible tutorials, especially videos and explanatory graphics.

In an ongoing series of annual DINAA workshops since 2014, the DINAA team has worked with potential partners about what DINAA is and how it operates, and to learn from each other and improve overall practice while growing DINAA. The 2019 DINAA Workshop in Berkeley, California, brought together DINAA team members, researchers, museum and library representatives, tribal heritage experts, and data managers from partner states as well as potential DINAA partner states. The goals were 1) to discuss strategies for adding states not currently participating in DINAA, 2) to establish opportunities for training (such as establishing data carpentry courses), 3) to explore greater integration with archaeology in cultural resources management (CRM), 4) to discuss longer-term management of data, specifically with regard to tribal governance, and 5) to plan future data acquisition strategies. One conclusion of the meeting was to expand DINAA's efforts by compiling data from a variety of online sources, including museum collections records, journal articles, and research databases. One idea being explored is to take bibliographic records generated with public funds in the National Archaeological Data Base (NADB) and link the bibliographic information to specific site records. We also concluded that major next steps for funding should involve establishing a longer-term plan for supporting 1) regular engagement with a (compensated) DINAA governance board of tribal representatives and 2) ongoing face-to-face workshops with state and tribal data managers in order to do trainings, get feedback for improvements to DINAA, and start to build a broader community of individuals who can work with Linked Open Data in US archaeology and cultural heritage.

## Learning From Implementation

As a data management tool, DINAA has the potential to transform the way we think about and conduct basic archaeological research, data and heritage management, and public education in the United States and beyond. The integration of site file data at continental scales in an open and readily accessible informational infrastructure allows, for the first time, the exploration of the North American archaeological record across multiple temporal periods and geographic regions. The utility of such a resource was directly seen in the 2017 *PLOS ONE* paper discussed previously, that showed the effect of sea-level rise on known archaeological sites and properties listed as eligible for inclusion on the National Register of Historic Places.[16] For the first time, the entire site database from a substantial portion of the Gulf and Atlantic coasts of the United States was examined, showing the scale of the problem, and making recommendations for management and mitigation (Figure 9.4).

However, DINAA's greatest value for museums, libraries and the public centers on Linked Open Data (LOD) applications. Open Context, like other LOD systems, emphasizes the use of stable Web Uniform Resource Identifiers (URIs, i.e., stable URLs that serve as universally unique "primary key" identifiers) to identify concepts and other entities so they can be easily and precisely referenced and related across different data collections on the Web. DINAA uses Open Context and the EZID service to mint persistent URIs for each site files record. In archaeology and historical geography, the "site" is a key organizational entity. Minting stable Web URIs and offering rich temporal, geographic, and cultural metadata (also available in machine-readable JSON-LD format) about sites will therefore create significant LOD resources essential for broadly integrating museum, library, and scientific datasets.

DINAA cross-references diverse museum, library, and archival resources. However, using DINAA involves several challenges for organizations that may have limited technical support and staffing. In our experience, efforts to reuse data offer some of the best ways to discover problems in data and data services.[19] To build experience needed to guide uses of DINAA, we have undertaken the following activities:

1. **Expansion through text-mining:** As a way of expanding DINAA's indexing capabilities, and adding site information from areas where we do not yet have site file data, we developed text mining software to find references to sites reported using a Smithsonian/River Basin Survey trinomial site numbering format in

published online journals and other data sources. An initial test found numerous references to archaeological sites in back issues of *American Antiquity*. In 2019, we began asking state information managers for lists of sites by county in their state, identified by their Smithsonian trinomial codes, so we can begin indexing items even in the absence of specific site information from their office. Ten states responded positively to this request, with the result that the DINAA team now has well over one million site numbers, of which 881,243 are already indexed.[20]

2. **Linking to external collections:** We established cross-referenced linkages between DINAA and the Federal Register, the primary source of US government regulatory determinations. The Federal Register references archaeological site records, and DINAA provides a powerful index illustrating the geographic and chronological scope of US government cultural heritage management. Similarly, in a collaboration with the Phoebe A. Hearst Museum of Anthropology, we indexed some 5,000 sites in California available in the Hearst's public online collection. DINAA can now more easily cross-reference with additional Hearst records because of improvements in the Hearst's collections management system resulting from this project. Furthermore, because the Hearst Museum's collections are documented in CollectionSpace, an open source museum information system that is used by several institutions that also curate significant North American collections (most notably the San Diego Museum of Man), extending CollectionSpace to use DINAA, as well as documenting cost-effective implementation methods, will help future institutional partnerships. While the technologies for connecting DINAA to other resources now function well, the greater challenge of determining what resources should be linked remains (see following).
3. **DINAA's Linked Open Data** approach through Open Context cross-references distributed collections on the Web, enables users to find and access relevant data in other online datasets using site numbers as the common identifier. To date, these linkages include:

   a. **Links to and from tDAR:** DINAA cross references site records with tDAR metadata records. Open Context uses this information to interface with tDAR's API to display links to tDAR-archived reports and data relevant to site records in DINAA. Additionally, tDAR recently enhanced its spatial metadata records to include DINAA site file record URIs. This marks a major development in interoperability between American archaeological information systems.
   b. **Eastern Woodlands Household Archaeology Database Project (EWHADP):**[21] Andrew White, a researcher investigating household structures for Woodland period sites in the Midwest and South, has incorporated DINAA identifiers in his online database. In doing so, his datasets are precisely related to DINAA data, and the DINAA search interface can be used to discover data about ancient households compiled and curated through his research efforts.
   c. **VertNet/GBIF:**[22] VertNet is a major contributor to the Global Biodiversity Information Facility (GBIF). VertNet has started to use DINAA identifiers and data as spatial metadata for zooarchaeological specimens. This means DINAA now helps support research and information management in a much broader world of bioinformatics systems.

d. **Pelagios:** Pelagios aggregates gazetteer data and annotations that link cultural heritage content to gazetteers. In order to more broadly disseminated DINAA data and annotation, Open Context implemented Pelagios-recommended Linked Open Data standards so that DINAA data are now discoverable via the Peripleo-Pelagios network.[23] This broadens the community of researchers and software developers working with DINAA data and annotations.

As discussed, DINAA primarily serves as an index and finding aid that helps make resources scattered across the web easier to discover. But even in this role, where DINAA relies upon partner institutions to properly and ethically curate collections, DINAA runs into ethical challenges. For example, at the workshop in August 2019, DINAA showed examples of linking site records to resources in the Hearst Museum's online collection. An Indigenous archaeologist workshop participant followed links from DINAA to the Hearst's online collection and saw materials inappropriate for public display. After raising this problem with the museum, the Hearst quickly took down the inappropriate materials. This incident highlights some issues. DINAA helped facilitate identification of materials that should not be public, leading to the improved curation of the Hearst collection. At the same time, however, this incident raises important questions about DINAA's role in linking to outside resources. What resources should DINAA point toward? DINAA can and should link to public online resources maintained by Tribal Nations, but many tribal communities do not have much of an online presence. How can DINAA help highlight Indigenous voices documenting their own heritage while avoiding inappropriate or even abusive resources? Furthermore, DINAA has a continental scope and scale. Most programs that can serve as a model for good ethical practices in working in partnership with Indigenous communities occur on a local scale. How can such models for partnership work to responsively meet the needs of several hundred diverse sovereign Tribal Nations across all of North America?

## Conclusion: Inclusive Stewardship of North America's Archaeological Heritage

The DINAA project recognizes significant challenges in ethical data management, especially given the often-tragic histories of colonialism and appropriation of Indigenous land, arts, and culture.[24] Recently, the National Endowment for the Humanities and the Institute of Museum and Library Services invested in projects like Mukurtu to address Indigenous information privacy needs. DINAA complements these prior investments and also looks to new efforts, such as the Collective Benefit, Authority to Control, Responsibility, and Ethics Principles for Indigenous Data Governance (CARE),[25] which builds on the widely-cited FAIR Guiding Principles for scientific data management and stewardship (Findable, Accessible, Interoperable, and Reusable)[26] by adding CARE's four key principles. Empowering communities with respect to digital cultural heritage involves a host of issues beyond access controls and intellectual property claims (the focus of Mukurtu). Native American communities must also interface with sometimes opaque and unresponsive government agencies that hold relevant cultural heritage data. DINAA makes key information used in the management and preservation of North America's rich cultural heritage available for inspection, evaluation, and use by descendant communities that have often been marginalized from administrative and political processes. In this sense, DINAA is an "open government" project that will make cultural heritage management more accountable to wider constituencies, especially descendant communities, and will improve the government-to-government relationships that are essential to cultural heritage management by sovereign tribal nations.

While DINAA itself is and will be freely accessible open data, it can empower tribal and related institutions managing sensitive, access-restricted data. Interoperability measures between tDAR and Open Context illustrate synergies between open data and access-restricted systems. Open Context's login-free and highly granular data facilitate access and use of site data with location information redacted, for museums, educators, and research applications built on the Open Web. Moving forward, DINAA is transitioning from a "proof-of-concept" project where collaborations focused on the specific needs of heritage professionals representing three of the hundreds of Tribal Nations in North America. The DIgital Index of North American Archaeology needs an ethical governance model appropriate for a continent-scale resource. This will require formation of a Native American Governance board where members that represent Tribal Nations can set policies for content, linking, notifications and takedowns, dispute resolution, and help identify ways that DINAA can maximize positive benefits in protecting, promoting and enriching Indigenous cultural heritage. There are many open questions on how to best form and finance this Governance Board, but the need for large scale digital resources responsive to the needs of Indigenous peoples will only grow.

## Notes

1 "Digital Index of North American Archaeology (DINAA)." Accessed May 22, 2020. http://ux.opencontext.org/archaeology-site-data/. DINAA has been funded by grants from the National Science Foundation and the Institute of Museum and Library Services.

2 Braun, Sebastian F., "Culture, Resource, Management, and Anthropology: Pipelines and the Wakan at the Standing Rock Sioux Reservation." *Plains Anthropologist* 65, no. 253 (2020): 7–24.

3 For example, Atalay, Sonya, "Indigenous Archaeology as Decolonizing Practice." *American Indian Quarterly* 30, no. 3/4 (2008): 280–310; Colwell-Chanthaphonh, Chip, T. J. Ferguson, Dorothy Lippert, Randall H. McGuire, George P. Nicholas, Joe E. Watkins and Larry J. Zimmerman, "The Premise and Promise of Indigenous Archaeology." *American Antiquity* 75, no. 2 (2010): 228–238; Margaret Conkey, "Dwelling at the Margins, Action at the Intersection? Feminist and Indigenous Archaeologies." *Archaeologies* 1, no. 1 (2005): 9–59; Henderson, James Youngblood. "The Context of the State of Nature." In Marie Ann Battiste (e.d.). Reclaiming Indigenous Voice and Vision. Vancouver: UBC Press, 2000, 11–38; Lightfoot, Kent G., Lee M. Panich, Tsim D. Schneider, Sara L. Gonzalez, Matthew A. Russell, Darren Modzelewski, Theresa Molino and Elliot H. Blair, "The Study of Indigenous Political Economies and Colonialism in Native California: Implications for Contemporary Tribal Groups and Federal Recognition." *American Antiquity* 78, no. 1 (2013): 89–103; Martinez, Doreen E., "Wrong Directions and New Maps of Voice, Representation, and Engagement: Theorizing Cultural Tourism, Indigenous Commodities, and the Intelligence of Participation." *American Indian Quarterly* 36, no. 4 (2012): 545–573; Watkins, Joe. *Indigenous Archaeology: American Indian Values and Scientific Practice*. Walnut Creek: Alta Mira Press, 2000; Colwell-Chanthaphonh, Chip and T. J. Ferguson. *History is in the Land: Multivocal Tribal Traditions in Arizona's San Pedro Valley*. Tempe: University of Arizona Press, 2006; Colwell-Chanthaphonh, Chip, and T. J. Ferguson (eds.). *Collaboration in*

*Archaeological Practice: Engaging Descendant Communities*. Walnut Creek: Alta Mira Press, 2008; Houde, Nicolas, "The Six Faces of Traditional Ecological Knowledge: Challenges and Opportunities for Canadian Co-Management Arrangements." *Ecology and Society* 12, no. 2 (2007); Irlbacher-Fox, Stephanie, "Traditional Knowledge, Co-existence and Co-resistance." *Decolonization: Indigeneity, Education & Society* 3, no. 3 (2014): 145–158; Menzies, Charles R. *Traditional Ecological Knowledge and Natural Resource Management*. Lincoln: University of Nebraska Press, 2006; Simpson, Leanne B., "Land as Pedagogy: Nishnaabeg Intelligence and Rebellious Transformation." *Decolonization: Indigeneity, Education & Society* 3, no. 3 (2014): 1–25; Wiethaus, Ulrike (ed.). *Foundations of First Peoples' Sovereignty: History, Education & Culture*. New York: Peter Lang, 2007; and Wildcat, Matthew, Mandee McDonald, Stephanie Irlbacher-Fox and Glen Coulthard, "Learning from the Land: Indigenous Land Based Pedagogy and Decolonization." *Decolonization: Indigeneity, Education & Society* 3, no. 3 (2014): I–XV.

4 Altschul, Jeff H. and Thomas C. Patterson. "Trends and Employment in American Archaeology." In Ashmore, W., D. Lippert and B. J. Mills (ed.). *Voices in American Archaeology*. Washington, D.C.: Society for American Archaeology, 2010, 291–316.

5 "National Archeological Database, Reports Module." Accessed November 1, 2019. https://www.nps.gov/archeology/tools/nadb.htm.

6 For example, Anderson, David G. and Virginia Horak (eds.). *Archaeological Site File Management: A Southeastern Perspective*. Atlanta: Interagency Archeological Services Division, National Park Service, Southeast Regional Office, 1995; and "NADB-Maps Archeological SiteCounts (State Historic Preservation Officers)." Archeology Program, National Park Service, Washington, D.C. Accessed November 1, 2019. http://wayback.archive-it.org/6471/20150825214608/. http://cast.uark.edu/other/nps/maplib/USsittot.1993.html.

7 Wells, Joshua J., Eric C. Kansa, Sarah W. Kansa, Stephen J. Yerka, David G. Anderson, Thaddeus G. Bissett, Kelsey N. Myers and Robert C. DeMuth. "Web-Based Discovery and Integration of Archaeological Historic Properties Inventory Data: The Digital Index of North American Archaeology (DINAA)." *Literary and Linguistic Computing* 29, no. 3 (2014): 349–360. https://doi.org/10.1093/llc/fqu028; and Kansa, Eric C., Sarah Whitcher Kansa, Joshua J. Wells, Stephen J. Yerka, Kelsey Noack Myers, Robert C. DeMuth, Thaddeus G. Bissett and David G. Anderson, "The Digital Index of North American Archaeology (DINAA): Networking Government Data to Navigate an Uncertain Future for the Past." *Antiquity* 92, no. 362 (2018): 490–506.

8 DINAA 2019. Digital Index of North American Archaeology (DINAA) Publication and Integration of North American archaeological site file records. Accessed September 3, 2019. http://ux.opencontext.org/archaeology-site-data/dinaamap//.

9 McManamon, Francis P., Keith W. Kintigh, Leigh Aenn Ellison and Adam Brin, "tDAR: A Cultural Heritage Archive for Twenty-First-Century Public Outreach, Research, and Resource Management." *Advances in Archaeological Practice* 5, no. 3 (2017): 238–249. doi.org/10.1017/aap.2017.18.

10 "Welcome to CARD 2.0." Accessed September 20, 2019. http://www.canadianarchaeology.ca/; Chaput, Michelle A., Björn Kriesche, Matthew Betts, Andrew Martindale, Rafal Kulik, Volker Schmidt and Konrad Gajewski, "Spatiotemporal distribution of Holocene populations in North America." *Proceedings of the National Academy of Sciences* 112, no. 39 (2015): 12127–12132; and Jones, Nicola. "World's largest hoard of carbon dates goes global." *Nature News* (11 July 2017). Accessed October 31, 2019. https://www.nature.com/news/world-s-largest-hoard-of-carbon- dates-goes-global-1.22287.

11 White, Andrew. "Eastern Woodlands Household Archaeology Data Project." Accessed September 20, 2019. https://www.householdarchaeology.org/.

12 Anderson, David G., David Echeverry, D. Shane Miller, Andrew A. White, Stephen J. Yerka, Eric C. Kansa, Sarah Whitcher Kansa, Christopher R. Moore, Kelsey Noack Myers, Joshua J. Wells, Thaddeus G. Bissett, and Ashley M. Smallwood. 2019 Paleoindian Settlement in the Southeastern United States: The Role of Large Databases. In *New Directions in the Search for the First Floridians*, edited by David Thulman and Irv Garrison, pp. 241–275. University Press of Florida, Gainesville.

13 Lynott, Mark J., "Ethical Principles and Archaeological Practice: Development of an Ethics Policy." *American Antiquity* 62, no. 4 (October 1997): 589–599.
14 "Code for America, Iterative Development." Accessed May 19, 2020. https://www.codeforamerica.org/practices/iterative-development; "Open Government Partnership: Third Open Government National Action Plan for the United States of America. Washington, D.C." Accessed May 19, 2020. https://www.opengovpartnership.org/wp-content/uploads/2017/05/final_us_open_government_national_action_plan_3_0_0_0.pdf; and Open Knowledge Foundation, "How to Open up Data." In *The Open Data Handbook*, 2019. https://opendatahandbook.org/guide/en/how-to-open-up-data/.
15 Anderson, David G., Thaddeus G. Bissett, Stephen J. Yerka, Joshua J. Wells, Eric C. Kansa, Sarah W. Kansa, Robert C. DeMuth, Kelsey Noack Myers and Devin White, "Climate Change and Archaeological Site Destruction: An Example from the Southeastern United States Using DINAA (Digital Index of North American Archaeology)." *PLoS One* 12, no. 11 (2017): e0188142. https://doi.org/10.1371/journal.pone.0188142.
16 Anderson, "Climate Change and Archaeological Site Destruction," 2017.
17 DINAA 2019. Digital Index of North American Archaeology (DINAA) Publication and integration of North American archaeological site file records. Accessed September 3, 2019. http://ux.opencontext.org/archaeology-site-data/dinaamap//.
18 From Anderson, "Climate Change and Archaeological Site Destruction," 2017.
19 Kansa, Eric C., Sarah W. Kansa and Benjamin Arbuckle, "Publishing and Pushing: Mixing Models for Communicating Research Data in Archaeology." *International Journal of Digital Curation* 9, no. 1 (2014): 57–70.
20 Digital Index of North American Archaeology (DINAA). Publication and integration of North American archaeological site file records. Accessed September 3, 2019. http://ux.opencontext.org/archaeology-site-data/dinaamap//.
21 White, Andrew. "Eastern Woodlands Household Archaeology Data Project." Accessed September 20, 2019. https://www.householdarchaeology.org/.
22 For example, LeFebvre, Michelle J., Laura Brenskelle, John Wieczorek, Sarah Whitcher Kansa, Eric C. Kansa, Neill J. Wallis, Jessica N. King, Kitty F. Emery and Robert Guralnick, "ZooArchNet: Connecting zooarchaeological specimens to the biodiversity and archaeology data networks." *PLoS One* 14, no. 4 (2019): e0215369, https://doi.org/10.1371/journal.pone.0215369.
23 "Open Context on Peripleo-Pelagios." Accessed May 26, 2020. https://peripleo.pelagios.org/ui#selected=http%3A%2F%2Fopencontext.org%2Fprojects.
24 Chander Anupam, and Madhavi Sunder, "The Romance of the Public Domain." *California Law Review* 92 (2004); Kansa, Eric C., Jason Schultz and Ahrash N. Bissell, "Protecting Traditional Knowledge and Expanding Access to Scientific Data: Juxtaposing Intellectual Property Agendas via a 'Some Rights Reserved' Model." *International Journal of Cultural Property* 12, no. 3 (2005): 285–314; Kansa, Eric C., "Openness and Archaeology's Information Ecosystem." *World Archaeology* 44, no. 4 (2012): 498–520. https://doi.org/10.1080/00438243.2012.737575; Christen, Kimberly A., "Does Information Really Want to Be Free? Indigenous Knowledge Systems and the Question of Openness." *International Journal of Communication* 6, (2012): 24; Christen, Kimberly A., "Tribal Archives, Traditional Knowledge, and Local Contexts: Why the 's' Matters." *Journal of Western Archives* 6, no. 1 (2015).
25 "CARE Principles for Indigenous Data Governance." Accessed May 22, 2020. https://www.gida-global.org/care.
26 "FAIR Principles," Accessed May 22, 2020. https://www.go-fair.org/fair-principles/.

## Bibliography

Altschul, Jeff H. and Thomas C. Patterson. "Trends and employment in American archaeology." In W. Ashmore, D. Lippert and B. J. Mills. Edited by *Voices in American Archaeology*. Washington, DC: Society for American Archaeology, 2010: 291–316.

Anderson, David G., Thaddeus G. Bissett, Stephen J. Yerka, Joshua J. Wells, Eric C. Kansa, Sarah W. Kansa, Robert C. DeMuth, Kelsey Noack Myers, and Devin White. "Climate Change and Archaeological Site Destruction: An Example from the Southeastern United

States Using DINAA (Digital Index of North American Archaeology)." *PLoS One 12*, no. 11 (2017): e0188142. https://doi.org/10.1371/journal.pone.0188142.

Anderson, David G. and Virginia Horak (eds.). *Archaeological Site File Management: A Southeastern Perspective*. Atlanta: Interagency Archeological Services Division, National Park Service, Southeast Regional Office, 1995.

Anderson, David G., and Shane Miller. "PIDBA (Paleoindian Database of the Americas): Call for Data." *PaleoAmerica 3*, no. 1 (2017): 1–5.

Anderson, David G., Shane Miller, Stephen J. Yerka, J. Christopher Gillam, Erik N. Johanson, Derek T. Anderson, Albert C. Goodyear, and Ashley M. Smallwood. "PIDBA (Paleoindian Database of the Americas) 2010: Current Status and Findings." *Archaeology of Eastern North America 38* (2018): 63–90.

Atalay, Sonya. "Indigenous Archaeology as Decolonizing Practice." *American Indian Quarterly 30*, no. 3/4 (2008): 280–310.

Braun, Sebastian F. "Culture, Resource, Management, and Anthropology: Pipelines and the Wakan at the Standing Rock Sioux Reservation." *Plains Anthropologist 65*, no. 253 (2020): 7–24.

"CARE Principles for Indigenous Data Governance." Accessed May 22, 2020. https://www.gida-global.org/care.

Chander, Anupam, and Madhavi Sunder. "The Romance of the Public Domain." *California Law Review 92* (2004).

Chaput, Michelle A., Björn Kriesche, Matthew Betts, Andrew Martindale, Rafal Kulik, Volker Schmidt, and Konrad Gajewski. "Spatiotemporal Distribution of Holocene Populations in North America." *Proceedings of the National Academy of Sciences 112*, no. 39 (2015): 12127–12132.

Christen, Kimberly A. "Does Information Really Want to Be Free? Indigenous Knowledge Systems and the Question of Openness." *International Journal of Communication 6*, (2012): 24.

Christen, Kimberly A. "Tribal Archives, Traditional Knowledge, and Local Contexts: Why the 's' Matters." *Journal of Western Archives* 6, no. 1 (2015).

"Code for America, Iterative Development." Accessed May 19, 2020. https://www.codeforamerica.org/practices/iterative-development.

Colwell-Chanthaphonh, Chip, and T. J. Ferguson. *History is in the Land: Multivocal Tribal Traditions in Arizona's San Pedro Valley*. Tempe: University of Arizona Press, 2006.

Colwell-Chanthaphonh, Chipe, and T. J. Ferguson (eds.). *Collaboration in Archaeological Practice: Engaging Descendant Communities*. Walnut Creek: Alta Mira Press, 2008.

Colwell-Chanthaphonh, Chip, T. J. Ferguson, Dorothy Lippert, Randall H. McGuire, George P. Nicholas, Joe E. Watkins, and Larry J. Zimmerman. "The Premise and Promise of Indigenous Archaeology." *American Antiquity 75*, no. 2 (2010): 228–238.

Conkey, Margaret. "Dwelling at the Margins, Action at the Intersection? Feminist and Indigenous Archaeologies." *Archaeologies 1*, no. 1 (2005): 9–59.

"Digital Index of North American Archaeology (DINAA)." Accessed May 22, 2020. http://ux.opencontext.org/archaeology-site-data/.

"FAIR Principles." Accessed May 22, 2020. https://www.go-fair.org/fair-principles/.

Henderson, James Youngblood. "The Context of the State of Nature." In*Reclaiming Indigenous Voice and Vision*. Edited by Marie Ann Battiste. Vancouver: UBC Press, 2000: 11–38.

Houde, Nicolas. "The Six Faces of Traditional Ecological Knowledge: Challenges and Opportunities for Canadian Co-Management Arrangements." *Ecology and Society 12*, no. 2 (2007).

Irlbacher-Fox, Stephanie. "Traditional Knowledge, Co-existence and Co-resistance." *Decolonization: Indigeneity, Education & Society 3*, no. 3 (2014): 145–158.

Jones, Nicola. "World's Largest Hoard of Carbon Dates Goes Global." *Nature News* (11 July

2017). Accessed October 31, 2019. https://www.nature.com/news/world-s-largest-hoard-of-carbon-dates-goes-global-1.22287.

Kansa, Eric C. "Openness and Archaeology's Information Ecosystem." *World Archaeology 44*, no. 4 (2012): 498–520. https://doi.org/10.1080/00438243.2012.737575.

Kansa, Eric C., Sarah W. Kansa, and Benjamin Arbuckle. "Publishing and Pushing: Mixing Models for Communicating Research Data in Archaeology." *International Journal of Digital Curation* 9, no. 1 (2014): 57–70.

Kansa, Eric C., Sarah Whitcher Kansa, Joshua J. Wells, Stephen J. Yerka, Kelsey Noack Myers, Robert C. DeMuth, Thaddeus G. Bissett, and David G. Anderson. "The Digital Index of North American Archaeology (DINAA): Networking Government Data to Navigate an Uncertain Future for the Past." *Antiquity 92*, no. 362 (2018): 490–506.

Kansa, Eric C., Jason Schultz, and Ahrash N. Bissell. "Protecting Traditional Knowledge and Expanding Access to Scientific Data: Juxtaposing Intellectual Property Agendas via a 'Some Rights Reserved' Model." *International Journal of Cultural Property 12*, no. 3 (2005): 285–314.

LeFebvre, Michelle J., Laura Brenskelle, John Wieczorek, Sarah Whitcher Kansa, Eric C. Kansa, Neill J. Wallis, Jessica N. King, Kitty F. Emery, and Robert Guralnick. "ZooArchNet: Connecting Zooarchaeological Specimens to the Biodiversity and Archaeology Data Networks." *PLoS One 14*, no. 4 (2019): e0215369. https://doi.org/10.1371/journal.pone.0215369.

Lightfoot, Kent G. Lee M. Panich, Tsim D. Schneider, Sara L. Gonzalez, Matthew A. Russell, Darren Modzelewski, Theresa Molino, and Elliot H. Blair. "The Study of Indigenous Political Economies and Colonialism in Native California: Implications for Contemporary Tribal Groups and Federal Recognition." *American Antiquity 78*, no. 1 (2013): 89–103.

Lynott, Mark J. "Ethical Principles and Archaeological Practice: Development of an Ethics Policy." *American Antiquity 62*, no. 4. (Oct. 1997): 589–599.

Martinez, Doreen E. "Wrong Directions and New Maps of Voice, Representation, and Engagement: Theorizing Cultural Tourism, Indigenous Commodities, and the Intelligence of Participation." *American Indian Quarterly 36*, no. 4 (2012): 545–573.

McManamon, Francis P., Keith W. Kintigh, Leigh Aenn Ellison, and Adam Brin. "tDAR: A Cultural Heritage Archive for Twenty-First-Century Public Outreach, Research, and Resource Management." *Advances in Archaeological Practice 5*, no, 3 (2017): 238–249. https://doi.org/10.1017/aap.2017.18.

Menzies, Charles R. *Traditional Ecological Knowledge and Natural Resource Management*. Lincoln: University of Nebraska Press, 2006.

"National Archeological Database, Reports Module." Accessed November1, 2019. https://www.nps.gov/archeology/tools/nadb.htm.

"NADB-Maps Archeological SiteCounts (State Historic Preservation Officers)." Washington, D.C., Archeology Program, National Park Service, Accessed November1, 2019. http://wayback.archive-it.org/6471/20150825214608/http://cast.uark.edu/other/nps/maplib/USsittot.1993.html.

"Open Context on Peripleo-Pelagios." Accessed May 26, 2020. https://peripleo.pelagios.org/ui#selected=http%3A%2F%2Fopencontext.org%2Fprojects.

"Open Government Partnership: Third Open Government National Action Plan for the United States of America. Washington, D.C." Accessed May 19, 2020. https://www.opengovpartnership.org/wp-content/uploads/2017/05/final_us_open_government_national_action_plan_3_0_0_0.pdf.

Open Knowledge Foundation. "*How to Open up Data*," in *The Open Data Handbook*, (2019). https://opendatahandbook.org/guide/en/how-to-open-up-data/.

Simpson, Leanne B. "Land as Pedagogy: Nishnaabeg Intelligence and Rebellious Transformation." *Decolonization: Indigeneity, Education & Society 3*, no. 3 (2014): 1–25.

Watkins, Joe. *Indigenous Archaeology: American Indian Values and Scientific Practice*. Walnut Creek: Alta Mira Press, 2000.

"Welcome to CARD 2.0." Accessed September20, 2019. http://www.canadianarchaeology.ca.
Wells, Joshua J., Eric C. Kansa, Sarah W. Kansa, Stephen J. Yerka, David G. Anderson, Thaddeus G. Bissett, Kelsey N. Myers, and Robert C. DeMuth. "Web-based discovery and integration of archaeological historic properties inventory data: The Digital Index of North American Archaeology (DINAA)."*Literary and Linguistic Computing29*, no. 3 (2014): 349–360. https://doi.org/10.1093/llc/fqu028.
White, Andrew. "*Eastern Woodlands Household Archaeology Data Project.*" Accessed September20, 2019. https://www.householdarchaeology.org/.
Wiethaus, Ulrike (editor). *Foundations of First Peoples' Sovereignty: History, Education & Culture* (New York: Peter Lang, 2007).
Wildcat, Matthew, Mandee McDonald, Stephanie Irlbacher-Fox, Glen Coulthard. "Learning from the Land: Indigenous Land Based Pedagogy and Decolonization." *Decolonization: Indigeneity, Education & Society* 3, no. 3 (2014): I–XV.

# 10 Finding Balance Between Development and Conservation

## The O'ahu Greenprint

*Holly Bostrom, Lea Hong, and Breece Robertson*

### Introduction

The island of O'ahu is diverse and multi-faceted. From fast-paced downtown Honolulu, to the agricultural plains of Central O'ahu, to heavily touristed Waikīkī beach, to the small communities along the wet Ko'olau Range and arid Wai'anae range, to the big surf of the North Shore, O'ahu is a study in contrasts. While the island and its people have experienced urbanization, residents and visitors love the island's cultural, historical, and natural landscapes. Outside of Honolulu's urban core, residents demand to "Keep the Country, Country." O'ahu is not the largest Hawaiian island, encompassing just 600 square miles of land, but with approximately one million residents it is the most populated and most dense. And its population continues to grow. The resident population of the City and County of Honolulu is projected to increase at an annual rate of 0.4% from 2010 to 2040.[1]

Along with this increased density, large areas of O'ahu are for sale, and development is planned or occurring on productive agricultural and scenic open space across the island. These changes are resulting in community tension, conflicts, and even costly litigation. How does O'ahu plan for future growth while preserving lands most vital to the island's past and future? The people of O'ahu are currently grappling with a number of issues related to increased development, congestion, and sprawl. Some of the pressing issues, as identified through interviews with community stakeholders, include:

- Managing population growth, which may include directing development toward urban centers, reducing congestion, providing the amenities needed for urban communities to thrive, ensuring affordable housing and maintaining quality-of-life for residents.
- Reducing divisiveness around development.
- Increasing food and energy security through the protection of lands for agriculture and renewable energy.
- Reducing invasive species in a fragile island environment.
- Adapting to climate change, particularly on the coast.
- Ensuring access to parks, trails and open space by residents.

It is clear that O'ahu's farms, sacred places, picturesque communities, pristine beaches, and verdant mountains weave an important part of the interconnected story of the island and its people. These landscapes hold *mo'olelo* (stories) that tie us to

Hawai'i's ancestral past and current practices; give us beautiful places to enjoy and explore; give us the opportunity to achieve food diversity and security; and keep us connected to a flourishing and vibrant Hawaiian culture.

In this context, much of The Trust for Public Land's work in Hawai'i focuses on returning culturally significant land to Native Hawaiians through nonprofit ownership, or structuring government agency land purchases with co-stewardship roles for Native Hawaiian organizations and agencies. To protect and support the return of culturally significant land to Native Hawaiians, The Trust for Public Land, the Office of Hawaiian Affairs, the Castle Foundation, and others, worked with local community and agency partners to map O'ahu's most culturally significant lands through the O'ahu Greenprint.[2,3] In the Native Hawaiian worldview, all land is culturally and spiritually important. The O'ahu Greenprint Cultural Resource layer is NOT intended to label any particular parcel land or site to be better or more culturally significant than any other. The layer is intended to help communities identify and prioritize those lands that have the greatest opportunity or likelihood of success using voluntary land conservation tools.

### *Why a Greenprint for O'ahu?*

The Greenprint for O'ahu is a way to address the environmental, cultural, and land use concerns facing the island today. The process began with a Greenprint for the North Shore (in partnership with the North Shore Community Land Trust) in 2012. The effort expanded to all of O'ahu in 2013 in partnership with the Office of Hawaiian Affairs (OHA) and others. Together, these entities worked to create a set of maps that identify high priority lands for voluntary conservation and develop strategies for implementation. In order to accomplish these goals, OHA and TPL, with support from Townscape, Inc., engaged with a broad cross-section of residents to quantify the subjective values that the people of O'ahu hold for conservation. These values were then translated into maps that highlight the special areas that embody the conservation values of residents. The result is a sustainable plan based on local priorities and grounded in science designed to meet community conservation goals of protecting the island's iconic cultural, historical, agricultural, natural, coastal, and scenic landscapes. Ultimately, this plan can guide voluntary land conservation with willing landowners in order to honor local values and culture as it incorporates hard science.

### *Mission and Objectives*

The O'ahu Greenprint kicked off with stakeholder meetings in which participants (collectively called the Island Leadership Team) developed and then confirmed a mission and set of objectives for the project. They are as follows:

### *Mission Statement*

To understand the values O'ahu residents associate with land and water resources and to use that community-based knowledge to develop a conservation plan that will help perpetuate those values for present and future generations by guiding purchases of threatened, privately owned land and resources through voluntary fee simple acquisition or conservation easements from willing landowners.

### *Objectives*

1. Establish the relative priority of values residents place on land and resources from the mountains into the ocean, including cultural, recreational, natural, historic, agricultural, subsistence, and other values.
2. Identify which important lands and water resources are most threatened or most in need of restoration and develop strategies to protect them and the associated values that make them important.
3. Strengthen and coordinate existing conservation networks island-wide.
4. Promote long-term conservation, restoration, and stewardship (*aloha 'āina*) that is strategic and responsive to community needs and values.
5. Increase community awareness of and ensure proactive community action related to:

    a. The value of and need to protect and revitalize Hawaiian cultural sites, places, landscapes, and historic properties.
    b. The importance of other conservation values such as agricultural, natural, and recreational values.
    c. The importance of creating synergy between all conservation values.
    d. The imperative for a balanced approach to growth.
    e. The tools for preserving critical lands.

6. Acknowledge and honor important resources that can't be mapped.

## The Greenprinting Process

### *Timeline at-a-Glance*

- Spring-Winter 2013: Community Engagement and Current Conditions Review
  In developing the Greenprint, TPL first completed a current conditions review of O'ahu. The TPL and OHA (project team) built on that research through community outreach including SpeakOuts, interviews, and Island Leadership Team meetings. Through this process, the project team was able to establish the value residents place on important cultural, recreational, and natural resources on O'ahu. A Technical Advisory Team (TAT) was created to guide the mapping process. Please see the Acknowledgments section for the full list of attendees at Island Leadership Team meetings and members of the Technical Advisory Team.
- Fall 2013–Winter 2014: Mapping Conservation Values
  GIS mapping experts from both TPL and OHA worked with the TAT to collect data, refine that data, and integrate it with the community-developed values to create maps identifying the best opportunities for land protection throughout O'ahu.
- Spring 2014: Island Leadership Team Creates an Action Plan for Realizing the Greenprint
  The project team compiled action ideas from the public outreach process and presented a draft action plan to the Island Leadership Team in March 2014. The ILT offered suggestions for refining the action plan for realizing the Greenprint in June 2014 and confirmed the Greenprint maps.
- Summer 2014: Finalize Greenprint Results

The project team created and placed the mapping site on-line and developed final materials.

## The O'ahu Greenprint

A. Current Conditions and Community Engagement

### *Current Conditions Review*

In order to develop the Current Conditions Report, The Trust for Public Land staff collected and synthesized background information. This information included related planning efforts such as local sustainable community and development plans, water management plans, state planning and economic development efforts, and demographic trends. The report provided context throughout the planning process and informed the project objectives, mapping analysis and action plan.

### *Community Engagement*

#### *SpeakOuts*

A SpeakOut is an interactive display where organizers can share informative materials on an issue and participants can express their views and provide their feedback in an informal and cooperative environment. In order to understand the community's conservation priorities, OHA and The Trust for Public Land worked with Townscape, Inc. to develop and host a booth at several community events throughout O'ahu in a "SpeakOut" style format. At these events, staff shared information on conservation and Greenprints with event participants and feedback was solicited on specific questions.

The O'ahu Greenprint team attended 12 events from May through December 2013, hosting booths in a "SpeakOut" style format. The goal was to solicit ideas and share information about the Greenprint with residents throughout the island's eight planning districts at community events, rather than through formal public hearings.

The project team solicited feedback in a variety of ways including through surveys (hard copy and on-line), mapping exercises, voting on conservation value priorities and even drawing pictures of favorite outdoor activities for keiki (children). A total of 910 surveys were completed, 784 at the SpeakOuts and 126 online. In the survey, participants were asked several questions to gain an understanding of the types of land most important to protect. Responses painted a picture of what residents value most about O'ahu, what lands are at risk of changing, and what people want for their island over the next 25 years.

#### *The Island Leadership Team*

The project team developed a stakeholder committee called the Island Leadership Team, comprised of more than 90 experts and interested residents. This team met in July 2013, October 2013, March 2014, and June 2014.

### *Interviews*

The project team completed 25 interviews by phone and in-person throughout O'ahu with stakeholders during the fall of 2013. The vast majority of these interviews took place in person in Honolulu from October 25-November 1, 2013. Through these interviews we received advice and candid feedback about what a successful Greenprint would entail and how to ensure this project and plan are implemented toward that goal.

In total, we estimate that that the O'ahu Greenprint Team reached more than 1,300 people through this public engagement effort.

### *Mapping Conservation Values*

Geographic Information Systems (GIS) analysis uses satellite photos, aerial photos, and on-the-ground verification to interpret the geography of a place. The conservation values expressed by residents through the public outreach and stakeholder engagement process informed the development of GIS maps. These maps, in turn, served as the basis for the online mapping tool that enables users to generate land conservation opportunity maps for the island.

In moving from the public outreach phase of the Greenprint to the Geographic Information System (GIS) mapping and modeling phase, a Technical Advisory Team (TAT) of local mapping experts provided strategic advice on data collection and data modeling. The TAT was responsible for making recommendations related to data. Their advice was invaluable in developing the criteria for each goal; identifying the best available data and their sources; and advising through the modeling process to ensure that that modeling assumptions were based on defensible science and that input data and model results were accurate.

### *The Greenprint Maps*

A set of seven conservation values emerged from the public outreach process. The TAT analyzed public input from surveys, Island Leadership Team guidance, and SpeakOut information to inform the development of a set of color coded maps of O'ahu that incorporate those land and water conservation values. By showcasing lands most in need of protection, these maps will allow organizations and agencies to coordinate planning and pursue projects where investments of limited dollars can yield the highest possible conservation impact. Today, 51,749 acres of land, nearly 18% of the study area, are already conserved. These lands include beach parks, botanical gardens, community parks, district parks, regional cultural areas, beach access walkways, steep slope areas, and state parks. Nearly a quarter of the study area—34,970 acres—is protected as a National Marine Sanctuary. Also, 51,501 acres (17.64% of the study area) are considered developed, whereas 10,157 acres are agricultural (Homer et al. 2015).

There is one map for each of the Greenprint values. These values are as follows: Protect Agricultural Lands, Preserve Cultural and Historic Places, Protect Coastal Regions, Protect Natural Habitats, Increase Recreation and Public Access Opportunities, Preserve and Enhance Viewplanes, and Protect Water Quality and Quantity.

While The Trust for Public Land and the Office of Hawaiian Affairs strive to provide the best data available, we depend on sources outside of our organizations for much of our information. We also acknowledge that there are likely data gaps in the mapped conservation values due to undocumented cultural sites, native natural habitat and other resources. Thus, there may be very special and significant places not yet reflected in the maps. Our maps are meant to provide information that can encourage various conservation efforts. The Trust for Public Land and the Office of Hawaiian Affairs are not responsible for any errors, omissions, or positional accuracy. This map is provided without warranties, expressed or implied, and will be updated and corrected over time as new data becomes available.

In the tables that follow (Tables 10.1 and 10.2), "study area" encompasses the 292,017 acres of land and 148,870 off-shore acres in the O'ahu Greenprint. These numbers were then combined with data derived from the North Shore Greenprint to get information for the whole island of O'ahu.

The seven maps describe the values as follows (in no particular order):

### *Protect Agricultural Lands*

O'ahu's subtropical climate and year round growing season provides ideal conditions for agriculture. However, O'ahu's topography, with the Wai'anae mountain range to the west and the Ko'olau mountain range running along its eastern coast, ensures that much of the island is steeply sloped. Agriculture is therefore focused in the fertile area between the ranges and within smaller coastal valleys.

A century ago, much of Hawai'i's agricultural lands were used for sugarcane and pineapple production. In central O'ahu, many of those lands are converting to diversified-crop agriculture or suburban use. Today, Hawai'i still imports 85–90% of its food (Office of Planning—Department of Business Economic Development & Tourism, and Department of Agriculture—State of Hawaii 2012, ii). Stakeholders consistently prioritized a move toward a more sustainable and self-reliant agricultural system on O'ahu in which farms cultivate locally consumed food crops.

Approximately 60,000 acres (15%) of the study area is zoned for agriculture and half of that is considered prime agricultural lands.

*Table 10.1* Conservation Opportunity Lands by Conservation Value

| *Greenprint Value* | *High Priority* Areas for Protection (% of Study Area)* | *High Priority Areas for Protection (% of O'ahu)* |
|---|---|---|
| Protect Agricultural Lands | 27,737 (10%) | 38,928 (10%) |
| Preserve Cultural and Historic Places | 88,097 (30%) | 98,760 (26%) |
| Protect Coastal Regions | 2,594 (1%) | 3,849 (1%) |
| Protect Natural Habitats** | 35,342 (12%) | 46,749 (12%) |
| Increase Recreation and Public Access Opportunities | 14,590 (5%) | 18,028 (5%) |
| Preserve and Enhance View Planes | 11,678 (4%) | 18,857 (5%) |
| Protect Water Quality and Quantity | 66,230 (23%) | 101,887 (27%) |

Notes

* "High-priority" areas reflect a score of "5" on a scale of 0–5 for the Greenprint. There may also be overlap between values.

** Includes North Shore Greenprint value "Protect Natural Habitats for Plants and Animals."

*Table 10.2* Conservation Opportunity Off-shore Areas by Conservation Value

| *Greenprint Value* | *High Priority* Areas for Protection (% of Off-Shore Study Area)* | *High Priority Areas for Protection (% of Off-Shore O'ahu**)* |
|---|---|---|
| Protect Cultural and Historic Places | n/a | 64 (0.03%) |
| Protect Coastal Regions | 36,376 (24%) | 40,916 (22%) |
| Protect Natural Habitats*** | 1,017 (1%) | 3,843 (2%) |
| Increase Recreation and Public Access Opportunities | 860 (1%) | 3,476 (2%) |
| Preserve and Enhance View Planes | 1,418 (1%) | 1462 (0.08%) |
| Protect Water Quality and Quantity | 1,531 (1%) | 1,611 (0.08%) |

*Notes*

* High-priority areas reflect a score of "5" on a scale of 0–5 for the Greenprint. There may also be overlap between values.

** "Off-shore O'ahu" includes 182,763 off-shore acres studied across the North Shore and O'ahu Greenprints.

*** Includes North Shore Greenprint value "Protect Natural Habitats for Plants and Animals."

This map reveals the results for the Protect Agricultural Lands value. Criteria considered included: protect lands to grow more taro and other traditional Hawaiian crops; protect farmlands; protect prime and important agricultural lands; identify lands zoned for agriculture; and protect large scale agriculture dedicated to year-round local consumption, including fish, produce and animal products. The greatest weight was applied to protecting prime and important agricultural lands, protecting farmlands, and protecting lands to grow traditional Hawaiian crops.

About 28,000 acres of land are identified as high priority (9.5% of the study area) for this value. Just over 1,200 acres have already been conserved (less than 5% of the study area). Across the island of O'ahu, nearly 40,000 acres have been identified as high priority for protecting agricultural lands. Outside of the developed areas of O'ahu, these conservation priority areas extend across much of the island, particularly in central O'ahu, North Shore, and the Leeward coast.

### *Preserve Cultural and Historic Places*

The Trust for Public Land map identifies areas identified as priority for the protection of cultural and historic places. The five criteria informing the development of the map were done through public outreach. Through the questionnaire, participants shared places throughout the island that they felt are in need of protection, including Hawaiian cultural and traditional agricultural sites such as lo'i, fishponds, and places necessary for cultural and religious practices. The resulting criteria were each weighted equally. These criteria include significant archeological, historic, sacred, religious sites and burial grounds; existing fishponds; historic ahupua'a (sustainable land units); cultural trails; and traditional Hawaiian agricultural zones. The equal weighting came from the belief that no cultural or historical site could be determined to be of a higher priority than another. For example, how can we determine that an ancient burial site is somehow more or less important than a current place of worship?

About 88,000 acres of land are identified as high priority (30% of the study area) for this value. Nearly 24,000 acres of land are identified as high priority for protecting traditional Hawaiian areas where cultural activities take place (8% of the study area). About 16,000 acres of high priority cultural and historical acres have already been conserved (18% of the study area). Across the island of Oʻahu, nearly 100,000 acres of land have been identified as high priority for cultural and historic preservation. The map includes all cultural sites, not just ancient cultural sites. All sites on the Hawaiʻi register of historic places are included in this map. Hawaiian cultural practices are dependent upon ʻāina, wai, and kai (land, fresh water, and ocean), resources. This pattern is evident in the high concentrations of priority lands identified within our watersheds, and near our streams, coastline, and fishponds. As a general premise, we recognize that our natural resources are also our cultural resources. However, for the purposes of these mapping exercises, it is helpful to place these resources in separate maps.

### *Protect Coastal Regions*

This map displays the results for the Protect Coastal Regions value. The criteria informing this value identify pristine shoreline; near shore waters, coral reefs, and fisheries; protect estuaries and input of fresh water into the ocean; and protect natural waterways with sufficient flow into the ocean. The greatest weight was applied to protecting pristine shoreline and protecting nearshore waters, coral reefs, and fisheries.

Approximately 2,500 acres of land and 36,000 acres off-shore have been identified as high priority for conservation across the study area for this value. These high priority coastal areas make up approximately a quarter of the study area. In total, 23% of high priority land and off-shore acres in the study area have already been conserved. Nearly 1,900 acres have been identified as "pristine shoreline" (areas with less than 10% impervious cover) and over 80% of these acres have been preserved. Across the island, more than 40,000 off-shore acres and 6,000 land acres and have been identified as high priority for coastal protection.

### *Protect Natural Habitats*

Oʻahu's unique island ecosystems have created important places for more than 50 endangered and threatened native species to grow and thrive. Examples of endangered or threatened species include the ʻŌpeʻapeʻa (Hawaiian hoary bat), ʻĪlioholoikauaua (Hawaiian monk seal), and the Honu (Green sea turtle). One of the greatest threats to the continued viability of these species is habitat loss. Throughout the Greenprint process, residents and stakeholders expressed a strong desire to use land conservation to protect these native plants and animals.

The TPL map displays the results of the Protect Natural Habitats value. The criteria included protection of native forests, native and natural habitat, vegetation along streams, endangered and rare native species habitat, spawning habitat, native animal populations, ridges and mountaintops, unfragmented areas and protect intact ahupuaʻa systems, protect and restore wetlands, provide wild and natural green areas, identify and protect endangered species areas for migration because of climate change, and protect karst systems. The greatest weight was applied to protecting native forests, native and natural habitat, and vegetation along streams.

More than 35,000 acres of land are identified as high priority for natural habitat protection within the study area. Nearly 18,000 acres of land have been identified as high priority for protecting unfragmented areas and intact ahupua'a systems and nearly 10,000 acres are high priority for protecting rare and endangered native species and habitats within the study area. Across the island, over 45,000 acres of land have been identified as high priority for natural habitats.[4] The highest priority areas for the protection of natural habitats are along the Ko'olau and Wai'anae mountain ranges and their watersheds.

### *Increase Recreation and Public Access Opportunities*

This map displays the results of the Increase Recreation and Public Access Opportunities value for the O'ahu Greenprint. The criteria informing this value included protection of access to mountains and ocean; beach access; ocean fishing areas; improve, protect, and increase hiking trails and trail heads; and create more protected bicycle lanes.

More than 14,500 acres of land and nearly 1,000 off-shore acres have been identified high priority for increasing recreation and public access opportunities within the study area. More than 9,000 high priority land and off-shore acres have already been protected.

Across the island, over 18,000 acres of land have been identified as high priority for recreation and public access opportunities. Many of the highest priority areas for increasing recreation and public access opportunities are along the O'ahu coastline where there is no coastal access within ¼ mile of the closest existing access point. The Mokapu peninsula and Pearl Harbor appear as high priority for increased access due to the fact that these military lands are currently off limits to the public. If all high priority bicycle trails/lanes were preserved within study area, 568 miles of new trails/lanes would be created.

### *Preserve and Enhance View Planes*

O'ahu's scenic landscapes, providing unparalleled views of both pristine beaches and dramatic mountains, are an important part of Hawai'i's character and identity. The TPL map displays the results of the Preserve and Enhance View Planes value. The criteria informing this value included the protection of views of the shoreline/ocean and to preserve views of the undeveloped mountains and scenic views from roads.

Across the island, nearly 19,000 acres of land have been identified as high priority for the preservation and enhancement of view planes. Less than 5,000 acres of the high priority land acres for this value have been conserved across the island. The highest priority areas for preserving and enhancing view planes are on the North Shore, along watersheds on the Leeward coast, and along the ridgelines of the Windward coast. Areas described as particularly important for protection during public outreach included the views—both to and from—the Ka Iwi coast (from Hanauma Bay to Makapu'u), views of the dramatic peaks of the Ko'olau range from the Pali and Kamehameha highways, and the views of Mt. Ka'ala from the Leeward coast and North Shore.

### *Protect Water Quality and Quantity*

Public water supplies in O'ahu are provided entirely by groundwater while streams provide irrigation and vital aquatic habitat (Anthony, Hunt, Brasher, Miller, and Tomlinson, 2004, 2). Municipal demands comprise more than 80% of total water use on the island and it is estimated that demand for municipal potable water will increase 33% between 2000 and 2030. Thus, the protection of water (both quantity and quality) is a pressing issue for the residents of O'ahu. This map displays the results of the Protect Water Quality and Quantity value for the O'ahu Greenprint. The public outreach that developed included input from stakeholders at the first Island Leadership Team meeting. At that meeting, stakeholders expressed a desire to preserve whole watersheds, protect aquifer recharge, create connected greenways and protect vegetation along streams. Ultimately, the criteria informing this value included: protect natural waterways, near shore waters, and fisheries; protect and restore entire stream corridors, including vegetation; protect and restore wetlands; protect aquifer recharge; identify areas to recreate in watersheds; preserve whole watersheds; preserve natural springs. The criteria given the most weight were protect natural waterways, nearshore waters, and fisheries; protect and restore entire stream corridors, including vegetation, and protect and restore wetlands.

About 67,000 acres of land within the study area are identified as high priority for this value. Nearly 17,000 of these acres have already been conserved. Across the island, more than 100,000 acres of land (27% of O`ahu), have been identified as high priority for the protection of water quality and quantity. The highest priority areas for protecting water quality and quantity are watershed lands throughout O'ahu. Throughout the Greenprint process, residents and stakeholders repeatedly stressed the importance of protecting water systems and land that can aide in creating continuity between mauka water sources, flowing streams, healthy estuaries, and the ocean.

## Action Plan

The project team, with stakeholder input, created a concise and focused action plan for the Greenprint. The Island Leadership Team asked TPL, OHA and other partners working on O'ahu to take the following steps:

### *Conserve*

1. Protect important, land, water and resources as identified on the Greenprint maps using voluntary land conservation tools (e.g. purchase, conservation easements, donations)
2. Develop a sustainable strategy for updating the Greenprint data. This strategy could include:
    a. Re-evaluate the maps yearly to determine if there is new data to be added.
    b. Do an updated survey every five to eight years and update the maps to reflect changes in knowledge/priorities by O'ahu residents.
3. Explore how the advancement of the Greenprint goals can reduce impacts to O'ahu from climate change.

4. Use Greenprint data to nominate lands for protection for existing city, state, and federal funding sources and mechanisms.

*Collaborate*

1. Empower partners to utilize the online mapping portal and maps. TPL to offer mapping portal training sessions.
2. Present the Greenprint at relevant forums, such as the Hawai'i Conservation Conference.
3. TPL to present Greenprint findings to governmental entities, commissions, and other conservation partners such as the City and County of Honolulu, Clean Water and Natural Land Commission, State Legacy Land Conservation Commission, and the Board of Land and Natural Resources Chair.
4. TPL to brief relevant agencies/commissions on how to use the maps through training sessions.
5. TPL and OHA to meet periodically to discuss ways to continue to implement the Greenprint.
6. Continue to develop and maintain healthy relationships with the owners and managers of threatened, privately owned land and resources, and identify willing landowners for voluntary fee simple acquisition or conservation easements.
7. Continue to consult with practitioners and cultural leaders regarding the conservation of lands identified on the Greenprint Cultural and Historic places map.
8. Educate and seek more buy-in from political/government leaders regarding Greenprint and voluntary land conservation tools.
9. Offer technical support to community groups so they can use and/or refine Greenprint maps for their communities.
10. Be prepared to respond flexibly to community land conservation initiatives, such as submitting letters of support for appropriate projects that will advance Greenprint goals.

*Raise Awareness*

Continue to conduct outreach to stakeholders and the public to raise awareness of the Greenprint; the importance of conservation to preserving our unique local heritage on O'ahu; and ensure proactive community action related to the value and need to protect and revitalize critical agricultural, cultural, historic, natural, and recreational lands and places.

1. Develop and deliver outreach toolkit around the Greenprint. The toolkit could include:

   a. Powerpoint presentation or video;
   b. Information about the benefits of land conservation, available tools for preserving critical lands, and tax incentives for private landowners;
   c. Information about the work of the Office of Hawaiian Affairs, The Trust for Public Land, and the role of voluntary land conservation on O'ahu; and
   d. Information about projects completed with willing landowners.

2. Engage specific communities and landowners where there are known land conservation and water protection struggles to educate and determine whether voluntary land conservation is an option.
3. Offer tours of the most threatened areas/sites where permitted by landowner.
4. Use social media to disseminate information related to the O'ahu Greenprint.
5. Spread the word about what individuals can do to assist in the implementation of the Greenprint including donating or presenting the Greenprint to community groups.
6. Where feasible, take maps to communities (such as individual ahupua'a or moku) to empower individuals and communities. The Greenprint at this scale may assist when communities review development plans, sustainable community plans, etc. Through this effort each community can collaborate around important lands for protection.

### *Increase Funding*

1. Actively encourage city, county, and state legislature to increase funding for conservation.
2. TPL to discuss funding resources to leverage with relevant smaller organizations.
3. Discuss and determine appropriate methods for financing conservation projects in the mapped priority areas. These methods may include:
   a. Local family and community foundation grants
   b. Local and state dedicated conservation funds (state and county)
   c. Federal grants
   d. Other sources
   e. Tax benefits of voluntary conservation of land

In addition to driving strategic and expanded conservation efforts throughout the island, the Greenprint can inspire residents to get more involved in local decision making about lands, join their local land trust, visit protected lands more often, or volunteer their time to help steward existing conserved land.

## Conclusion

In light of the development pressures facing the island today, this Greenprint comes at an important turning point for the island. The people of O'ahu have spoken, and this Greenprint can express their collective desire to protect culturally, historically, or environmentally sensitive areas; increase awareness about resources in need of protection; and strengthen the special relationship residents have with these island landscapes.

## Notes

1 This text was adapted from an original report authored by: Kelley Hart, Director of Planning, Conservation Vision, The Trust for Public Land; Bob Heuer, Associate GIS Director, The Trust for Public Land; Sherri Hiraoka, Senior Planner, Townscape, Inc.; Lea Hong, Hawaiian Islands State Director, The Trust for Public Land; Katherine Jones, Program Manager, Conservation Vision, The Trust for Public Land; Laura Kaakua, Native Lands Project Manager, The Trust for Public Land; Koa Kaulukukui, Pohaku Kihi, Kanawai

Pili Nohona a me Na Pono 'Oiwi Counsel for Environmental Law and Native Rights, Office of Hawaiian Affairs; Mitchel Hannon, GIS Program Manager, The Trust for Public Land; Kaimo Muhlestein, 'Aho Kuahui Pu'ulu Huliamahi Community Outreach Coordinator, Office of Hawaiian Affairs; Kamoa Quitevas, Land, Ka Pou Kakou Noi'i 'Ike Kūpuna, Paia Kāne, Land Culture and History Research Manager, Office of Hawaiian Affairs; Steve Rafferty, Project Manager, The Trust for Public Land; Leslie Uptain, State Director of Philanthropy, Hawai'i, The Trust for Public Land.

2 The Trust for Public Land (TPL) and the Office of Hawaiian Affairs (OHA) gratefully acknowledge the individuals and organizations that contributed their time, energy, and ideas toward the creation of the Greenprint for O'ahu. Approximately 1200 people visited the Greenprint booths at SpeakOuts, more than 900 people completed the Greenprint survey, and more than 25 people participated in interviews. Those who attended the Island Leadership Team meetings supervised and guided the Greenprint. The Technical Advisory Team provided scientific and technical expertise to assist the Island Leadership Team in development of the Greenprint.

3 The Trust for Public Land gratefully acknowledges the following supporters for their generous contributions to this project: Atherton Family Foundation, Harold K.L. Castle Foundation, Marisla Fund of the Hawaii Community Foundation, The Office of Hawaiian Affairs.

4 Note: in the North Shore Greenprint, this value was called "Protect Natural Habitats for Plants and Animals."

## References


Anthony, Stephen S., Charles D. Hunt, Jr., Anne M.D. Brasher, Lisa D. Miller, and Michael S. Tomlinson 2004. "Water Quality on the Island of Oahu, Hawaii, 1999–2001—Circular 1239": 2. https://pubs.usgs.gov/circ/2004/1239/pdf/circular1239.pdf.

Department of Business, Economic Development and Tourism, Research and Economic Analysis Division 2012. "Population and Economic Projections for the State of Hawaii to 2040": 1. Accessed January 22, 2020, http://www.oahumpo.org/wp-content/uploads/2013/02/2040-long-range-forecast.pdf.

Hart, Kelley, Bob Heuer, Sherri Hiraoka, Lea Hong, Katherine Jones, Laura Kaakua, Koa Kaulukukui, Pohaku Kihi, Kanawai Pili Nohona a me Na Pono 'Oiwi, Mitchel Hannon, Kaimo Muhlestein, 'Aho Kuahui Pu'ulu Huliamahi, Kamoa Quitevas, Ka Pou Kakou Noi'i 'Ike Kūpuna, Paia Kāne, Steve Rafferty, and Leslie Uptain. 2015. "Finding Balance between Development and Conservation | The O'ahu Greenprint": 8. https://www.tpl.org/sites/default/files/files_upload/OGPR_WEB_PRF3_FINAL.pdf.

Homer, C., J. Dewitz, L. Yang, S. Jin, P. Danielson, G. Xian, J. Coulston, N. Herold, J. Wickham, and K. Megown 2015. "Completion of the 2011 National Land Cover Database for the conterminous United States—representing a decade of land cover change information, Photogrammetric Engineering and Remote Sensing," *Photogrammetric Engineering and Remote Sensing*, American Society for Photogrammetry and Remote Sensing, Bethesda, MD, 81: 345–353. https://www.mrlc.gov/data/references/national-land-cover-database-2011-nlcd2011.

Office of Planning—Department of Business, Economic Development & Tourism, in cooperation with the Department of Agriculture, State of Hawaii 2012. "Increased Food Security and Food Self-Sufficiency Strategy": ii. http://files.hawaii.gov/dbedt/op/spb/INCREASED_FOOD_SECURITY_AND_FOOD_SELF_SUFFICIENCY_STRATEGY.pdf.

# 11 Native Land

## Social Media Education and Community Voices

*Victor Temprano*

### Introduction

In this chapter, I will spend some time discussing the specifics of the Native Land project. I will touch on a wide range of points: Native Land's creation; its scope and challenges; criticisms of the project; its multifaceted impacts; its future goals. Mapping Indigenous territories, treaties, and languages is a challenge in many respects, and doing it in a highly visible online public forum, with different technologies at play, adds complexity that is worth breaking down and explaining in writing.

However, I have a more important goal here: to encourage you to use your skills, whatever those skills may be, and to help you to find honest ways that you believe will make your communities more whole than they currently are. I don't claim that Native Land has solved any big, real problems in the world, but it has made an impact, in some way, and can serve as an example to others of the value of working on something you believe in.

My name is Victor Temprano, and I am a settler born on Katzie territory. I grew up in Okanagan territory, in unceded land now called British Columbia, Canada. I had never used map technology or geographic information systems (GIS) before working on Native Land, although mapping has now become a special skill set. The core of why I used a map at all was simple: I wanted to tell a story, a story that would lodge itself inside people who encountered it. So as you read, please consider the ways in which your own skills might be put to use in something that touches others' hearts. We should take criticism, we should be skeptical, but we should still aim for something higher, amidst all the difficulty and confusion. I hope that you will find the following words useful for your own projects and life.

### What Is Native Land?

I usually introduce Native Land (Figure 11.1) as a "site that attempts to map Indigenous territories, treaties, and languages." In truth, it is a hobby project that got out of hand. I didn't start it with any great vision, or any funding or institutional support. It was a spin-off of another project that I was working on at the time, based around pipeline protests and education regarding resource development.

Native Land is a simple idea. You can go to the site (or the app), and enter your address. You'll get zoomed to a location. Results will pop up on the side with the territories, treaties, or languages that are relevant to that area. You can click these for

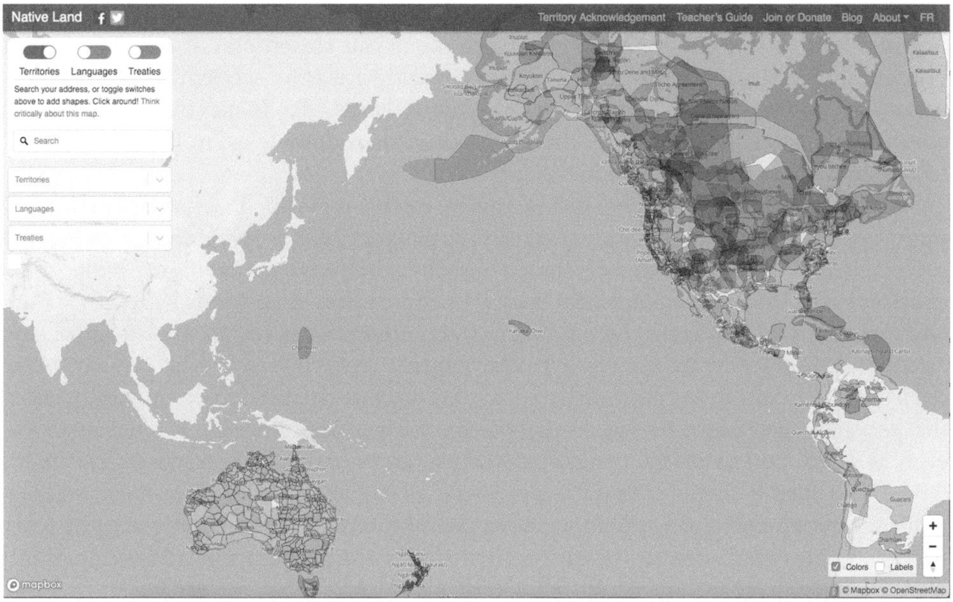


*Figure 11.1* The full scope of Native Land as of summer 2019. Victor Temprano, Native Land Digital, 2019.

more information, and then visit individual websites related to nations, tribes, languages, or treaty documents.

There's very little additional information given: no years, no extensive sources (you can click for source information, though), no reservation locations, no pronunciations. You can turn the "colonial" labels (modern borders, modern towns) on and off, and mouse around to quickly get a sense of the different nations around your area. Given this very basic functionality, there are a surprising number of concerns and problems when it comes to making the actual project work, especially on an international or global scale. I will discuss these a bit further on. But in order to understand the general way that I have approached these problems, it is important to talk a little about how the purpose of the map evolved as I got into building and publicizing the site.

I often describe Native Land as a "rhetorical tool." That is, the map and the site overall makes an argument, and that argument is embedded as much in the design of the map as in the language around it, the sources used, and the theoretical approach. So what is this argument?

Most simply put, the argument of Native Land is that Indigenous people have been, and continue to be, present across colonized territories, and that they should be recognized as having an important ancestral connection to colonized lands. This is actually a very tame argument: Native Land does not (explicitly) argue that lands are stolen, that modern borders should be redrawn, that colonialism is bad, or any such confrontational point. The argument it does make—that Indigenous people are from these lands—is so self-evident that it is virtually impossible to refute. Yet, just because it is obviously true doesn't mean that the majority of those born and raised within colonial societies accept it or think about it in daily life. And so, this is the goal of Native Land as I originally conceived it: to reach out to settlers, or non-Indigenous people, and to immediately bring to their consciousness the scope, depth, and complexity of Indigenous connection to colonized lands. To make them realize this connection as they walk around, drive, go about their daily lives; to bring them back to the land, its people, and its living history.

I will explore more aspects of this argument throughout this chapter, but a few more brief words before I dive in. It was important to me to reach out to what I perceived as "my" community—that is, settlers—rather than trying to make a resource "for" Indigenous people. Indigenous people know their lands better than I do, and I do not feel qualified to help Indigenous communities in some way without being asked and directed as to what is actually helpful.

However, I *did* grow up in settler society as a white-identified person, and I know the pervasive ignorance and racism that can exist in that society. It frustrated me, even as a kid, and made me cynical and angry, and I wanted to change *something*. I do feel qualified to criticize "my" community, as vague as it is to define such a thing, and I wanted to tackle the ignorance head-on. This has always felt like the right path: help my community, before I try to go about fixing the problems in others. It can feel like navel-gazing at times, but the truth is that it is often *much* more difficult to try to teach your friends something than to go to some far-off place and "help" the locals. I don't intend to disparage other ways of doing work, but this has been my approach, and I believe it is an approach that has often been minimized in Western culture at the expense of a kind of "white man's guilt" that imposes solutions on far-off places.

Now that I have discussed some of the ways in which I approached this project, let's move on to a few of the specific challenges I faced in building and maintaining the site.

## Challenges

### *Source Maps*

One of the first challenges that I encountered, and that people ask about, is data. How does Native Land source its data, and what are some of the issues there? Obviously, we look for existing maps, but there is a massive disparity in the quality of maps across different municipalities, states, provinces, countries. There are big differences in how indigeneity itself is conceived around the world, and by no means do all Indigenous groups think of themselves as "nations" in ways that are common in Canada and the United States.

I started with British Columbia, Canada, because that is where I am from and where I was most interested in at first (Figure 11.2). By a stroke of (ironic) luck, British Columbia has great "traditional territory" maps. I believe that this is because the land is almost totally unceded. Due to this lack of treaties, the Canadian government is extremely interested in having Indigenous nations define their home territories for the ultimate purpose of being able to sign them away. Thus, many grants have been provided and much work done in establishing the traditional territories of different nations. This was good for my project. For other parts of Canada east of the Rockies, generally speaking, the existence of traditional territory maps gets much less

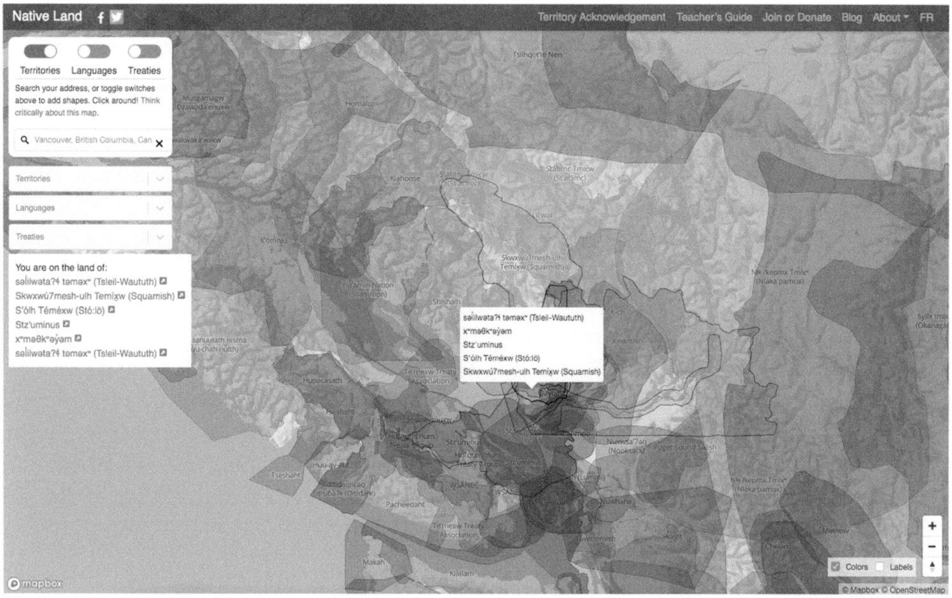


*Figure 11.2* Native Land in Vancouver and British Columbia. Victor Temprano, Native Land Digital, 2019.

defined and reliable. There are some exceptions for very large nations, such as the Anishinaabe.

I started, therefore, with a basic idea: If a nation has a territory map available on their own website, that would be the ideal source. This worked well in British Columbia, but fell apart as I started mapping the rest of Canada. As a second-level source, I would then look for any academic studies that have been done on a given nation, and if any maps were available. If that could not be found, then I would fall back to large "conglomerate" maps; maps of many nations that attempted to cover the country or continent. And, failing that, I would just use absolutely any map that I could find, even old, disputable maps from early colonial times.

These four source levels have their drawbacks, especially if one approaches them from an academic perspective or from the perspective of looking for "correct" GIS data. There are also issues with the original intentions of these maps, especially the conglomerate maps or old colonial maps. Many times, these maps were made without any regard for the actual Indigenous people they were representing, or with the active intention of occupying land. Many times names were written incorrectly or without care, borders were made to match colonial intentions, and more. But still, I regarded it as important to start *somewhere*. And that it would be better to have something on the map for a nation rather than nothing. This is a tenuous justification, but it stands again when one considers the overall goal of the project.

It has been important to get a nation on the Native Land map and to eliminate "empty" spaces, because the truth of the matter is that *every* part of Canada and the United States was part of the traditional territories of one nation or another, at some point in time. Even places that today are considered "uninhabited" were thoroughly explored by Indigenous people across the many-thousand-year history of their living on the land. Many nations utilized spaces in different ways: some peoples were migratory, others highly settled. It is not to say that every spot of ground was actively lived on but, just like modern-day maps of Canada or the United States, people do not actively inhabit every inch contained by the borders. Boundaries can indicate broad areas of interest and occupation. And to have "empty" spaces is to suggest that the land was not occupied or regarded as homeland by anyone, and this is not really accurate to Indigenous history.

As Native Land has grown in popularity, we have updated the shapes on the map often, mostly with the help of user input. Sometimes this means expanding traditional territory to include areas where people live today, sometimes it means refining the borders to be more accurate on a geographical basis, and sometimes it means removing or renaming nations that are more colonial fictions than real, living peoples. This process is complicated and would deserve its own paper to explain in detail, and I will touch on it a little further in the section on user feedback. Inevitably, as the creator of the site, I am in a position of authority in deciding the legitimacy of maps—a problem that I will discuss later as I explain why I felt the need to form a non-profit with a board of directors to oversee the project.

### *Defining Indigeneity*

Another major problem that came up early was deciding *where* to map. Obviously, I started in Canada and then wanted to move to mapping the United States. This was largely because I knew that by making an impact in the United States, the website would grow much larger globally. But people started to send me some interesting questions that made me consider how to approach mapping things on a global scale, and which nations or groups should be included in such a map.

An early email that I often recount was from a person of Irish descent, who explained that the Irish were an Indigenous people who had suffered extensive colonialism, and should have their homelands mapped. This posed an interesting issue for me. Certainly it is true that the Irish were colonized in every sense of the word: language repressed, lands stolen, and more. But yet, there is a country "Ireland" that exists today, and there are many people of Irish descent spread around the world with the spread of British colonialism. So—and this is a decision that I am still questioning—I decided to put aside mapping the Irish on the map, for a few reasons. Foremost, I was concerned that people of Irish descent in North America would find that a reason to disregard colonialism of Indigenous people; I could imagine someone saying "well, we were colonized, they were colonized, so what's the difference?" and thereby erasing the active history of colonization in North America. Furthermore, mapping Ireland would turn the map's focus to Europe, which would inevitably raise more questions that I would not be able to handle: Who are the Indigenous people of Britain? Of Germany? Of Italy? And, since these countries are ruled today by people from their own countries, is colonialism still ongoing in the same way that Canada and the United States are actively ruled by people of European descent? These problems were enough to have me put aside mapping Europe until I had more input from others. I had enough work, anyway, with continuing to map Canada and the United States (and Australia and South America), without trying to include Europe into the mix.

When it comes to placing new peoples on the map, how does one decide who is Indigenous and who is not? This may seem like a fairly obvious delineation in certain cases, such as in much of North America, where a line between settlers and those who were here before them seems clear. But even here, there are difficult questions to be parsed. For instance, in Canada, one of the more recently government-recognized groups of Indigenous people are the Métis, whose heritage consists of a mixture of European and North American lines. I have placed them on the map, but have received a variety of feedback: Some people say that "Métis" territory should cover all of North and South America, while others say that they should be removed from the map entirely because they are not "pure" Indigenous. This is a question we are trying to explore now that the non-profit has wider expertise to draw from, but still, we don't necessarily know what is the right way to map things in this case.

An early suggestion made was to use UN definitions of Indigenous. Some of these are extremely vague. For example, Indigenous people are: "inheritors and practitioners of unique cultures and ways of relating to people and the environment," or Indigenous peoples each have unique and distinctive cultures, languages, legal systems and histories. Most Indigenous peoples have a strong connection to the environment and their traditional lands and territories. They also often share legacies of removal from traditional lands and territories, subjugation, destruction of their cultures, discrimination and widespread violations of their human rights.

These definitions are sufficiently vague that arguments could be made that almost any human culture on earth could fit into the definition. And perhaps that's right—some people say that "everyone is Indigenous." But then, what am I mapping? This returns us to the project's goals.

In addition, how does one determine which cultures in Africa, Iran, or India are Indigenous and which are not? Does a culture stop being Indigenous when it begins to colonize other cultures, or when it gains control of its own lands? These questions are far beyond the scope of Native Land and would have paralyzed the project.

As a result, I chose to stick to areas of the world where defining Indigenous was simpler: that is to say, clearly colonized countries where the descendants of the colonizers are still in power: North and South America and Oceania are the most obvious examples. It is not that I claim that I have a perfect definition and that this is an easy thing to discuss: but, in order to move ahead, it has been important to start somewhere, and at least in many of these countries there is a governmental or functional definition at work already that I could lean on.

### *Use and Misuse of the Map*

Another major challenge is in trying to assess and anticipate how Native Land would be used—and misused. I have already outlined my goals in terms of how I hoped the resource would be used: That is, to inspire a thought process and discussions around colonialism, with an aim to opening settlers' eyes to the ongoing existence of (and injustice done to) Indigenous people. But I am not so optimistic as to assume that all who would use the site would do so with a generous eye.

My biggest concerns in the area of site misuse centered around corporations and governments. While individuals themselves may bicker back and forth over the validity of the site, its meaning, and so on, this is all within the scope of the goal of Native Land to start a conversation. Large organizations, however, may attempt to use the site for their own goals. Indeed, I have received a few emails over the course of the site's history that points to the resource being misused in this way.

Most of this stems from the tendency of both people and organizations to take resources on the internet at face value. While for individual people this is not so serious, corporations and governments have considerable power and need to take more responsibility when considering how they are interacting with Indigenous people. I have received emails from corporations asking me who they should be consulting with before a project, or Indigenous Affairs in Canada attempting to use the site to determine which nation's interests should be regarded when allocating funding. These uses are absurd and speak to a lack of respect and effort when it comes to engaging with Indigenous people.

Corporations and governments have immense resources at hand: financial, staff, technological, and more. Governments have their own records of different nations, all the phone numbers and contact information, and a history of interaction. Corporations have a duty to do their own research and to work with nations and governments directly in consultation. Using an online resource is not appropriate for making real legal decisions. As such, I put up a disclaimer on the site hoping to discourage any kind of "official" usage of the site, reminding users that these are not legal boundaries and are not to be used in any kind of legal situation.

Nevertheless, the site will, undoubtedly, continue to be used as a shortcut by governments and corporations. No number of disclaimers can stop this. At times, I considered removing the site because of this potential misuse; but, in weighing out the good and bad of the site, and consulting with colleagues and users, I determined that it was still worth keeping online. This negative element of the site, however, remains and should not be forgotten.

## User Feedback

In this chapter so far, I've touched on a number of key theoretical or practical issues in doing mapping on the site. In this section, I wanted to explore the impact of the site itself on users based on more of that feedback. Hopefully, this will also help you in your own projects; to handle your feedback in a way that strengthens you, but also engages the hard questions about your work.

### *Thank-Yous and Positive Feedback*

Some of the emails I've received are a great help in giving inspiration and validation in continuing the project. These range from simple "Thank you for making this site" emails, to emails from Indigenous people who express that this is the first time they have ever seen their own ancestral territory mapped. Teachers may email to thank us for providing a resource for them and their students, or others for giving them a way to discuss these issues with friends and family.

One email in this category that sticks in my mind was received in the first year of the site's existence. The site went down due to high bandwidth around Thanksgiving (a popular time of year for the site, due to its implications for Indigenous relations and history). A person emailed me to ask when the site would be back online, because he was planning to present it to his family at Thanksgiving that year and start a conversation about Indigenous history and the history of settling on the land.

This email made a big impact on me. It made me realize that the site had effects far beyond what I could have imagined, and it inspired me that people were taking this on and using it in their own ways to discuss these issues with friends and family. It gave me, like many of these positive emails, the inspiration to keep working despite the complications of the project.

In general, it's important to remember that we tend to discount positive feedback in favor of negative comments. So savor the good ones when they come, and allow them to strengthen you.

### *Suggestions*

Suggestions (and critiques) fall into a few categories. The most common would be emails that suggest a new nation that I had missed, or provide a resource for mapping a new area of the world that I was not yet able to tackle, such as northern Europe or Taiwan. I couldn't always follow some of these (due to workload), but they are important because they tell me what people are expecting to see and what they want to see mapped in the future.

Suggestion emails are crucial for the functioning and improvement of the site: without user input, the crowdsourcing model crumbles and the site would stagnate. People often dig up better resources, whether from their own community knowledge or from a part of the internet I missed. Some examples might be: an email from a band office with the latest updated shapefile of their territory; an anonymous message telling me that I am using an incorrect name for a nation; or an email from someone expressing confusion about why a certain nation is not included. I filed these emails away and slowly worked at them over time, adding the shapes, doing additional research, or asking for clarification from the user.

Other suggestions can include things like improving the map design, removing settler state boundaries, or adding new features such as the territory acknowledgment section. Inviting and responding thoughtfully to this feedback—even if I ultimately turn down the new ideas—is crucial in having a site that is collaborative. This takes time and energy away from the technological improvements or new map fixes, but it's important to building a community and a sense that people are involved. We now have a Slack channel populated with numerous volunteers who have helped us immensely over the years in many areas, and more importantly, they are able to connect with each other and discuss ideas together.

It is important, in my opinion, to make giving input extremely easy for users. This means no big complex user forms: just a place for people to put an email and a message. If you create barriers or try to make people fill in too much information, you will lose people along the way. Make it easy, and make sure you follow up with those who make suggestions, even if they don't provide quite enough information for what you need.

### *Critical and Difficult Feedback*

Finally, we come to critical emails. These have often made me feel panicked and turn my attention very quickly to areas of the site that are lacking and potential harm being done by the site. I must note here that I am not concerned at all with emails that are racist or destructive: that is, emails that say that I am doing this out of white guilt, slam me as stupid, or insist that Indigenous people don't deserve to be mapped, etc. These emails (or Facebook conversations), in my opinion, are already generating dialogue, and I do not need to change anything to modify the site for them. They are hardly worth the time.

The critical emails that do matter the most tend to come from Indigenous people whose nations have been left off the map. Sometimes people feel erased, removed from history, left out of a broader narrative—which feels like a replication of the entire history of settler colonialism. I invite and embrace these emails and apologize as well as I can, and attempt to rectify the issue as soon as possible, with the help of the person who emailed, if they wish to be part. Sometimes they do not. They do not owe me their time or knowledge, especially after the hurt inflicted, and I do my best to repair things regardless. Of course, I never intend to make such errors of omission or of naming, but nevertheless I do make such mistakes and they have to be taken seriously.

These emails are an opportunity, actually, to realize how important the site is and what an impact it makes. The project is important because it actually *affects* people—good and bad—and that meant I had to work hard to try to get things right. While I had limitations due to lack of funding, personnel, and expertise, I did try to do my best while also getting a broad swathe of territories, languages, and treaties. I tried not to give excuses, but rather aim to help people understand why I might have missed their nation and that it was not from malice (if that matters to them at all, which it may not). At the end of the day, though, I want to fix the issue, not argue about whether I was right or wrong or malicious or ignorant.

Other critiques that are more difficult to handle or respond to may touch on my place as a settler running the site, or on theoretical problems, such as the site's very existence replicating "government sanctioned" ideas of tribal history and nationhood, or the inclusion of groups that some feel are questionable in terms of being defined as Indigenous. Because of the complex history of colonialism, many nations are seen as having been "created" by governments, and not as "original" inhabitants.

Overall, these critical and intelligent emails made me question whether the site should exist at all. A number of times I nearly took it down in response, but I had to admit that it is also important to take on board the positive feedback. After engaging with many of these hard critiques, it grew increasingly clear to me that I could not really be running this big and this impactful of a site all by myself, without help from people beyond the occasional email.

I alone could not responsibly be in charge of this site. I needed help. I needed more voices to be included in positions of power in the project. I needed a board of directors.

## Building a Non-profit

There is a saying out there: "If you want to go fast, go alone. If you want to go far, go with others." This might not apply to your style of working, but as an entrepreneurial self-starter, this applied to me. It was good to start Native Land on my own, because I was able to work quickly and move through the theoretical problems using my own sense of right and wrong. But, as the project grew and those problems started to become more apparent (largely through criticisms and feedback), I realized that this resource and project was much bigger than me or what I could handle alone.

This meant that some sort of organization needed to be formed that could help me, or, even better, take over running the project entirely one day. Ideally, this would be a non-profit, with a board of Indigenous people with expertise and knowledge in the areas of mapping and social impact. The organization could pursue funding, deal with staff, explore new opportunities, interconnect with Indigenous communities around the world: all things that were beyond my own time, initiative, and wallet.

And so I began to reach out, emailing some of the people who had helped me over the years, trying to put together a few staff members and hoping to find some people to help. Asking for help is important, and accepting it can be even harder, especially when it seems to come "too slow" or sometimes conflicts with your own ideas of what's good and bad. But this is part of the "going far," and in truth, it's been very exciting to see the growth of the organization. Difficult, and at times frustrating, but rewarding.

Thus, the past year or so has seen the growth of Native Land Digital, the Canadian non-profit that is now in charge of maintaining and running the website and its associated initiatives, such as Education Guides and media work. We have five board members: Leena Minifie, Shauna Johnson, Mesiah Burciaga-Hameed, Lee Timutimu, and Rudo Kemper. These people come from all around the world and have been huge helpers and guides. We have been able to secure some initial funding, and are working on finding more so that we can continue to sustain the project as it scales in size and complexity as we go forward and continue growing.

It is important for me to hand this project off, as much as possible, to Indigenous leadership in one form or another. This is difficult, because, in forming the board initially, I had to make decisions on who was a suitable member and who was not. These decisions were not easy and had to be informed as much by who was available as by things like land base, educational expertise, or passion and willingness to help. At times, I have wanted to abandon the project and go back to doing other things. But, the more that Native Land sticks around and grows, the more it is clear to me how important this project is, not even as much to the broader world as to me personally. I will be happy if, one day, my responsibilities are able to decrease to the point where I am only involved in a small way; but in the meantime, I resolve to take on whatever needs to be done in order to help the organization get its legs under it.

## Conclusion

At the end of the day, what you are putting your time and energy into matters to you. It is important to take the challenges, criticisms, and problems seriously, but it is also important to allow yourself some space for validation and praise from those who you affect positively. Native Land has been a space of personal struggle and personal triumph for me, and most of all, a place to learn my own boundaries and my own need for collaboration and help.

If you have a project in mind, or a project underway, I encourage you to stick with it and to run with your strengths. Ask for help from others when you don't know what to do anymore—people all have their own work, but even a few minutes or a short conversation can revitalize you to keep going.

The future of Native Land, now Native Land Digital, is unclear, but I have faith in myself and in the team we've assembled to continue the work as long as it makes sense to do so. We have much learning to come, much research to do, and many more people to reach. I want to thank all the members of the board and the many people who have emailed and helped over the years—your support, your criticisms, and your words mean the world to me. Good luck with your endeavors, and remember not to give up.

## References

"Carte Linguistic du Canada," MuturZukin, September 2019, http://www.muturzikin.com/cartesusa/canada.htm.

"Early Indian West," University of Texas, September 2019, https://legacy.lib.utexas.edu/maps/united_states/early_indian_west.jpg.

"Indigenous Peoples at the UN," United Nations, September 2019, https://www.un.org/development/desa/indigenouspeoples/about-us.html.

"Language Map of British Columbia," First Peoples' Cultural Council, September 2019, https://maps.fpcc.ca/.

"UNDRIP Manual for NHRIs," United Nations, September 2019, https://www.ohchr.org/Documents/Issues/IPeoples/UNDRIPManualForNHRIs.pdf.

# 12 Mapping Indigenous American Cultures and Living Histories (MIAC-LH)

## A Gathering Place

*Janet Berry Hess*

## Introduction

A study conducted in 2015 found that 87% of state-mandated K–12 education standards placed Native Americans in a solely pre-1900 context only.[1] A 2019 report by the National Congress of American Indians found that only 11 states in the U.S. mandate "instruction in tribal culture, history, and/or government to all students in K-12 public schools,"[2] and "27 states make no mention of a single Native American in their K–12 curriculum."[3] States which require education about tribal history often fail to support their mandate. In California, for example, despite a 2017 bill which requires public schools to adopt and teach curriculum addressing Native American culture and history, no funding has been provided for training, development, or even textbooks addressing the topic.[4] Despite the efforts of Native scholars, no California state oversight of efforts to implement the bill exists.[5] The future K–12 teachers I teach frequently ask for guidance about teaching Native American culture and history; some do not teach Indigenous history or culture at all, for fear of doing so wrongly or disrespectfully. This silence impacts Native children directly: "more than 90 percent of all Native American students nationwide attend regular public schools or so-called mainstream schools side-by-side with non-Natives."[6] It also deprives children of traditional ecological knowledge disrupted by colonialism.[7] As Jeanette Haynes Writer has argued, "if schools' citizenship education does not address the historical and contemporary context of Native American citizenship, then this is evidence of ongoing colonization and the practice of cultural imperialism."[8]

Many scholars have worked on digital resources to address the exclusion of contemporary Native American history,[9] but a comprehensive, public-facing "gathering place" of place-based, tribally approved narratives, with links to other resources, has yet to be constructed.[10] Inspired by a digital map of the continent of Africa (created by my advisor at Harvard University, Suzanne Blier), and with the assistance of Pbonchai Tallman, the support of the Chairman of the Federated Indians of Graton Rancheria Greg Sarris, and the creator of Native Land, Victor Temprano, I received a National Endowment for the Humanities Digital Advancement grant to create a prototype for a publicly accessible educational map with tribally approved content, "Mapping Indigenous American Cultures and Living Histories," or MIAC-LH (http://indigenousmap.org).

## The Vision

In the original proposal for the National Endowment for the Humanities, we suggested that MIAC-LH (originally titled IndigenousMap)

> would connect the study of humanities by scholars and the general public to the dynamic conditions of social and cultural life, and document cultural heritage materials, by creating a prototype digital map related to three pre- and post-contact Native tribal regions and cultures in the U.S.: the Osage, Modoc, and the Federated Indians of Graton Rancheria.[11]

MIAC-LH, at the time the grant was awarded, proposed an online resource for the general public and K–college educators to access narratives, images, and other tribally approved information. MIAC-LH was intended as the prototype for a "gathering place" of digital projects and tribally approved information. In light of the imperiled state of Indigenous sovereignty, as well as the lack of understanding and curriculum related to Native cultures, the objective of MIAC-LH was therefore twofold: 1) to lower barriers to accessing basic information related to Native histories and cultures for people of all identities/affiliations and educational levels as a prototype for a broader resource in the future, and 2) to consolidate tribally approved information related to Native cultures and histories in America, including links to related digital projects addressing languages, art forms, and histories.

Although information about Native nations is widely available, accurate and tribally approved scholarship—including resources related to the distribution of languages, shifting histories of land ownership and occupation, historical and cultural narratives, and other information—is widely dispersed among natural history and art museums, private collections, and academic holdings, as well as digital projects and archives. It is often difficult to discern from publicly available information what is tribally approved for public and educational use. Obtaining even a comprehensive and current list of Native and non-Native archives, libraries, and museums that house material pertinent to regional Native American life is challenging. Materials purporting to teach children about tribal life often neglect traditional ecological knowledge (TEK) and the diverse, frequently intersecting nature of knowledge sets that stretch across Nations due to trade, travel, and dislocation. Pbonchai Tallman, for example, a key figure in the creation of MIAC-LH, is an Elder of Blackfeet ancestry, brought up on a Paiute reservation, and trained by a Yaqui mentor, who now lives near the Pit River Nation. He states of his intersectional background and environment,

> in the Paiute Nation there were bands of people living near the lake [Pyramid Lake in Nevada]. They fished. Pine nut people lived in the forest and collected nuts and hunted deer, elk, and antelope. The Tule people made many things from cattails. The flower bulbs were eaten, the stems made tule boats, decoys for ducks, and clothing for warmer months. Sage people rounded up jackrabbits and got meat and a special type of robe used to keep elders warm. I now live about 15 miles east of Mt. Lassen in northern California. I collect mullen leaves, sage, night moss, long needle pine buds, and hammerstone for making arrow points... In the wilderness I like to sit still and wait for animals to come out. I've seen deer,

> elk, bears, eagles, wolves, coyotes, badgers, bobcats, racoons, porcupines, different birds, and small creatures such as squirrels and mice, as well as insects such as bees, yellow jackets, grasshoppers, bumblebees, butterflies and many more. From what I observe I take messages and that helps me make right choices in my life... This area [the Pit River region] used to be a gathering place for tribes pre-European to gather for trade and ceremony. There are artifacts left from those times, but I leave them because they don't belong to me. Modoc, Northern Paiute, Maidu, and Pit River people all used to meet about four times a year and share or trade items. This is just one area, so you can imagine how diverse the needs and materials across our country must be. This mapping project is an attempt to show the diversity and show old and current boundaries of our people, with further knowledge available to those wanting to learn more about our ancestors and see the changes in our land.[12]

It is this kind of rich, environmentally informed, and intersectional narrative that MIAC-LH seeks to include (after approval by a Tribal Council or Tribal Heritage Protection Officer [THPO]), along with historical, linguistic, and artistic information.

## The Process

In constructing MIAC-LH, Pbonchai Tallman's spiritual perspective and experience, and Victor Temprano's technical expertise, insight, and friendship, swiftly emerged as the anchoring base for the project. MIAC-LH became in a sense the next layer of Native Land: a linguistic map of the region of the United States with tribally approved cultural and historical narratives. As we attended National Endowment for the Humanities meetings, shared at conferences, and met with scholars and THPOs, it became clear that the map, initially conceptualized in data tree-like terms, would function best as a collection of tribally approved accounts, overlaid with region shapes and languages, emphasizing what individual Nations wanted highlighted in the public realm. MIAC-LH contains a Google earth depiction of the continent—eliminating colonial boundaries—with regions, names, and languages of Indigenous nations, a directory of research/tribal resources and resources for children, and narratives and images specifically chosen by THPOs and tribal members (see Figure 12.1)

The process of constructing the map involved learning and challenges. Our friend Paula Farid, a member of the Five Woman Council in Pawhuska who supported the vision that Pbonchai Tallman and I had shared of describing the Osage Nation, passed away after the grant was written; further collaboration with the Osage Nation stalled. The Modoc Nation, too, proved not to be in the position to pursue collaboration as a Council, despite the interest of individual members. As a settler of European descent, I could not ethically move forward without tribal input. I was prepared to end the project if more tribal guidance did not appear. But as Pbonchai Tallman states, "Grandfather chooses our paths, and we follow them."[13]

While attending a session of the Association of Tribal Archives, Libraries, and Museums (ATALM), I heard a talk by the CEO, Walter Echo-Hawk (now Tribal Chairman of the Pawnee Nation), describing a family biography he had written. I raised my hand and mentioned that I, too, had written a family autobiography set in the Pawnee region, based on the records kept by my Uncle Everett Berry.[14] Chairman

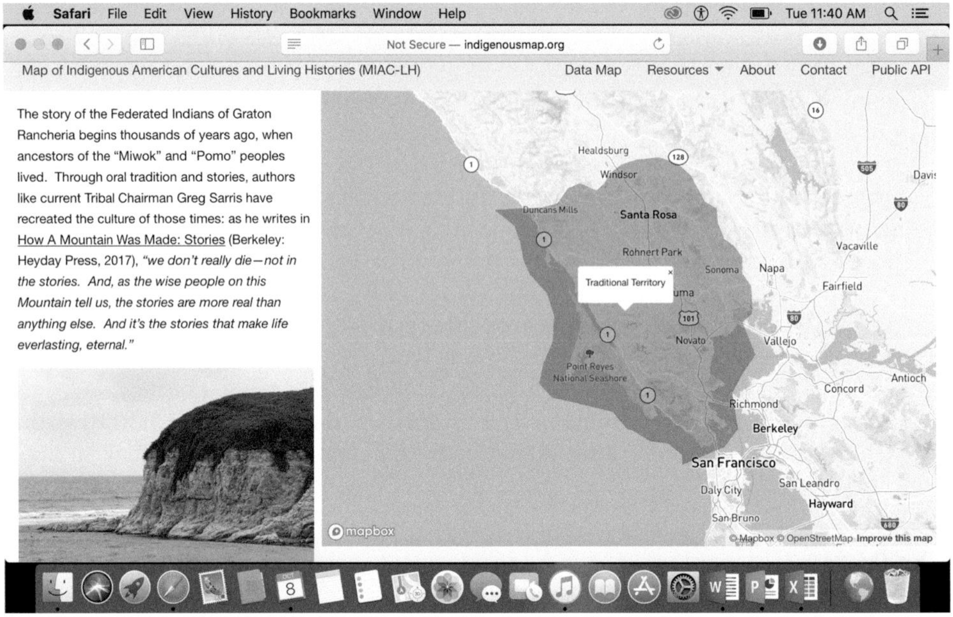


*Figure 12.1* Graton Rancheria First Contact maps and narrative from Mapping Indigenous American Cultures and Living Histories. Narrative, screenshot from Mapping Indigenous American Cultures and Living Histories, 2020.

Echo-Hawk stated, "I knew your Uncle," and marvelous events unfolded. I was honored to visit the Chairman in his home, and to be invited to a Pawnee dance held in June 2019. Mr. Echo-Hawk approved the story I wrote about the Pawnee, which highlighted the history of Pawnee art, including the art of his uncle, Brummett Echo-Hawk, as well as the Echo-Hawks' legal advocacy (see Figure 12.2). We began receiving information from the Tribal Heritage Protection Officer of the Federated Indians of Graton Rancheria, Buffy McQuillen, and constructed the narrative she and the Tribal Council wished to tell, with images they provided: the story of first contact on the west coast, and the reclamation of that story through the establishment of a monument located at what is called "Drake's Landing" in northern California, as well as stories from Tribal Chairman Gregory Sarris. Through ATALM's Facebook presence, we connected with Alicia Marie Recountre-da Silva, a longtime participant in ATALM. Alicia conducted outreach to add to the map information that would be useful to ATALM, specifically the location of archives, libraries, and museums, as approved by ATALM President Susan Feller. At the time of this writing, we are working on the Yaqui narrative—of particular importance, as it involves a Nation stretching across colonial borders. Most importantly, we have planned a listening session at ATALM to discover what participants—particularly members of Tribal Councils—would like to see represented in MIAC-LH in the future. We envision a map with the capacity for Native organizations to add narratives, images, and websites, in addition to the locations of Native and non-Native archives, libraries, and museums.

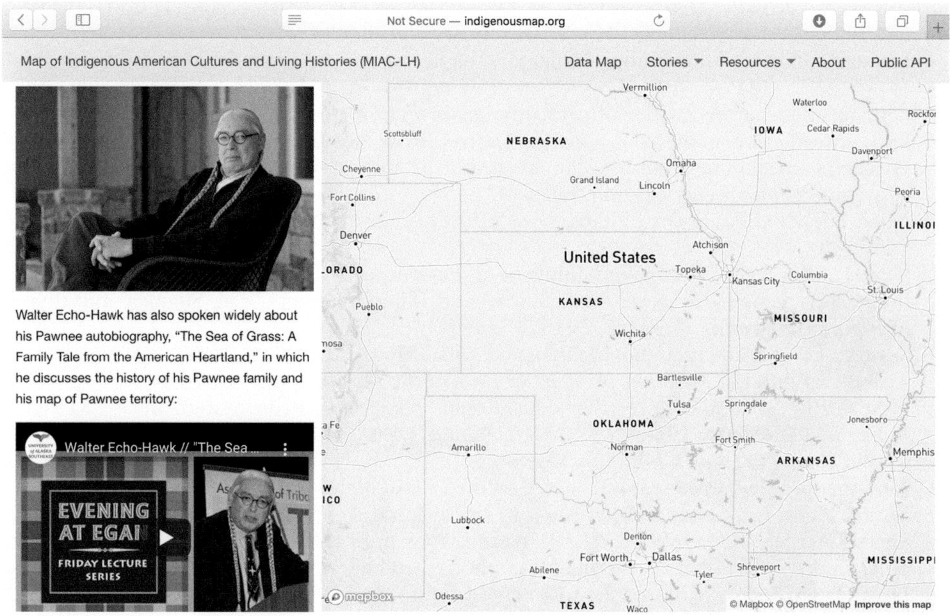


*Figure 12.2* Conclusion to Pawnee maps and narrative, screenshot from Mapping Indigenous American Cultures and Living Histories.

Digital resources often exist briefly, only to make way for new and enriched visions. If one has the ability to listen and follow the advice of tribal members, as well as humility and perseverance (or the ability to acquire these things), however—if one can follow the "light" of collaboration rather than the "darkness" of academic or professional self-promotion—remarkable things may happen. If you wish to create a digital resource, begin by studying existing digital projects, attending the ATALM Conference, and investigating institutional, state, and federal funding opportunities (see the Appendix at the end of this book), particularly the National Endowment for the Humanities and the Institute of Museum and Library Services. Ask for feedback, reshape your project with feedback in mind, and seek collaboration from the outset and along the way. I invite any tribal member or creator of a digital project reading this chapter who wishes to work with MIAC-LH's "gathering place" to contact us through its website. I wish you the best!

## Notes

1 Shear S. B., R. T. Knowles, G. J. Soden and A. J. Castro, "Manifesting Destiny: Re/presentations of Indigenous Peoples in K-12 U.S. History Standards." *Theory and Research in Social Education* 43, no. 1 (2015): 68–101.
2 Ngai, Phyllis, Unn-Doris Karlsen Baek and Gry Paulgaard. "Indigenous Education in the Norwegian and U.S. Contexts." In Kathryn W. Shanley and Bjorg Evjen (eds.). *Mapping Indigenous Presence: North Scandanavian and North American Perspectives*. Tucson: University of Arizona Press, 2015, 80.

3 National Congress of American Indians, Becoming Visible: A Landscape Analysis of State Efforts to Provide Native American Education for All, Washington, D.C. (September 2019), 7.
4 Herrera, Allison, "Indigenous Educators Fight for an Accurate History of California" (April 29, 2019). https://www.hcn.org/issues/51.7/tribal-affairs-indigenous-educators-fight-for-an-accurate-history-of-california-missions.
5 Ibid. Information gathered by Gregg Castro, Rose Borunda and the California Indian History Curriculum Coalition can be found at https://www.csus.edu/college/education/engagement/indian-curriculum.html.
6 Ngai, Baek and Paulgaard, "Indigenous Education in the Norwegian and U.S. Contexts," 2015, 80.
7 Shanley, Kathryn. "'Mapping' Indigenous Presence: The Declaration on the Rights of Indigenous Peoples at Rhetorical Turns and Tipping Points," in Shanley and Evjen, *Mapping Indigenous Presence*, 2015, 13.
8 Writer, J., "Broadening the Meaning of Citizenship Education: Native Americans and Tribal Nationhood." *Action in Teacher Education* 32, no. 2 (2010): 70–81.
9 Noteworthy examples are the educational resources incorporated into "Mapping Indigenous L.A." (see Chapter 1), "Native Land" (see Chapter 11), and the resources listed in the Appendix (see "Children's Resources").
10 A further problem in educational access is the lack of high-speed internet on tribal land. See Wang, Hansi Lo. "Native Americans on Tribal Land Are 'The Least Connected' to High-Speed Internet" (December 6, 2018). https://www.npr.org/2018/12/06/673364305/native-americans-on-tribal-land-are-the-least-connected-to-high-speed-internet.
11 Hess, Janet Berry. National Endowment for the Humanities Digital Advancement proposal, 2018.
12 Letter from Pbonchai Tallman to Janet Hess, June 1, 2020.
13 Conversation with Pbonchai Tallman, July 5, 2020.
14 Hess, Janet Berry. *Osage and Settler: Reconstructing Shared History through an Oklahoma Family Archive*. Jefferson, NC: McFarland & Co., 2015.

## References

AfricaMap, https://worldmap.harvard.edu/africamap/.

California Indian History Curriculum Coalition, https://www.csus.edu/college/education/engagement/indian-curriculum.html.

Herrera, Allison. "Indigenous Educators Fight for an Accurate History of California," April 29, 2019, https://www.hcn.org/issues/51.7/tribal-affairs-indigenous-educators-fight-for-an-accurate-history-of-california-missions.

Hess, Janet Berry. *Osage and Settler: Reconstructing Shared History through an Oklahoma Family Archive*. Jefferson, NC: McFarland & Co., (2015).

Hess, Janet Berry. *National Endowment for the Humanities Digital Advancement Proposal*, (2018).

National Congress of American Indians. *Becoming Visible: A Landscape Analysis of State Efforts to Provide Native American Education for All*, Washington, D.C. (September 2019).

Ngai, Phyllis, Unn-Doris Karlsen Baek, and Gry Paulgaard, "Indigenous Education in the Norwegian and U.S. Contexts," in Kathryn W. Shanley and Bjorg Evjen, (ed.), *Mapping Indigenous Presence: North Scandanavian and North American Perspectives* Tucson: University of Arizona Press, (2015): 78–119.

Shanley, Kathryn, "'Mapping' Indigenous Presence: The Declaration on the Rights of Indigenous Peoples at Rhetorical Turns and Tipping Points," in Kathryn W. Shanley and Bjorg Evjen, (ed.), *Mapping Indigenous Presence: North Scandanavian and North American Perspectives* Tucson: University of Arizona Press, (2015): 5–26.

Shear, S.B., R.T. Knowles, G.J. Soden and A.J. Castro, "Manifesting Destiny: Re/presentations of Indigenous Peoples in K-12 U.S. History Standards," *Theory and Research in Social Education 43*: 1 (2015): 68–101.

Wang, Hansi Lo. "Native Americans on Tribal Land Are 'The Least Connected to High-Speed Internet." December 6, 2018 https://www.npr.org/2018/12/06/673364305/native-americans-on-tribal-land-are-the-least-connected-to-high-speed-internet.

Writer, Jeanette Haynes. "Broadening the Meaning of Citizenship Education: Native Americans and Tribal Nationhood," *Action in Teacher Education 32*: 2 (2010): 70–81.

# 13 William Commanda, Oral Wampum Storytelling, Digital Technology and Remapping Indigenous Presence Across North America

*Romola V. Thumbadoo, and D. R. Fraser Taylor*

## Introduction

Who is William Commanda? How can an Indigenous man born in 1913 possibly find a place in a scientific academic publication on contemporary mapping, especially in the digital era? This is a challenge, and it requires an innovative conceptual framework to position his geo-cybernetic mapping within the evolving domain of cartography, cybercartography, and digital mapping. We suggest that he did not map Indigenous territory in a traditional place-based manner as a standard geographic information system (GIS) product, but rather by intention, Mamiwinini movement, ceremony, thought-driven orality, storytelling and Ginawaydaganuc (digital atlas creation) relationality, across the continent. Thus, the central question that this paper seeks to examine is the nature, relevance, and effectiveness of the geo-cybernetic journeys and discourse of late Indigenous Elder William Commanda in his approach to mapping Indigeneity across North America.

Exploring *Why Ancient Wisdom Matters in the Modern World* in *The Wayfinders,*[1] Wade notes in discussing Polynesian mapping that

> the issues and misconceptions … both inspire and haunt us to this day: the sheer courage that true exploration implies, the brilliance of human adaptation, the dark impact of conquest and colonialism. It reminds us, too, of the need always to be skeptical about the tenacious grip of academic orthodoxy. Knowledge is rarely completely divorced from power, and interpretation is too often an expression of convenience.[2]

He quotes Albert Einstein:

> The intuitive mind is a sacred gift and the rational mind is a faithful servant. We have created a society that honours the servant and has forgotten the gift.[3]

In discussing the Songlines of Indigenous Peoples of Australia he writes, they

> are not straight or linear. They do not even exist in three-dimensional space. In their numbers, they weave a web across an entire continent. For a civilization that lacked the written word, they became a record of the past, a promise of the future, and a network that in movement bound together all of the people.[4]

These ideas open the landscape to William Commanda's mental mapping.

John Edward Huth takes the discussion further in this direction, observing in his book, *The Lost Art of Finding Your Way,* that in our times of change, our need for metaphorical wayfinding has never been greater; and as we struggle to get our bearings and find our way across this unfamiliar terrain, navigation is a useful tool, and not just for visualizing a path in the wild, but also in the workplace and elsewhere; a navigator's mental processes are the same. Successful navigators are keen observers, incorporate what they experience, and use this data to create the mental map they need for orientation. In the *Globe* article he coauthored with Julia Moulden,[5] the work of neuroscientist Eleanor Maguire is explored next, and the writers point out that brain research on the hippocampus indicates that, while observation helps us figure out where we are and fluid thinking helps widen our view, scenario planning (creating stories about how things might unfold) extends our map into the future. Further, the mental processes required for successful navigation, especially in the most challenging situations, include an emotional component, and inquisitiveness lies at the heart of emotional intelligence.

Also known by his Algonquin name, the motional signifier, Ojigkwanong, Morning Star, William Commanda was born on November 11, 1913, on the eve of the first World War, on the first Native reserve created in Canada; he died on August 3, 2011, almost 98 years old. His people were the Mamiwinini, the ones who traveled all the time. For 40 years, he was the Keeper and Carrier of ancient Algonquin wampum shell belts, mnemonic artifacts of spiritual, historical record-keeping and storytelling importance. He is Canada's best-known Indigenous birch bark canoe maker—this marking another motional motif of his identity—was a passionate advocate for respect and responsibility to Mother Earth, a tireless peace maker, and founder of a global social justice eco-peace community, the *Circle of All Nations.* His singular stature earned him two honorary doctoral degrees, further ratified in the doctoral thesis entitled, *Ginawaydaganuc and the Circle of All Nations: The Remarkable Environmental Legacy of Elder William Commanda* (Thumbadoo),[6] and appointment as Officer of the Order of Canada, the country's penultimate honor. He was a brilliant linguist, entrenched in the orality of his land-based Algonquin language, fluent also in multiple Algonquin dialects, and French. Self-proclaimed as a kindergarten dropout, he learned to read and write as an adult, embraced communications and digital technology in concert with its development, and, as he entered the millennium, marked his place on the web and in social media, where he remains encore vivant as storyteller emeritus. He engaged in a vast and diverse body of work over the course of almost a century.

This chapter examines his digital mapping of presence across the continent in select examples of wampum, language, and journeying agency. It explores how he used oral and digital storytelling to map his presence across North America. It examines theoretical approaches that contribute to scrutinizing this, consistent with the spiraling medicine wheel conceptual tool developed in the *Circle of All Nations* work (www.circleofallnations.ca). Thus, it examines the theme of movement located in the ideological foundation of his world view; this is animated through physical journeys over land and water; it is negotiated in the cybernetic agency underpinning his approach; and is articulated in land-based knowledge generation and digital storytelling.

The emerging discipline of cybercartography provides the framework to explore the multiple expressions of William Commanda's geo-cybernetic mapping. In his work, the map is not an object, or physical demarcation on the land. Syntropic movement, physical, ephemeral, and performative and processual, are the constitutive elements of his mapping; and diversity and relationality produce an infinite number of interfaces, each enacted with individual agency yet also integrated and intermeshed. The concept of syntropy introduces the temporal dimension into the spatial, consistent with the epic Seven Fires Prophecy. The *Circle of All Nations* global eco community is the ephemeral manifestation of the Commanda map in his Facebook pages; the *Circle of All Nations* digital atlas accesses the innovative Nunaliit data management framework: the digital atlas, as discussed in the volume by Taylor, is "quite different from a conventional atlas and is a metaphor for all kinds of qualitative and quantitative information linked by location. These are interactive, multimedia and multisensory on-line products."[7] Cybernetics and cybercartography serve as the methodological tools to both scrutinize and animate this phenomenon, since in William Commanda's discourse, map is not primarily a cartographical or material product, defined by delimitations and physicality, but rather map is the animation he mediated across time and space. This paper examines the processes that he used to advance the comprehensive manifestation of the core elements of his vision and work, and to contribute to an understanding of the surefootedness of his mapping of Indigenous presence across the continent.

## Discussion of Select Terminology

This section provides a brief discussion of some conceptual ideas and Indigenous theoretical frameworks critical to an understanding of the Commanda digital mapping process and discourse.

### *Wampum*

A close examination of William Commanda's public practice, performance mapped at multiple levels, over the course of several decades, and more intensively over the past 40, reveals power generation via interaction with the unseen force. His chosen public word for this unknown force is Creator. Here he asserts connection with the procreative, evolving, emergent, multidirectional and temporal dynamic. William Commanda became Keeper and Carrier of three sacred wampum belts of spiritual, historical, and political significance in 1970, the Seven Fires Prophecy, Three Figure Welcoming, and the Jay Treaty Border Crossing belts. William Commanda is the last and longest known guardian of the ancient mnemonic Wampum prophetic oral storytelling and animative record-keeping heritage of his Mamiwinini/Nomad Algonquian ancestors; he held the sacred wampum belts for 40 years. The following excerpt from the book, *Learning from a Kindergarten Dropout*, provides a contextual description: "From the earliest of times, William's ancestors along the eastern seaboard of North America created belts woven with beads made of the purple and white shells of the marine sea creatures. The word wampum emerges from the Algonquin word wampumpeage which described white shells and or strings of shells; its roots link also with the words waban and wabanpaog, referring to the dawn or the first light of the east, and the eastern people, the Algonquians. The belts that William carries are predominantly purple; the beads were made from the shell of the quahog. It is not clear how they fashioned the thousands of tiny beads before the age of industrialization, but it is well known that they created many beautiful belts that served as devices to record their stories, legends, prophecies and agreements. These served both as a living record of a commitment and also a means to recall the messages."[8] The wampum belts were the mnemonic devices or memory awakeners, ideographic rather than phonetic. They complemented language and oral delivery in communications. Thus, it is suggested that Wampum was *spirit text*, only accessed and understood by certain people. It is interesting that the Seven Fires Prophecy itself warns about false prophets. Part of the reason William Commanda's message resonated for many diverse peoples emerges from the authenticity of his voice as prophet, even more evident almost a decade after his death.

### *Orality and Language*

William Commanda's understanding of land emerged from his orality and language emergent of his cosmic world. Language was critically important to him. Countless people have affirmed how seamlessly Algonquin, French, and English languages were used by him and his wife in their daily communications. Algonquin is a land-based language, wherein, according to him in his analysis and articulation, the land itself, its sounds, its physical, geological and biological realities (rocks, stone, animal, plant and the elements) shaped and determined the evolution of the language. The generic language which emerged with the Algonquins of the Ottawa River Watershed was transmitted across the country by his nomadic ancestors; he emphasized that there were 84 Algonquin Nations across North American that spoke dialects of the mother tongue. William Commanda was raised on the ancient Algonquin language. While confined to French and English in his communications with others, he nonetheless performance-communicated his own much vaster understandings of truth and meaning to complement his words. The import of his words and other dimensions of communications warrant careful examination to grasp an understanding of the depth of his thoughts and by extension, this applies to the understanding of Indigenous peoples and their communication styles. He, however, was a virtuoso performer and, as is still evidenced in social media, drew a diversity of people into his web of communications.

### *Mamiwinini*

William Commanda identified himself and his people as Mamiwinini; the word itself incorporates the critical motional element integrative of both time and space and brings a unique element to the study of his Indigenous mapping. As discussed in *Learning from a Kindergarten Dropout*,[9] "The Mamiwinini journeyed over the waterways of Turtle Island, spinning a web of protection and prayer over the vast continent for thousands of years, passing over lightly, leaving an indelible trace of the brunches of the great family tree that comprised 84 nations linked by both language and a deep connection to the land."[10] Motional spatial understanding was determined by and emergent from the name and identity was commensurate with the idea of map in motion. Nomad is a restrictive translation of the word because it implies transience; William Commanda affirmed cyclical permanence of presence on the land. His temporal and spatial understanding incorporated elemental and motional Law of Nature knowledge.[11]

### *Cybernetics*

With respect to the academic analysis of geo-cybernetics, in the context of the emergent field of cybercartography, Reyes presents several angles to facilitate research in three chapters in *Cybercartography: Theory and Practice.*[12] By way of background, she notes that Norbert Wiener (1894–1964), a mathematician, used the word cybernetics [derived from the Greek kybernetes, which means steersman, and shares its etymological root with the word governing—our note] to describe the science of communication and control in the organism, the machine, humans, and society. He noted that society can only be understood through a study of messages and the communication facilities to which it belongs, thus emphasizing the relevance of the study of communications and the role of information in human life and for the user via a feedback process.[13] The explicit involvement of the user in the reflection on the feedback is of interest in the Commanda geo-cybernetic research. Wiener notes that communications and language are inseparable. The relevance of language issues in cybernetics are not only for people or living creatures, but also for communications between humans and machines or among machines themselves. This sets the context for social control cybernetics, algorithm, and artificial intelligence. However, as Taylor points out, Weiner conceptualizes a scientifically objective first-order cybernetic system with an observer who is not seen as part of the system; "In Cybercartography, the user is an integral part of the creative process."[14]

Reyes also discusses modeling in the context of cybernetics, noting that maps are visual models of the geographical landscape where a geometrical language is used to represent geospatial information.[15] She elaborates on Russell's analysis of Lobachevshy's demonstration that the truth of Euclidean geometry (grounded in axiom, a statement taken to be true) could only be established by observation, not by reasoning; i.e. geometry is an empirical science based on observations. She points out further, that "Once knowledge is formalized in either mathematical or other terms, a reader might deduce wrongly that the frameworks have emanated from pure reasoning processes, rather than from empirical observation."[16] This critical deduction leads us to a grasping of the *science* and *mathematical* precision foundational to Indigenous knowledge.

Reyes raises two other research items of interest: 1) Von Bertalanffy (1950) developed the systems theory to transcend the limitation of the reductionist approach to studying nature and this recognized wholistic structures and interrelationship of system and context; this aligns with William Commanda's central theoretical principle of Ginawaydaganuc—that we are all connected with each other and with nature; and 2) Jackendoff (1994) raises the question of the relationships between natural language and knowledge, and spatial images and knowledge. He notes further that the theory of spatial representation is less developed than that of the encoding of linguistic meaning and he points also to the synergetic effect of integrating various models and languages.[17] Again, this has implications for wampum storytelling that compounds spatial and linguistic dimensions, and that incorporates and animates synergy and syntropy. Reyes argues that "the interactivity inherent to cybercartography allows the user to manipulate messages so that new geospatial information is created. This feedback process between the prototype and the users is an essential characteristic of cybernetics and allowed the advancement of the Helix trajectory."[18] This presents a context to understand the *Sacred Cane/Axis Mundi* motif intrinsic to many Indigenous belief systems, including the *Circle of All Nations* Logo tree, around which the Commanda thesis is structured.[19]

With her discussion of Integral Theory, Reyes notes that Wilber draws an approximate contour associating formal-operational cognitive capacities with modern rationalism, and he uses the term *vision-logic* to denote higher and deeper level cognitive capacities that transcend (and include) rationalism, and modernity-into-postmodern forms of representation, discourse, epistemology, expression and being. Vision-logic is essentially transrational and therefore, more apt to better contextualize levels of meaning beyond strictly factual or codified knowledge and is presently emerging as the next wave in the study of human consciousness. Formalisms, as represented in various texts, narratives, statistics, and conventional maps, all have limited capacity to respond to the increasing need to put very large amounts of information into forms that are meaningful to society.[20] This creates an opportunity for cybercartography to advance itself as an important tool for this emerging vision-logic capacity, as a means to help integrate, synthesize, visualize, and communicate very complex information in ways that go much beyond the conventional use of the geographic map, thereby giving planning and mapping a new level of application and meaning in the information era. Without the cybertechnological tools, William Commanda accesses this domain of vision-logic; the cyber zone enables the transference of his transrational knowledge and consciousness, as is evident in his active presence on Facebook and social media and the web. The mere inclusion of his photograph on Facebook on June 21, 2019, almost eight years after his death, drew over 8,000 hits in 24 hours, intimating this cyber reach over time (see Figure 13.1).

### *Cybercartography*

While cybernetics permits a temporal/spatial examination of the Commanda discourse, particularly in digital social media, cybercartography opens doors to the creation of a living archive in a digital atlas platform. Taylor had introduced the term cybercartography at the 1997 International Cartographic Conference in Stockholm to describe the transformative innovations taking place in the field, stating, "Cybercartography will see cartography applied to a much wider range of topics than has traditionally been the case […] It will also utilize an increasing range of emerging media forms and telecommunications networks such as the Internet and the World Wide Web. It will be multidimensional cartography using multimedia formats and is more likely to be an integral part of an information package than a stand-alone product. Cybercartography will also be highly interactive and engage the user in new ways. In organizational terms, it will see new partnerships being created between national mapping organizations, the private sector and educational institutions and the products of cybercartography are likely to be compiled by individuals from very different disciplines and professional perspectives working together."[21] By 2003, Taylor had articulated cybercartography as "the organization, presentation, analysis and communication of spatially referenced information on a wide range of topics of interest and use to society in an interactive, dynamic, multimedia, multi-sensory format, with the use of multimedia and multimodal interfaces."[22] In 2019, cybercartography was redefined as "… a complex, holistic, user centred process which applies location-based technologies to the analysis of all types of topics of interest to

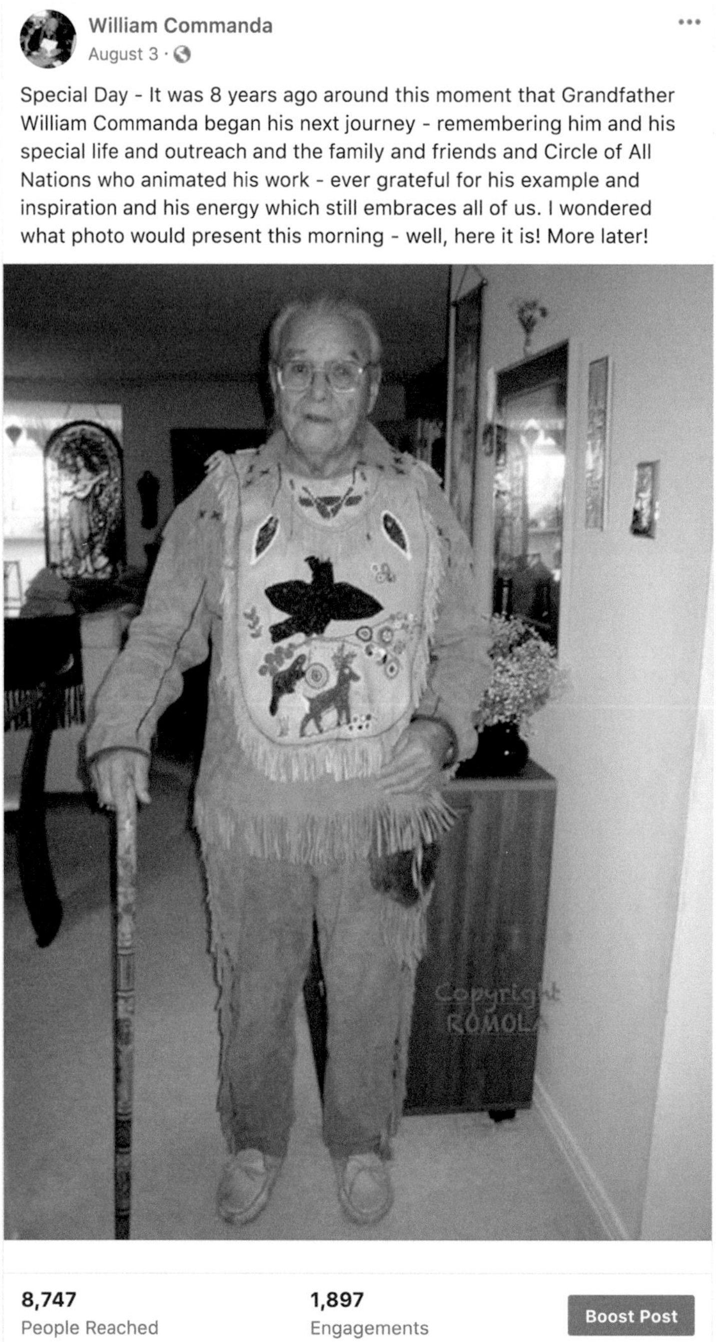


*Figure 13.1* William Commanda (1913–2011) Facebook screenshot 2019. Source: Thumbadoo.

society and the presentation of the results in innovative ways through cybercartographic atlases. A cybercartographic atlas is a metaphor for all kinds of qualitative and quantitative information linked by location and displayed in innovative, interactive multimodal and multisensory formats."[23]

## Cyber Land-Based Knowledge Generation

*Kwe Beckian! Hello Stranger!*—that, as William Commanda informed countless people, was how his Algonquin ancestors greeted Cartier when he first set foot in North America in 1534. That then is the root for the name of the province of Quebec. In clarifying the roots of the meaning of the colonial name, William Commanda was both performing his decolonizing intent, and at the same time affirming a welcome. In his dialectics, William Commanda used his language to assert truths rendered invisible by colonial place naming and history telling.

The following map history analysis trail is another example of the William Commanda discourse and illustrates its relevance to understanding his performance mapping approach.

In his re-mapping of relationships, he not only challenged the post-colonial discourse, but also extended performance mapping to the transformational, and in his *Circle of All Nations* work, his performance mapped the inter-cultural zone of interface, to deconstruct the land-based colonial manifesto and re-constitute it in multi-directional Indigenous-emergent mapping. Thus, when he was presented with the Key to the City of Ottawa at Victoria Island, Asinabka Sacred Chaudiere Site (the ancient gathering place of his ancestors) on June 21, 2006, he said, "I see three levels of significance that make me happy. The City of Ottawa is acknowledging someone from the City of Maniwaki; the Province of Ontario is reaching across to the Province of Quebec; and the National Capital Region, representative of the entire country, is opening its heart to someone seen generally as belonging within the confines of the reserve, the Kitigan Zibi Anishinabeg."[24] He presented his vision for the Asinabka Sacred Chaudiere Site on a birch bark map (see Figure 13.2; this was later recreated digitally). "Thus he deconstructed colonial maps (Harley), exercised border crossing discursive power (Border Crossing Wampum Belt and Foucault), assigned meaning to metaphoric devices like his Wampum belts (Rundstrom 1991, Turnbull 2007, Casti 2016) and with acceptance of the Key to the City, empowered act and artifact. With his presentation of his birch bark 'perpetually-live' map of the site (Crampton 2009), he produced territory endowed with ontogenetic potential (Kitchen and Dodge 2007), indicative of emergent mapping (Turnbull 2007, Pearce 2008), anticipatory counter-mapping (Rundstrom 2007) and interactive visualization (Crampton 2009), and ignited hybrid spaces and interactive networks of cybercartography and social media (Taylor 1997, 2003, 2005, 2008, and 2009)."[25]

He was also immersed in communications technology from the time of the creation of almost every new device: he was given a Wagner typewriter by its manufacturer, J. T. Underwood, and taught himself to type. A world-renowned canoe builder, he used the earliest commercial cameras and a slide projector to register and share his canoe-making images. He had the first television set on his reserve. He accumulated a singular library of documentaries and enthusiastically embraced VHS and later DVD players as teaching tools; he acquired the first video recorders and the earliest overhead projector. In addition to the visual acuity demonstrated in his iconic birch bark

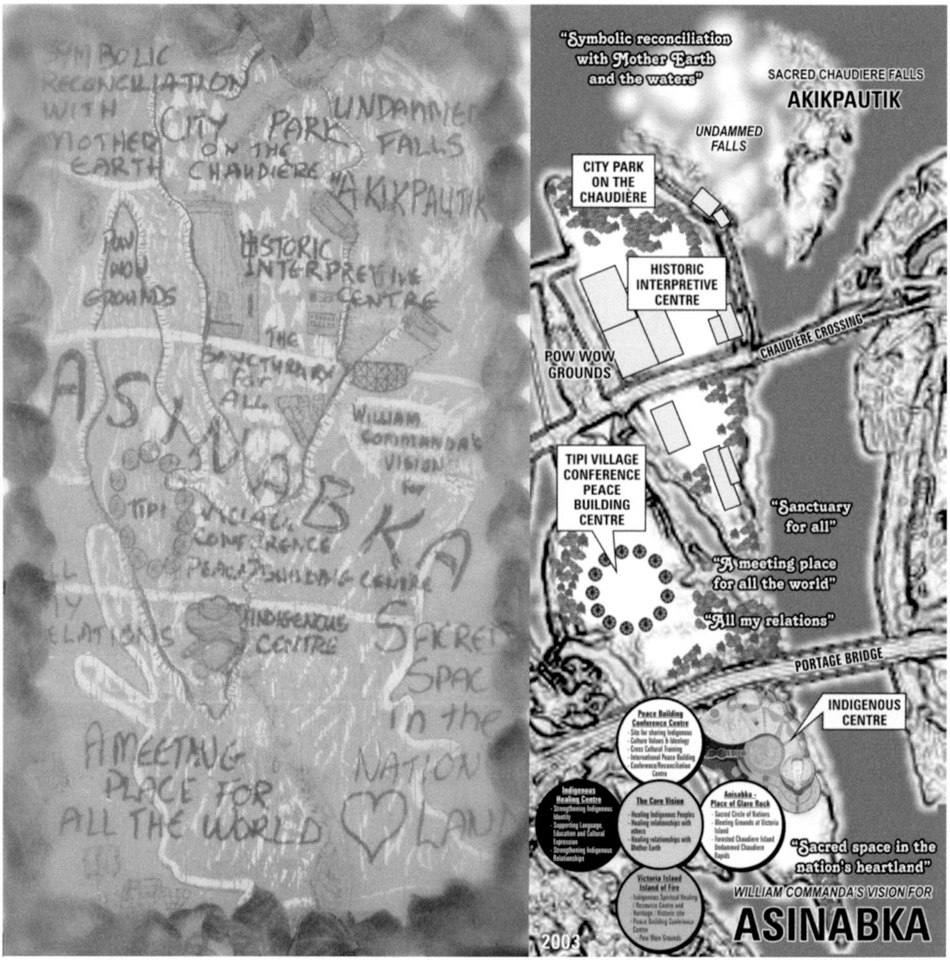


*Figure 13.2* Asinabka Chaudiere Map: on birch bark and in Digital Representation. Source: Thumbadoo.

canoe etchings, he had a fine ear for sound; he drummed, read music, sang a cappella, played the violin and valued a diversity of music for its energetic configurations. He was interviewed and recorded on multiple occasions. He was a formal presence on the world wide web following the national Aboriginal Constitutional Conference of 1987 (Canada), the 1993 United Nations *Cry of the Earth* Indigenous conference (New York), and the 2000 initiation of the *Circle of All Nations* websites and blogs.[26] In 2009, he engaged with Facebook, where he remains a meaningful and influential presence nine years after his death, with any *Circle of All Nations* reflection on his historical views on critical contemporary news items in five Facebook pages attracting thousands of hits in hours. His cybercartographic atlas created in the Nunaliit framework developed by the Geomatics and Cartographic Research Centre[27] (within the new domain of integrated cybercartographic theory, research, practice and technology developed by its Director, D. R. Fraser Taylor) begins the rich interactive living atlas archival compilation of his work, thinking, and discourse, inclusive of the contemporary reflections and reciprocal contributions of his ever-evolving *Circle of All Nations* geo-eco peace community.

We reiterate that William Commanda came to the domain of cybercartographic communications under his own volition. Identifying himself as Mamiwinini/Nomad, someone who moves all the time, he mapped his presence across his motherland, the continent of North America, and in particular, the American North East, in countless journeys, and in the hearts and minds of countless diverse peoples. He also traveled globally, east, south, west and north (Japan, Africa, Mexico and Europe). As the carrier of the oral Wampum storytelling heritage of his people he held a tremendous responsibility to share its messages, inclusive of a passionate concern about the destructive human impact on a living Mother Earth, and the concomitant global social injustices. The fact that his *Circle of All Nations* was entirely unfunded attests further to the unique power of this geo-cybernetic mapping and inclusive discourse.

## Conclusion

This chapter has explored digital mapping from a new conceptual paradigm. It reveals the operation of a cybernetic system emergent from a completely different ontological and epistemological foundation than is commonly researched in digital mapping. There are many lessons to be learned from the William Commanda story, his geo-cybernetic legacy, and his mental and physical mapping practice.

Cybercartographic atlases can be viewed as geo-spatial models of communication containing information and knowledge framed in a communication paradigm designed to impact social action and decision making. Reyes suggests that cybercartographic atlases are social products fed by society in their development, but when embodied in technology, the feedback loop impacts society. This leads to the question of the potential role of locational technology in social steering. Consistent with his understanding of Ginawaydaganuc, and his passionate concern for environment in the age of the Anthropocene, William Commanda seized upon digital telecommunication and information technology to impact global society in some microcosmic way via his animation of a *Circle of All Nations*. While websites are now viewed as static repositories of information, online social media platforms like Facebook represent a digital zone of motional fast shifting information sharing, spontaneous commentary, and individual and collective reflexivity, and they engage diverse communities in live conversation. William Commanda remains a dynamic and influential presence here. As Taylor has demonstrated in his research of the past two decades, cybercartographic atlases are complex digital mapping sites, and serve as repositories for libraries of digitized locational information; they are interactive and dynamic, inclusive of text, photos, videos, semiotic elements and graphics, reflecting journey and movement. Cybercartographic storytelling atlases further extend digital agency into academic research and knowledge generation, integrating theory, practice and technology. William Commanda's presence on the land is already registered via photo, slides, videos and the internet, and his unique discourse lends itself to the use of a digital atlas as a powerful methodological tool for qualitative and quantitative analysis and research. The evolving Commanda atlas ingests a vast body of interactive multimedia place-based archival and emergent social media and Facebook material, and the development and examination of his digitally mapped *Circle of All Nations* discourse will contribute to new practices in knowledge generation.

## Notes

1 Davis, Wade. *The Wayfinders: CBC Massey Lectures*. Toronto: House of Anansi Inc., 2009.
2 Davis, *The Wayfinders*, 2009, 64.
3 Albert Einstein, quoted in Davis, *The Wayfinders*, 2009, 116.
4 Davis, *The Wayfinders*, 2009, 157.
5 Huff, John Edward and Julia Moulden. "Your Car Has GPS. Your Brain Needs It, Too." *Globe and Mail* (June 21, 2019). https://www.theglobeandmail.com/opinion/article-your-car-has-gps-your-brain-needs-it-too/.
6 Thumbadoo, R. V. "Ginawaydaganuc and the Circle of All Nations: The Remarkable Environmental Legacy of Elder William Commanda." Ph.D. thesis, Carleton University (2017). https://curve.carleton.ca/aa4e3cbb-5b83-464d-8286-a901fcd77b06.
7 Taylor, D. R. F., Erik Anonby and Kumiko Murasugi (eds.). *Further Developments in the Theory and Practice of Cybercartography*. San Diego: Elsevier, 2019, 21.
8 Thumbadoo, R. V., *Learning from a Kindergarten Dropout*. Ottawa, Ontario: Circle of all Nations, 2005, 49.
9 Thumbadoo, *Learning From a Kindergarten Dropout*, 2005.
10 Thumbadoo, *Learning From a Kindergarten Dropout*, 2005, 13.
11 Thumbadoo, R. V., "Ginawaydaganuc and the Circle of All Nations," 2017, 61
12 Taylor, D. R. F. (ed.). *Cybercartography: Theory and Practice*. Amsterdam: Elsevier, 2005.
13 M. del C. Reyes, Cybercartography from a Modelling Perspective. In D. R. F. Taylor, (ed.). *Cybercartography: Theory and Practice*. Amsterdam: Elsevier, 2005, 71.
14 Taylor, *Cybercartography*, 2005, 5.
15 M. del C. Reyes, Cybercartography from a Modelling Perspective. In D. R. F. Taylor (ed.). *Cybercartography: Theory and Practice*. Amsterdam: Elsevier, 2005, 68.
16 M. del C. Reyes, Cybercartography from a Modelling Perspective. In D. R. F. Taylor (ed.). *Cybercartography: Theory and Practice*. Amsterdam: Elsevier, 2005, 68.
17 M. del C. Reyes, Cybercartography from a Modelling Perspective. In D. R. F. Taylor (ed.). *Cybercartography: Theory and Practice*. Amsterdam: Elsevier, 2005, 74.
18 M. del C. Reyes, Cybercartography from a Modelling Perspective. In D. R. F. Taylor (ed.). *Cybercartography: Theory and Practice*. Amsterdam: Elsevier, 2005, 85.
19 Thumbadoo, "Ginawaydaganuc and the Circle of All Nations," 2017, 61.
20 M. del C. Reyes, Cybercartography from a Modelling Perspective. In D. R. F. Taylor (ed.). *Cybercartography: Theory and Practice*. Amsterdam: Elsevier, 2005, 58.
21 Taylor, D. R. F. "Maps and Mapping in the Information Era." In L. Ottoson (ed.). Proceedings of the 18th ICA International Cartographic Conference (ICC 97), Stockholm, Sweden, Vol. 1. Gavle: Swedish Cartographic Society, 1997, 3–10.
22 Taylor, D. R. F. "The Concept of Cybercartography." In M.P. Peterson (ed.). *Maps and the Internet*. Amsterdam: Elsevier, 2003, 404.
23 Taylor, Anonby, Murasugi*Further Developments in the Theory*, , 2019, 20–21.
24 Thumbadoo, R.V. "Asinabka." Last modified June 2019. www.asinabka.com.
25 Thumbadoo, R. V. and D. R. F. Taylor. "Storytelling with Cybercartography: The William Commanda Story." In D. R. F. Taylor, Erik Anonby and Kumiko Murasugi (eds.). *Further Developments in the Theory and Practice of Cybercartography*. San Diego: Elsevier, 2019, 265.
26 Thumbadoo, R. V. "Circle of All Nations." Last modified June 2017. www.circleofallnations.ca (archival sites); Thumbadoo, "Asinabka."
27 "Geomatics and Cartographic Research Centre." GCRC. Accessed June 2019. https://gcrc.carleton.ca/index.html.

## References


Burgess, Neil, Maguire, Eleanor A. and John O'Keefe. "The Human Hippocampus and Spatial and Episodic Memory." *Neuron 35* (2002).

Casti, Emaneula and D. R. F. Taylor, Series Editor. *Reflexive Cartography: A New Perspective on Mapping*. Amsterdam: Elsevier, 2015.

Crampton, Jeremy W. "Cartography: Mapping 2.0." *Progress in Human Geography 33*, no. 1 (2009): 91–100.

Cresswell, Tim. "Discourse." In *International Encyclopaedia of Human Geography*. Nigel Thrift and Rob Kitchen (Eds.). Oxford: Elsevier, 2009: 211–214.

Davis, Wade. *The Wayfinders: CBC Massey Lectures*. Toronto: House of Anansi Press Inc., 2009.

"Geomatics and Cartographic Research Centre." GCRC. Accessed June 2019. https://gcrc.carleton.ca/index.html

Huff, John Edward and Julia Moulden. "Your Car Has GPS. Your Brain Needs It, Too." *Globe and Mail*, June 21, 2019. https://www.theglobeandmail.com/opinion/article-your-car-has-gps-your-brain-needs-it-too/.

Kitchin, Rob and Martin Dodge. "Rethinking Maps." *Progress in Human Geography 31*, no. 3 (2007): 331–344.

Pearce, Margaret Wickens. "Mapping, Non-Western." In *International Encyclopedia of Human Geography*. Rob Kitchin and Nigel Thrift (Eds.). Oxford: Elsevier, 2009: 372 -384.

Rundstrom, Robert "The Changing Direction of North American Cartography." *Cartographica 28*, no. 2 (1991): 1–12.

Rundstrom, Robert. "Counter-Mapping." In *International Encyclopedia of Human Geography*. Rob Kitchin and Nigel Thrift (Eds.). Oxford: Elsevier, 2009: 1–5.

Smith, Linda Tuhiwai. *Decolonizing Methodologies: Research and Indigenous Peoples*. London, New York: Zed Books, 2012.

Taylor, D. R. F. "The Concept of Cybercartography." In *Maps and the Internet*. M. P. Peterson (Ed.). Amsterdam: Elsevier, 2003.

Taylor, D. R. F. "Cybercartography: Theory and Practice, Volume 4." In *Modern Cartography Series*. Amsterdam: Elsevier, 2005.

Taylor, D. R. F., and T. Lauriault, (eds.). *Developments in the Theory and Practice of Cybercartography: Applications and Indigenous Mapping*. Amsterdam: Elsevier, 2014.

Taylor, D. R. F., Anonby, Erik and Kumiko Murasugi (Eds.). *Further Developments in the Theory and Practice of Cybercartography*. San Diego: Elsevier, 2019.

Thumbadoo, R. V. *Learning from a Kindergarten Dropout*. Ottawa, Ontario: Circle of All Nations, 2005.

Thumbadoo, R. V. *"Ginawaydaganuc and the Circle of All Nations: The Remarkable Environmental Legacy of Elder William Commanda."* Ph.D. thesis, Carleton University, 2017. https://curve.carleton.ca/aa4e3cbb-5b83-464d-8286-a901fcd77b06.

Thumbadoo, R. V. *"Circle of All Nations." Last modified* June 2017. www.circleofallnations.ca.

Thumbadoo, R. V. *"Asinabka." Last modified* June 2019. www.asinabka.com.

Thumbadoo, R. V. and D. R. F. Taylor. "Storytelling with Cybercartography: The William Commanda Story." In *Further Developments in the Theory and Practice of Cybercartography*. D. R. F. Taylor, Erik Anonby and Kumiko Murasugi (Eds.). San Diego: Elsevier, 2019.

Turnbull, David. "Maps Narratives and Trails: Performativity, Hodology and Distributed Knowledges in Complex Adaptive Systems—an Approach to Emergent Mapping." *Geographical Research 45*, no. 2 (2007).

# 14 Indigenous Place Names as Visualizations of Indigenous Knowledge

*Rebekah R. Ingram*

## Introduction: Why Are Place Names Important?

A place name is a symbol which conveys spatial information through language. Put more simply, place names are a way of using language to transmit landscape, geographical, and environmental information through space and time. Through the medium of language, they represent people's relationships with a location or landscape. Why and how we name place, as well as which places are named or nameless allows insight into many of the different aspects of life. The giving of a place name may occur as a result of an event in the area worthy of recognition, such as Council Bluffs, Iowa, which was named for a council meeting between the Lewis and Clark expedition and the Otoe and Missouri Indigenous bands (National Park Service, 2004). Anglo-Saxon (sometimes called "Old English") place names, such as Birmingham, Arlesey, and Bakewell all utilize personal names to indicate possession or control.[1] Finally, the importance of religion and the influence of the Catholic Church[2] is demonstrated by the common use of "saint" names, which number over 2,200 in Quebec.[3] The referents of naming and the act of naming itself cover a wide variety of topics including navigation, historical settlement and ecological knowledge. Examining the salience, frequency of use, and semantic (meaningful) concepts used within place names allows those studying them to infer what is important to the group of people doing the naming and what kinds of information they are aware of.

Names descriptive of some specific feature of the landscape, whether regarding a physical feature, such as the White Mountains (New Hampshire), a source of food, such as Cranberry Isles (Maine), or important sources of materials, such as Ash Creek (Arizona, Oregon, California, South Dakota, etc.) often demonstrate people's knowledge of and relationship to a specific location. Many Indigenous place names demonstrate Indigenous Knowledge (IK) or Traditional Environmental Knowledge (TEK), "[t]he understandings, skills and philosophies developed by societies with long histories of interaction with their natural surroundings"[4], which is acquired during travel and trade, through acts of cultural significance, and in the process of gathering sustenance and materials. The Assembly of First Nations elaborates that this is "knowledge informed by aboriginal paradigms as applied to skills, understandings, expertise, facts, familiarities, beliefs, revelations and observations ... it is location specific and reflects the particular conditions of unique cultures and peoples in specific geographic locations."[5] In short, Indigenous Knowledge requires an intimate knowledge of the landscape which is passed intergenerationally.

*Table 14.1* Place names of the Kwakwa̱ka̱'wak demonstrate Indigenous Knowledge[7]

| Place Name | Meaning |
|---|---|
| *mᴇlā'dēʿ* | having sockeye salmon |
| *sā'dzadēʿ* | having spring salmon |
| *wa'yadēʿ* | having herring spawn |
| *dō'yadēʿ* | having poisonous clams |
| *sᴇ'ldzadēʿ* | having blueberries |
| *q!waᵋlā'dēʿ* | having salmon berry shoots |
| *de'wadēʿ* | having yellow cedar |
| *lō'gwēdzas* | halibut fishing beach place |
| *ts!āg̨as* | mountain goat place |

Boas' (1934) study "Place names of the Kwakiutl"[6] provides many excellent examples of IK as embedded within place names. Some of these are given in Table 14.1, which can be found below.

These examples outline knowledge of different plants and animals important to (or, in the case of poisonous clams, dangerous to) the Kwakwala'wakw as food and material resources. In addition, they also demonstrate long-term, seasonal knowledge as evidenced through the specifications of "spring" salmon, the sights of fish spawning, and the appearance of blueberries and berry shoots, events which occur only at particular times of year.

Many Indigenous place names are in current use in North America. Familiar examples of these include names like Quebec, from the Algonquin language, meaning "narrows"[8] and Skanehtati (Schenectady, NY), which describes the extensive area of pine bush between Albany, New York, and the location of the present-day city of Schenectady. These names originated within their own **naming stratum**, where the naming stratum is the set of names within a particular language as used by that sociolinguistic group. For example, a number of Indigenous peoples including the Natchez and the Caddo peoples named places in what is today called Louisiana[9] according to their own sociolinguistic **naming conventions**, or their cognitive and semantic rules for what features were important enough to name and how they should be named. After European arrival in North America, the French subsequently named, or renamed places within the same area according to their naming conventions, thus creating a French naming stratum. Language patterns from each wave of new arrivals to the landscape were layered over each other as each group of people renamed and documented those names, or utilized the old name, often changing the name to conform to the language of the new namers, or simply to make the name easier to say. Each of these layers comprises its own naming stratum and, depending upon the movement and location of the namers, these strata may take up vast stretches of land (for example, the American English naming stratum which arguably consists of all of the names located within the administrative boundaries of the United States) or a relatively smaller geographic area (for example, the Pennsylvania Dutch naming convention of Lancaster County, PA). Strata may therefore be further defined as what I call macrostrata (a large area of layers that have been named by what may be considered a dominant language family), or microstrata (layers of smaller areas, or layers that have been named by what may be considered minority languages). Since macrostrata are comprised of such a large geographic area, microstrata may be contained within them, and they may also overlap with each other. While the macrostrata may be responsible for the renaming of places, those that are not renamed become part of a substratum. Thus, Indigenous place names that are in use today originated from one stratum and remained in use despite the fact that, in many cases, the naming language became less frequently used, in some cases even becoming endangered or even dormant.

## Indigenous Naming Strata and Mapping

Indigenous naming strata can be accessed by utilizing present-day Indigenous place names, those that can be found within oral histories and held by community members, and also through archival research within colonial administrative documents, ethnologies, or previous place name studies. Given an appropriate mapping framework, these place names can be used to create visualizations of place which essentially recreate an Indigenous naming stratum or Indigenous geography thus imparting historical knowledge of the landscape, marking important locations of resources and physical features. Furthermore, mapping with a framework which enables the use of multimedia also enables the continued preservation of the IK embedded within the landscape itself through the medium of place name mapping. This information can then go on to be used in language and knowledge revitalization efforts through the ability to share a historical or restructured conceptualization of space with others.

The documentation and preservation of Indigenous place names, and the IK/TEK embedded in place, has generally been confined (with reason) to paper, which imposed limitations in terms of focus. In written format, place name studies focused on the linguistic aspects, while paper maps, or collection of maps which formed atlases, focused on spatial and geographical aspects. Today's digital atlases allow for greater flexibility and manipulation in terms of data visualizations and collection of spatial data. In this section, I present three different techniques that can be used to map Indigenous place names. The spatial patterns which emerge upon a map when using these techniques can help us to think about place differently by presenting them visually, a literal viewpoint which is not offered by linguistic analysis alone. Mapping place names can also be used to better understand historical landscapes and environmental changes as will be outlined in sections 1, 2, and 3 below.

Indigenous place names have been a topic of interest for millennia (see, for example, Ptolemy's *Geografika*[10]); in North America, numerous sources have sought to document and translate Indigenous place names. Some of the maps within this chapter have been made using the Indigenous place names collected in volumes like Bright's *Native American Placenames of the United States*, and informed by others like Beauchamp's *Aboriginal Place Names of New York*, and Ruttenber's *Indian geographical names in the valley of Hudson's river, the valley of the Mohawk, and on the Delaware*[11]. Ideally, mapping place names should occur as part of an overall community mapping project in conjunction with the naming group in question, or as part of an overall methodology which includes work with knowledge holders from the naming group. In my case, the techniques outlined here were largely the result of the discussions over several years with members of the Kanyen'kehá:ka (Mohawk People) on place names and Indigenous place naming conventions. These relationships informed my dissertation work, but also led to the *Atlas of Kanyen'kéha Space*, a community digital mapping initiative currently in production with Carleton's Geomatics and Cartography Research Centre under the guidance of Dr. Kahente Horn-Miller (Kanyen'kehá:ka from

Kahnawà:ke) which seeks to document and preserve Kanyen'kéha place names together with Kanyen'kehá:ka oral history, IK, TEK, and language.[12] Thus, it has been my own experience that the insights of those who originally named place are invaluable in providing historical information from their own perspectives, as well as context in terms of landscape, environment, and culture. While I am fortunate to be able to work with Kanyen'kehá:ka in helping to document and preserve their place names and geography, it will certainly take several lifetimes to do so. As a result, Section 1 discusses one small aspect of that study, and the information in Sections 2 and 3 are offered only as examples for the potential applications of these techniques rather than as the final result of such a study. It is my hope that these techniques can be used by Indigenous naming groups themselves in order to provide spatial visualizations and preserve their own Indigenous Knowledge and understandings not only of their place names, but also their own cognitive geography.

### *1. Mapping Morphemes*

Place names follow the linguistic rules of the individual language in which they were conceived.[13] All place names are composed of **morphemes** according to those linguistic rules, where a morpheme is a basic element of meaning. In English, this can take the form of a single word, like "hill," but it could also be as small as the suffix "-s" on the word "hills," since this would indicate the meaning "plural." Some Indigenous languages use many morphemes to construct what could be considered an entire sentence in English, as in Example (1), below:

| (1) | |
|---|---|
| English: | I like you |
| Anishnaabemowin: | Giminwenimin[14] |

What is expressed by three separate words in English ("I," "like" and "you") is expressed as one entity in Anishnaabemowin (called "Ojibwe" in English). The technique described in this section makes use of these different elements of meaning, whether they are individual words, like "you" in "I like you" or individual morphemes like the "-in" suffix in "Giminwenimin."[15] Descriptive place names are often based upon landscape features and the words for these features are made up of morphemes. For example, the descriptive name "Springfield" is made up of the morphemes "spring" and "field." The place name Springfield in English demonstrates that it is not uncommon for morphemes or entire place names to be used repeatedly. It is precisely because these elements are used repeatedly that they can provide the basis for a visualization in the form of input data.

The first step in this process is to create a set of data to be mapped. Essentially, this can take the form of a list of place names which have been (or may be) linguistically analyzed and the individual morphemes have been identified. The names in question can be place names used in the present day, historical names, or both. It is also not necessarily the case that the entirety of the place name be morphologically clear: As long as the morpheme which is to be mapped is clear, a place name can be used to make a visualization. Next, a morpheme is chosen and all names which utilize that morpheme are noted. The location of each name which utilizes the morpheme is then marked on a map, and a visual pattern emerges.

In Ingram (2018), I analyzed the place name Kahnawà:ke, the name of the present-day Kanyen'kehá:ka Reserve located across from the city of Montreal in Quebec, Canada. Although used as a present-day place name, this name, meaning approximately "at the rapids"[16] has also been used historically throughout Haudenosaunee (often called the Six Nations or Iroquois Confederacy) territory in related Haudenosaunee languages.[17] The name is made up of three morphemes outlined in Example (2), below:

(2)

| | | |
|---|---|---|
| *Ka* | *hnaw* | *à:ke* |
| "it" | "rapids" | "at" |

Maracle (2001) suggests that *-hnaw-*, the morpheme which provides meaning for this name,

> "describes a situation where water (or some liquid) is moving quickly in a particular way or direction, or against the intended direction and therefore is often interpreted as a 'current,' 'rapids,' or as in reference to a 'spring' or 'well' where the water is being forced out from where it is."[18]

Instances of place names which utilize this morpheme demonstrate Indigenous Knowledge: those locations were of enough importance to encode them into language such that they might be shared between people in various different locations as well as intergenerationally.

In addition to the name Kahnawà:ke, *-hnaw-* is also found in other place names. Thus, mapping all names which use this morpheme will reveal the places describing this element of meaning. Mapping these instances produced the spatial patterns in the map in Ingram (2018)[19] which illustrates Haudenosaunee place names utilizing the root *-hnaw-*. Each point on this map represents a place name using the morpheme *-hnaw-*, which transmits knowledge of the location of "rapids." This is especially important given that the waterways have been modified in some of the areas where these names were used. For example, both the Erie Canal and St. Lawrence Seaway affected places named Kahnawà:ke[20] in that, rapids are no longer found at those locations; neither place can now be considered "at the rapids." However, the Indigenous Knowledge preserved within these place names can be used to create a visualization by mapping places utilizing the morpheme *-hnaw-*. Although the original referents of these names, the rapids, have been lost in some instances, their memory remains anchored to the landscape through these place names.

While even a static map can yield insight, digital atlases may feature tools which allow the user to compare and contrast visualizations created by mapping different morphemes or combinations of morphemes. This creates the opportunity to compare visualizations of the linguistic information contained within specific place names at specific locations.

## 2. *Mapping Semantic Concepts*

Similar to mapping morphological elements as outlined in the previous section, another technique involves mapping the topic of the name itself together with the

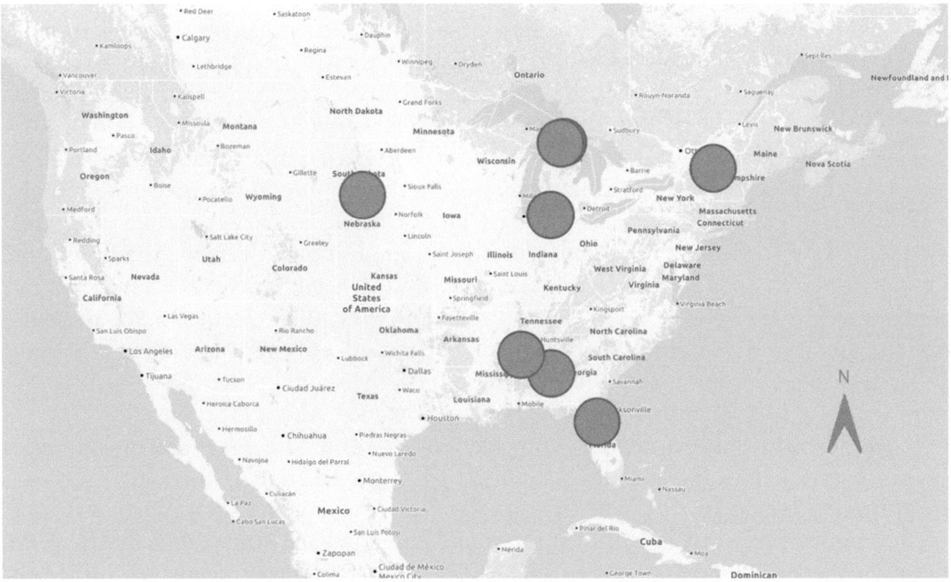


*Figure 14.1* Locations of places named for turtles. © Rebekah R. Ingram 2020.

location, but across several different naming groups. As outlined in the Introduction, Indigenous place names often encode materials or species of importance to place namers; not surprisingly, certain plants, animals, and materials found over much of North America are of importance to more than one group and thus appear in more than one group's naming convention. Bright's *Native American Placenames of the United States* reveals many places named for bears, beavers, kinds of fish or fishing spots, berries, different species of trees, and minerals such as iron and copper.[21] The map in Figure 14.1, below, was made by locating all instances of places named for turtles within that text and then mapping their approximate locations.

Again, it is important to verify the accuracy of each translation through linguistic analysis before utilizing this technique. Another issue which may arise is that of whether the location is that of the original name, or if it has been "transplanted" from somewhere else. Yet again, this outlines the need to work with the original naming group who may be able to verify both the meaning of a name as well as the original location or referent of the name. However, this technique may prove to be particularly useful in terms of environmental mapping. In this example, I have used turtles that "are indicator species, which means they can reflect the health of the ecosystems in which they live—if turtles aren't doing so well, there might be something wrong in that habitat or environment, [such as a high concentration] of fertilizers or other toxins."[22] This technique, when combined with a digital map, which enables the collection of other environmental information, could be used to view a landscape that has been altered (similar to the case with Kahnawà:ke outlined in Section 1), or to facilitate the sharing of Western knowledge and Indigenous Knowledge concerning the environment.

### *3. Mapping Sounds*

The final technique shown here utilizes the sounds found within place names to make maps. This technique helps to define the locations that different linguistic groups inhabited or where they interacted with each other on the landscape. Every human language ascribes some sort of meaning to a specific subset of sounds out of all possible human sounds. Many of these sounds are shared between many languages (such as, for example, the vowels represented in the words "ah" and "bee"). However, there is also wide variation in which sounds are used or not used. For example, the consonant symbolized by the letter "r" sounds different in English and German. Similarly, both French as well as Kanyen'kéha make use of nasalized vowels such as the one represented in the word "non." Making a map indicating whether a name uses a specific sound or does not use that sound can help to identify the place name according to languages which do or do not use those sounds; the spatial patterns revealed by this map will help to indicate which language group named which locations.

I demonstrate this example in this section by using the Algonquian languages of northeastern North America (Mahican, Mohegan, Lenape, Abenaki)[23] and the neighboring Iroquoian language family, also found in northeastern North America. First, an inventory must be taken of the sounds used in each language and differences between the two language groups identified. The Algonquian languages utilize the sounds represented by the letters "b," "p" and "m"; the Iroquoian languages can be distinguished from Algonquian by a complete absence of these sounds. Instances of Algonquian place names can then be identified by the appearance of "b," "p" or "m." Therefore, while it is possible that a name like "Papaconck" is Iroquoian, it is highly unlikely due to the presence of the sound "p."

Again, a list of place names is compiled belonging within the geographic area in question. In order to make the map given in Figure 14.2, I utilized some of the Iroquoian names from work with Kanyen'kehá:ka communities[24] as well as Bright's *Native American Placenames of the United States* (2004).[25] In this case, I was specifically interested in the areas of jurisdictional interaction between the two linguistic groups which, according to historical sources, occurred in an area along the Hudson Valley corridor, Lake George and Lake Champlain, and branches of the Susquehanna River.[26] These areas comprise the modern-day states of Pennsylvania, New Jersey, New York, Connecticut, Massachusetts, and Vermont. Using this data, I plotted over 225 place names onto a map. For this technique, the place names using the appropriate contrasting sounds must be distinguished in some way. This could be through the use of different symbols or colors, for example. In the map presented in Figure 14.3, those place names using "b," "p" or "m" (making them Algonquian) are represented by a black square, while those that do not are represented with a white square. Some places have more than one place name—one which uses a "b," "p" or "m" and one that does not. These are represented with a gray square.

Using this technique, a spatial pattern emerges which can help to visualize areas of land use between Algonquian groups and Iroquoian groups.[27] The delineation shows that the Hudson River does, indeed, appear to serve as the division between the two groups. The southern delineation requires further detailed analysis, but both Iroquoian and Algonquian place names can be found in four different places on the map (indicated by a gray square), showing where the two groups both had names for the same place.

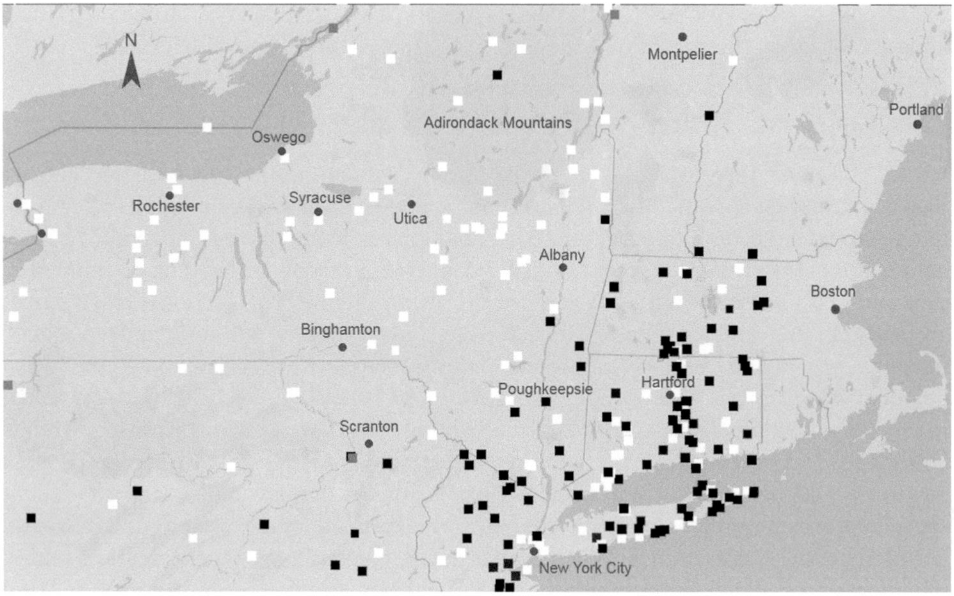


*Figure 14.2* Mapping the sounds of place names. Basemap © Mapbox 2020 © OpenStreetMap; Map © Rebekah R. Ingram (2020).

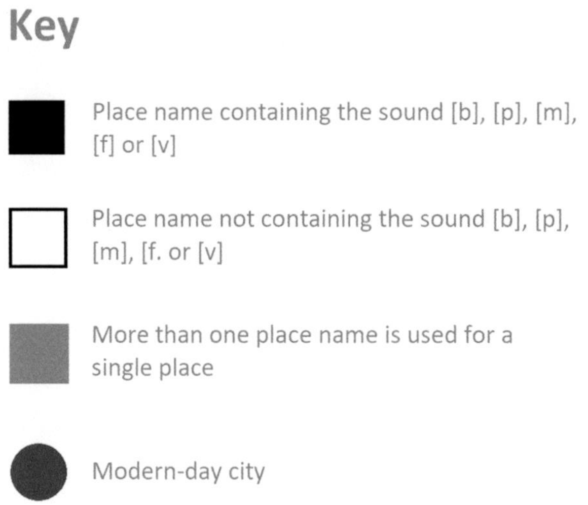


*Figure 14.3* Key for Figure 14.2.

The map presented here again serves only as a demonstration of the technique. Modern place names can be confirmed by language speakers as belonging to a specific naming convention; however other place names (for example, Cohoes, NY) could be of several different linguistic origins, including Kanyen'kéha, Dutch, or an Algonquian language. As stated previously, work with Indigenous speakers will again provide insight regarding linguistic analysis, the original locations of place names, and cultural relevance.

## Conclusion

Jett (2001) states that "placenames reveal how particular cultures perceive and classify their environments: what they see as significant—economically, religiously, and so forth—about how they differentiate particular places from space in general."[28] Place names are often viewed as secondary or extraneous in terms of language documentation and grammar creation; however, as outlined in the previous sections, place names are a significant source of Indigenous Knowledge, including semantic conceptualizations of space and TEK—Traditional Environmental Knowledge, both of which demonstrate deep ties to a geographic location or landscape. The methods presented here offer ways of creating spatial visualizations of language data in order to view landscape in different ways. Because of differences in ontology and landscape delineation[29] and because current mapping platforms are largely built using models of space based on European language concepts, this method is best used in collaboration with Indigenous communities. When used as a complement to linguistic analysis within the context of place name studies, these methods provide opportunities to help researchers, educators, and Indigenous communities bridge the gaps in conceptual, cultural, and linguistic space.

## Notes

1 University of Nottingham, "Key to English Place Names" (2019). http://kepn.nottingham.ac.uk/#.
2 Couture, Claude. "Quebec." In *The Canadian Encyclopedia*, last revised January 22, 2019. https://www.thecanadianencyclopedia.ca/en/article/quebec.
3 Rayburn, Alan. "Place Names." In *The Canadian Encyclopedia*, last revised October 21, 2015. https://www.thecanadianencyclopedia.ca/en/article/place-names.
4 UNESCO, "What Is Local and Indigenous Knowledge?" In: *Local and Indigenous Knowledge Systems*, 2017, http://www.unesco.org/new/en/naturalsciences/.
5 Assembly of First Nations. *First Nations Ethics Guide on Research and Aboriginal Traditional Knowledge*. Ottawa, ON: Assembly of First Nations, 2015, 4.
6 The term Kwakwaka'wakw has subsumed the term "Kwakiutl" as used by Boas (First Peoples' Cultural Council. "About the Kwakwaka'wakw people." *First Voices* (2020). http://www.firstvoices.com/en/Kwakwala); the language used by the Kwakwaka'wakw is known as Kwak'wala (ibid).
7 Following Boas, Franz. *Geographical Names of the Kwakiutl Indians*. New York, NY: Columbia University Press, 1934, p. 11.
8 Gouvernement du Québec, "Québec." *Commission de Toponymie* (2012). http://www.toponymie.gouv.qc.ca/ct/toposweb/recherche.aspx.
9 Swanton, John Reed. *Indian Tribes of the Lower Mississippi Valley and Adjacent Coast of the Gulf of Mexico*. Washington: Government Printing Office, 1911, 1.
10 Ptolemy, Claudius. *Geography*. New York: Cosimo Inc., 2011.
11 Beauchamp, William. *Aboriginal Place Names of New York*. Albany, NY: New York State Department, 1907; Ruttenber, Edward Manning. *Footprints of the Red Man: Indian Geographical Names in the Valley of Hudson's River, the Valley of the Mohawk, and on the Delaware, Their Location, and the Probable Meaning of Some of Them*. Newburgh, NY: New York State Historical Association, 1906.
12 Dr. Kahente Horn-Miller, p.c.; Dr. D. R. Fraser Taylor, p.c.

13 However, when borrowed into another naming stratum, place names are often changed in terms of structure and sound to conform to the linguistic rules and sounds of the borrowing language.
14 Chacaby, Maya, Facebook message to author, March 31, 2020.
15 Ibid.
16 Mithun, Marianne, "Principles of Naming in Mohawk." *Proceedings of the American Ethnological Society* no. 4 (1984).
17 Ingram, Rebekah R., Erik Anonby and D. R. Fraser Taylor. "Mapping Kanyen'kéha (Mohawk) Ethnophysiographical Knowledge." In D. R. Fraser Taylor, Erik Anonby and Kumiko Murasugi (eds.). *Further Developments in the Theory and Practice of Cybercartography* Cambridge: Elsevier, 2019.
18 Maracle, David. Kanatawakhan, *Karoron ne Owennahshonha: A Mohawk Thematic Dictionary*. London, ON: Kenyen'keha Books, 2001, 148.
19 Ingram, Rebekah R. "Mapping Indigenous Landscape Perceptions." In *Endangered Languages and the Land: Mapping Landscapes of Multilingualism*. Proceedings of FEL XXII/2018, Reykjavík, Iceland. London: EL & EL Publishing, 2018.
20 Langbein, Walter B. *Hydrology and Environmental Aspects of Erie Canal (1817–1899)*. (Washington, D.C.: United States Government Printing Office, 1976, 9; Phillips, Stephanie K. "The Kahnawá:ke Mohawks and the St. Lawrence Seaway." Master's Thesis, Montreal: McGill University, 2000. http://digitool.library.mcgill.ca/R/?func=dbin-jump-full&object_id=33310&local_base=GEN01-MCG02.
21 Bright, William. *Native American Placenames of the United States*. Norman, OK: University of Oklahoma Press, 2004.
22 McCallum, Jenn. "In Peterborough and Across Ontario, Turtles are at Risk." *The Peterborough Examiner (Peterborough, ON)* (July 11, 2018). https://www.thepeterboroughexaminer.com/community-story/8735609-in-peterborough-and-across-ontario-turtles-are-at-risk/.
23 See, for example, Cuoq, Jean André. *Lexique de la Langue Algonquine*. Montreal: J. Chapleau, 1886; Brinton, Daniel Garrison. *A Lenâpé-English Dictionary: From an Anonymous Ms. in the Archives of the Moravian Church at Bethlehem, PA*. Philadelphia: The Historical Society of Pennsylvania, 1888.
24 See Ingram, Rebekah R. "Naming Place in Kanyen'ké:ha: A Study Using the O'nonna Three-Sided Model." PhD diss., Carleton University, 2020).
25 Bright, William. *Native American Placenames of the United States*. Norman, OK: University of Oklahoma Press, 2004.
26 See Snow, Dean. *The Iroquois*. Malden, MA: Blackwell Publishing, 1996; Parmenter, Jon. *The Edge of the Woods: Iroquoia, 1534–1701*. East Lansing: Michigan State University Press, 2010; Ingram, Rebekah R. "Naming Place in Kanyen'ké:ha: A Study Using the O'nonna Three-Sided Model." PhD diss., Carleton University, 2020.
27 For the purposes of this exercise, I only utilized those names which Bright specifies as being Indigenous, i.e. I did not take into account places named for people as these may sometimes be contrary to Indigenous languages conventions (Ingram, Rebekah R. "Naming Place in Kanyen'ké:ha: A Study Using the O'nonna Three-Sided Model." PhD diss., Carleton University, 2020, 72) or for groups of people, since these are often based upon exonyms, or names that are given to a group by those from outside the group. I also avoided names which were of ambiguous origin.
28 Jett, Stephen C. *Navajo Placenames and Trails of the Canyon de Chelly System, Arizona*. New York, NY: Peter Lang, Inc., 2001.
29 See Ingram, Rebekah R. "Naming Place in Kanyen'ké:ha: A Study Using the O'nonna Three-Sided Model." PhD diss., Carleton University, 2020.

## Bibliography

Assembly of First Nations. *First Nations Ethics Guide on Research and Aboriginal Traditional Knowledge*. Ottawa, ON: Assembly of First Nations, 2015.

Beauchamp, William. *Aboriginal Place Names of New York*. Albany, NY: New York State Department, 1907.

Boas, Franz. *Geographical Names of the Kwakiutl Indians*. New York, NY: Columbia University Press, 1934. https://babel.hathitrust.org/cgi/pt?id=inu.39000005878918;view=1up;seq=7.

Bright, William. *Native American Placenames of the United States*. Norman, OK: University of Oklahoma Press, 2004.

Brinton, Daniel Garrison. *A Lenâpé-English Dictionary: From an Anonymous Ms. in the archives of the Moravian church at Bethlehem*, Philadelphia, PA: The Historical Society of Pennsylvania, 1888.

Couture, Claude. "Quebec". *The Canadian Encyclopedia*. Last revised January 22, 2019. https://www.thecanadianencyclopedia.ca/en/article/quebec.

Cuoq, Jean André. *Lexique de la Langue Algonquine*. Montreal: J. Chapleau, 1886.

First Peoples' Cultural Council. "About the Kwakwaka'wakw people." *First Voices*. Last revised May 22, 2020. http://www.firstvoices.com/en/Kwakwala.

Gouvernement du Québec. "Québec". *Commission de Toponymie*, 2012. http://www.toponymie.gouv.qc.ca/ct/toposweb/recherche.aspx.

Ingram, Rebekah R. "Mapping Indigenous landscape perceptions". In *Endangered languages and the land: Mapping landscapes of multilingualism*; Proceedings of FEL XXII/2018 (Reykjavík, Iceland), 6–32. London: EL & EL Publishing, 2018.

Ingram, Rebekah R. "*Naming Place in Kanyen'kéːha: a study using the O'nonna three-sided model.*" PhD diss., Ottawa, Ontario, Canada: Carleton University, 2020.

Ingram, Rebekah R., Anonby, Erik, & Taylor, D. Fraser. "Mapping Kanyen'kéha (Mohawk) Ethnophysiographical Knowledge." In *Further Developments in the Theory and Practice of Cybercartography* D. Fraser Taylor, Erik Anonby & Kumiko Murasugi (Eds.), Cambridge: Elsevier, 2019.

Jett, Stephen C. *Navajo Placenames and Trails of the Canyon de Chelly System, Arizona*.New York, NY: Peter Lang, Inc., 2001.

Langbein, Walter B. *Hydrology and Environmental Aspects of Erie Canal (1817–1899)*. Washington, D.C.: United States Government Printing Office, 1976.

Maracle, David Kanatawakhan. *Karoron ne Owennahshonha: A Mohawk Thematic Dictionary*. London, ON: Kenyen'keha Books, 2001.

McCallum, Jenn. "In Peterborough and Across Ontario, Turtles Are at Risk". *The Peterborough Examiner*, July 11, 2018. https://www.thepeterboroughexaminer.com/community-story/8735609-in-peterborough-and-across-ontario-turtles-are-at-risk/.

Mithun, Marianne. "Principles of Naming in Mohawk." *Proceedings of the American Ethnological Society, no. 4* (1984): 40–53.

Parmenter, Jon. *The Edge of the Woods: Iroquoia, 1534–1701*. East Lansing: Michigan State University Press, 2010.

Phillips, Stephanie K. *The Kahnawá:ke Mohawks and the St. Lawrence Seaway*. Master's Thesis, Montreal: McGill University, 2000. http://digitool.library.mcgill.ca/R/?func=dbin-jump-full&object_id=33310&local_base=GEN01-MCG02.

Ptolomy, Claudius. *Geography*. New York: Cosimo, Inc, 2011.

Rayburn, Alan. "Place Names." *The Canadian Encyclopedia*.Last revised October 21, 2015. https://www.thecanadianencyclopedia.ca/en/article/place-names.

Ruttenber, Edward Manning. *Footprints of the Red Man: Indian Geographical Names in the Valley of Hudson's River, the Valley of the Mohawk, and on the Delaware, Their Location, and the Probable Meaning of Some of Them*. Newburgh, NY: New York State Historical Association, 1906.

Snow, Dean. *The Iroquois*. Malden, MA: Blackwell Publishing, 1996.

Swanton, John Reed. *Indian Tribes of the Lower Mississippi Valley and Adjacent Coast of the Gulf of Mexico*. Washington: Government Printing Office, 1911.

UNESCO. "What Is local and Indigenous Knowledge?" *Local and Indigenous Knowledge Systems*, 2017. http://www.unesco.org/new/en/naturalsciences/.

University of Nottingham. *Key to English Place Names*, 2019. http://kepn.nottingham.ac.uk/#.

# Appendix

*Samantha Steindel-Cymer and Janet Berry Hess*

Below is a partial list of resources related to Native American, First Nations Alaskan Native, and Indigenous Hawaiian Nations. Consulting across resources, and conducting research in person at archives where/when possible, is recommended.

## Digital Mapping

Among the digital mapping resources established and under development are the maps and resources listed below.

American Indian Library Association Map of Tribal Libraries, https://ailanet.org

Coming Home to Indigenous Place Names in Canada

> Created to honor the 150th anniversary of the Confederation of Canada, the "Coming Home" map aims to "...preserve the cultural significance of First Nations, Métis, and Inuit communities of Canada by displaying landmarks, points of interest and areas of Canada in the native languages of said groups with accompanying English translations."
>
> https://umaine.edu/canam/publications/coming-home-map/

Database of Indigenous Slavery in the Americas

> "This project, housed at Brown University, is a collaborative effort to build a database of enslaved Indigenous people throughout time all across the Americas."
>
> https://indigenousslavery.org

Digital Index of North American Archaeology

> "The Digital Index of North American Archaeology (DINAA) aggregates archaeological and historical data sets developed over the past century from numerous sources, especially state and federal government agencies. DINAA provides public and research communities with a uniquely comprehensive window into human settlement across North America."
>
> http://ux.opencontext.org/archaeology-site-data/

Early California Cultural Atlas

"Early California Cultural Atlas (ECCA) is developing a digital atlas to integrate and manage historical resources and enable analysis of historical data related to the colonization and settlement of early California. Through development of various educational tools the atlas will contribute to the education of students from elementary school to the university classroom."

http://ecai.org/nehecca/

LandMark: Global Platform of Indigenous and Community Lands

"LandMark is the first online, interactive global platform to provide maps and other critical information on lands that are collectively held and used by Indigenous Peoples and local communities. The global platform is designed to help Indigenous Peoples and communities protect their land rights and secure tenure over their lands. LandMark provides several categories of data to show the land tenure situation for Indigenous Peoples and communities, as well as potential pressures on their lands, changes in land cover over time, and their contributions to protecting the environment."

http://www.landmarkmap.org

Living Nations, Living Words: A Map of First Peoples Poetry (under development by U.S. Poet Laureate Joy Harjo).

Lḵóot - Jilḵáat Storyboard

In association with the Haines Borough Public Library, the Storyboard is an interactive map which charts Native place names.

https://cvstoryboard.org/

Loss of Indian Land Interactive Map, http://www.slate.com/blogs/the_vault/2014/06/17/interactive_map_loss_of_indian_land.html

Mapping Indigenous L.A.

Mapping Indigenous L.A. is a project that seeks to put Indigenous place names, stories, memories, and demographics and to uncover "…multiple layers of indigenous Los Angeles through digital storytelling & oral history with community leaders, youth and elders from indigenous communities throughout the city."

https://mila.ss.ucla.edu

Mapping Indigenous American Cultures and Living Histories

Mapping Indigenous American Cultures and Living Histories, a project funded by the National Endowment for the Humanities, gathers in one site tribal locations, languages, resources, and selected National narratives approved by tribal leaders and Tribal Heritage Protection Officers. It also includes the location of tribal archives, libraries, and museums and selected non-tribal collections. MIAC-LH serves as a "gathering place" for public-facing information intended to educate the public and K-12 students about the presence and ongoing vitality of tribal Nations in their words. Indigenousmap.com

Native Land—Map and Teaching Guide

Native Land, constructed by Victor Temprano, contains a teaching guide. Native Land is an interactive map of traditional Indigenous territories around the world. See the guide: "This version includes detailed instructions on how to use Native Land, as well as exercises for use by teachers of different levels, from kids to adults. The Guide discusses the pros and cons of the map itself, the importance of learning more about colonialism, and provides resources for teachers to learn more."

https://native-land.ca/teachers-guide/

Pan Inuit Trails

"The Atlas provides a synoptic view (although certainly incomplete) of Inuit mobility and occupancy of Arctic waters, coasts and lands, including its icescapes, as documented in written historical records (maps of trails and place names). Delineations of trails and place names play a critical role in documenting the Inuit spatial narratives about their homelands. To show where these trails lead and connect to other trails, the historical records used in making this Atlas are being relationally linked, referenced geospatially, and displayed on a base map."

http://www.paninuittrails.org/index.html

PBS Native America Interactive Map, https://www.pbs.org/native-america/extras/interactive-map/

Residential Schools Land Memory Mapping Project, https://residentialschoolsatlas.org/index.html

The Invasion of America (https://www.arcgis.com/apps/webappviewer/index.html?id=eb6ca76e008543a89349ff2517db47e6)

The Ways Great Lakes Native Culture and Language, https://theways.org/map
Tribal Nations Maps

Tribal Nations Maps offers a series of maps for purchase featuring the historical territories of Indigenous Nations through the United States, Canada, Mexico, and the world. Further maps feature U.S. Native American Reservations, pipelines, and more.

http://www.tribalnationsmaps.com/

Trust for Public Land: Indigenous Hawaii

"Our Hawaiian Native Lands program works to preserve and promote the unique land-based cultures of native peoples. By protecting sites of traditional value and ensuring public access to the land, and, frequently, by placing property directly under native stewardship, we help native communities preserve their spiritual, cultural and economic relationship to the land."

https://www.tpl.org/our-work/hawaiian-heritage-lands#sm.00000sl5urp1dtdq1tcxrja6fyffw

Whose Land Are You On, https://library.chatham.edu/whoseland

## Repatriation, Reconciliation, and Healing Digital Resources

Among the digital mapping resources related to repatriation, reconciliation and healing are the maps and resources listed next.

Carlisle Indian School Digital Resource Center

> "The Carlisle Indian Industrial School is a major site of memory for many Native peoples, as well as a source of study for students and scholars around the globe. This website represents an effort to aid the research process by bringing together, in digital format, a variety of resources that are physically preserved in various locations around the country. Through these resources, we seek to increase knowledge and understanding of the school and its complex legacy, while also facilitating efforts to tell the stories of the many thousands of students who were sent there."
>
> http://carlisleindian.dickinson.edu/

Genoa Indian School Digital Reconciliation Project

> "The Genoa Indian School Digital Reconciliation Project is digitizing and describing government records from various archives, which are often difficult to locate and to access. Returning these records to American Indian families and tribes is an act of archival reconciliation—bringing history home. We are also gathering oral histories, community narratives, and privately owned items to tell a more complete story of Genoa and other Indian boarding schools. The Project's American Indian Community Advisors are providing guidance to the project so that it represents the various perspectives and experiences of those who attended the school. The Project will enable American Indian communities and families to review materials, establish protocols for how materials are to be shared, and contribute their own digital content and knowledge to the documentary record."
>
> https://genoaindianschool.org/

Hopi Music Repatriation Project, http://ethnocenter.org/HopiMusicRepatriation-ProjectFirstReport

Marshall Islands/Burke Museum Project

> A project initially founded by six University of Washington students, this endeavor, created in partnership with the Burke Museum, aims to share the story of the Marshallese community and acknowledge their history, their hardships, and their need for medical reform and greater access to mainland Medicaid health services.

https://www.burkemuseum.org/blog/healing-and-solidarity-marshallese-community

National Native American Boarding School Healing Coalition

> "NABS was created to develop and implement a national strategy that increases public awareness and cultivates healing for the profound trauma experienced by individuals, families, communities, American Indian and Alaska Native Nations resulting from the U.S. adoption and implementation of the Boarding School Policy of 1869. To lead in the pursuit of understanding and addressing the ongoing trauma created by the U.S. Indian Boarding School policy."
>
> https://boardingschoolhealing.org/

Project Naming, https://www.bac-lac.gc.ca/eng/discover/aboriginal-heritage/project-naming/Pages/introduction.aspx

Sherman Indian Museum

> "The Sherman Indian Museum collection documents the history of the Perris Indian School, Sherman Institute, and Sherman Indian High School, as well as the Native American experience in the U.S. and within government-run American Indian boarding schools. The collection spans more than a century and richly documents the experiences of students, representing more than 50 tribal nations, who attended the school since its inception in 1892. The collection is being digitized during 2017–2019, and materials will continue to be added and updated on Calisphere over the course of the project."
>
> https://calisphere.org/collections/27124/

Stewart Indian School, https://stewartindianschool.com

Sovereign Bodies Institute—The Missing and Murdered Indigenous Women Database, https://www.sovereign-bodies.org/; https://www.muckrock.com/project/the-missing-and-murdered-indigenous-women-database-293/

## Research Resources

### *Federal Resources*

Among the federal resources related to Native American Nations are the following maps and resources.

American Folklife Center

> "The American Folklife Center Archive, established in the Library of Congress Music Division in 1928, is now one of the largest archives of ethnographic materials from the United States and around the world, encompassing millions of items of ethnographic and historical documentation recorded from the nineteenth century to the present. These collections, which include extensive audiovisual

documentation of traditional arts, cultural expressions, and oral histories, offer researchers access to the songs, stories, and other creative expressions of people from diverse communities."

http://www.loc.gov/folklife/

Bureau of Indian Affairs map of BIA leadership

An interactive map that regionally lists Bureau of Indian Affairs (BIA) office leadership, tribal leadership, their organizations and contact information as well as related agencies throughout the United States.

https://www.bia.gov/bia/ois/tribal-leaders-directory/

Bureau of Land Management Cultural Heritage Program

The Bureau of Land Management (BLM) offers digital access to a Tribal Relations Manual and Tribal Relations Handbook, in adding to resources regarding tribal consultation and outreach. "The new *BLM 1780 Tribal Relations Manual and Handbook* represents the culmination of years of outreach and coordination between the BLM and American Indian tribes, and has been developed to complement the direction of the Administration and the Department. Beginning in August 2008, the BLM initiated comprehensive outreach to the tribes that garnered valuable input for improving BLM tribal consultation policy and practice. Tribes also provided insights regarding tribal consultation required by the National Historic Preservation Act (NHPA) and National Environmental Policy Act (NEPA). The decision to create a comprehensive manual resulted from tribal feedback."

https://www.blm.gov/programs/cultural-heritage-and-paleontology

Department of the Interior resources, https://www.doi.gov/tribes

Holdings of the Library of Congress

"The Library of Congress is the largest library in the world, with millions of books, recordings, photographs, newspapers, maps and manuscripts in its collections. The Library is the main research arm of the U.S. Congress and the home of the U.S. Copyright Office."

The Library of Congress' research areas, including computer catalog centers and Copyright Office public service areas, requires reader identification cards, which are free and require pre-registration for use. These cards are issued to researchers aged 16 years and older, and must be obtained by completing the registration process by presenting a valid driver's license, passport, or state identification card.

https://wwws.loc.gov/readerreg/remote/

National Archives and Records Administration

The following link features the results of the Native American collection search in the National Archives catalog, including Tribal files, NAGPRA files, grants, letters, and more.

https://catalog.archives.gov/

Bureau of Indian Affairs in the National Archives

> "The National Archives (NARA) maintains historically significant and permanently valuable records created by the Bureau of Indian Affairs (BIA) and its predecessor agencies from as early as 1793. Tasked with federal oversight of American Indians, the BIA interacted with many individuals and families. Most American Indian-related records held by the NARA primarily relate to administrative matters and the management of tribal and individual resources by the Federal government. Records created by the BIA can be found at many of the NARA facilities throughout the country. There is no comprehensive index to these records. It is important to know the Indian tribe and/or BIA agency to locate potentially relevant records."
>
> https://www.archives.gov/research/native-americans/bia

Photographs in the National Archives

> "The pictures listed in this leaflet portray Native Americans, their homes, and activities. They have been selected from pictorial records deposited in the National Archives by 15 Government agencies, principally the Bureau of Indian Affairs, the Bureau of American Ethnology, and the United States Army."
>
> https://www.archives.gov/research/native-americans/pictures

Native American Graves Protection and Repatriation Act Resources

Smithsonian Institution Archives

> "The Smithsonian holds thousands of museum collections, archives, books, and more related to the history of Native Americans and Indigenous communities of the world. Many of these materials were created or collected by the United States National Museum's (now the National Museum of Natural History) Bureau of American Ethnology and by the George Heye Foundation during the nineteenth and twentieth centuries. These collections comprise the largest set of Indigenous and Native materials within the Smithsonian, now held in the National Anthropological Archives and the National Museum of the American Indian. Other Smithsonian museums, libraries, and archival units, also hold collections documenting Native history. Browse the topics and images below to explore some of the Smithsonian's digitized collections relating to Native Americans and Indigenous Communities."
>
> https://collections.si.edu/search/gallery.htm?og=native-americans

*Contact information for collections and archives of the Smithsonian. The two collections are not merged*:

National Museum of the American Indian (NMAI)

> NMAI Archive Center: NMAIArchives@si.edu
> NMAI Collections Report Request:
> https://nmai.si.edu/explore/collections/accessing/

National Museum of Natural History
National Anthropological Archives: naa@si.edu
Anthropology Collections: rosenthd@si.edu or beauchac@si.edu
https://naturalhistory.si.edu/

### *Language Resources*

Among the many university and museum collections and tribal efforts related to languages are the resources listed here. Many Tribal Nations are engaged in digital language and oral history preservation projects. Resources for language preservation also include the OHMS (Oral History Metadata Synchronizer) and the Sustainable Heritage Network.

Alaska Native Language Archive, https://www.uaf.edu/anla/

American Philosophical Society, https://www.amphilsoc.org/library/guides/indigenous-peoples-and-languages

Arizona State Museum Library and Archives, https://libguides.asu.edu/c.php?g=263762&p=1761975

Autry Museum of the American West, https://oac.cdlib.org/institutions/Autry+Museum+of+the+American+West?descriptions=show

California Language Archive—University of California, Berkeley

> "The California Language Archive is an online catalog of indigenous language materials in archives at the University of California, Berkeley. It includes physical and digital materials held by the Survey of California and Other Indian Languages, the Phoebe A. Hearst Museum of Anthropology, and the Bancroft Library."
> http://cla.berkeley.edu

Center of Southwest Studies at Fort Lewis College, https://swcenter.fortlewis.edu/research-resources/archives

Digitizing Crow History, https://sustainableheritagenetwork.org/digital-heritage/digitizing-crow-oral-history-trials-tribulations-and-successes

Digitizing Zuni Pueblo Language Materials, https://econtent.unm.edu/digital/collection/zla

Native American Languages Collection—Sam Noble Oklahoma Museum of Natural History, University of Oklahoma

> "Founded in 2002, the Native American Languages collection at the Sam Noble Museum provides invaluable resources to researchers, educators and students. The collection includes audio and video recordings, manuscripts, books, journals, ephemera and teaching curricula, including lesson plans, from more than 175 Native North American languages. This collection is intended to be a resource

center where scholars and community members develop mutually beneficial relationships by preserving language resources, conducting research, providing services to Native American communities and educating the community about the importance of Native American languages and cultures."

https://samnoblemuseum.ou.edu/collections-and-research/native-american-languages/

Recovering Voices

"Founded in 2009, Recovering Voices (RV) is a collaborative program of the National Museum of Natural History, the National Museum of the American Indian, and the Center for Folklife and Cultural Heritage that partners with communities around the world to revitalize and sustain endangered languages and knowledge. Through interdisciplinary research, community collaboration, and public outreach, we strive to develop effective responses to language and knowledge loss. Our research program strives to understand the dynamics of intergenerational knowledge transfer and recognizes that language communities and scholars have a mutual interest in documenting, revitalizing and sustaining languages and the knowledge embedded in them. The Community Research Program (CRP) seeks to improve access to the Smithsonian's diverse collections—archival, biological, and cultural—and to support diverse approaches to the work of knowledge revitalization."

https://naturalhistory.si.edu/research/anthropology/programs/recovering-voices

RecoveringVoices@si.edu
twitter@RecoverVoices
facebook.com/recoveringvoices
(See also the American Philosophical Society, https://www.amphilsoc.org/library/CNAIR, and the holdings of the Smithsonian)

## Organizations and Museums

Among the organizations and museums with resources related to Native American Nations are those listed here. Please consult your regional and state museums and institutions.

Administration for Native Americans

"Established in 1974 through the Native American Programs Act (NAPA), the Administration for Native Americans (ANA) serves all Native Americans, including federally recognized tribes, American Indian and Alaska Native organizations, Native Hawaiian organizations and Native populations throughout the Pacific Basin (including American Samoa, Guam, and the Commonwealth of the Northern Mariana Islands)." ANA offers funding under the following topics:

Native American Language Preservation and Maintenance (P&M)
Native American Language Preservation and Maintenance—Esther Martinez Immersion (EMI)
Social and Economic Development Strategies (SEDS)
Environmental Regulatory Enhancement (ERE)

Social and Economic Development Strategies for Alaska (SEDS-AK)
https://www.acf.hhs.gov/ana/grants/funding-opportunities?fbclid=IwAR2ScRmz6L2PgtUF3bPKpgOsyY5av8jYqdhN87acxJhELVtJfbaJ1CFWLL4

American Philosophical Society, Center for Native American and Indigenous Research

The Native American Scholars Initiative at the American Philosophical Society offers fellowships for university- and community-based scholars working on digital projects that connect archives and Indigenous communities. "The Center for Native American and Indigenous Research (CNAIR) works with indigenous communities throughout North America and with scholars in many disciplines. The goal of CNAIR is to assist people in discovering and utilizing the American Philosophical Society Library's extensive archival collections in innovative ways that honor indigenous knowledge, cultivate scholarship, and strengthen languages and cultural traditions."
Contact: Alyssa Mt. Pleasant at amtpleasant@amphilsoc.org
http://amphilsoc.org/CNAIR

Amherst College Kim-Wait Eisenberg Collection, https://www.amherst.edu/library/archives/holdings/nativeamericanlit

Archives of Traditional Music, Indiana University, https://libraries.indiana.edu/archives-traditional-music

Association of College and Research Libraries (Native American Resources), https://crln.acrl.org/index.php/crlnews/article/view/8424/8636

Autry Museum of the American West, https://theautry.org/research-collections/online-research-tools

Burke Museum of Natural History and Culture
(*See also The Marshall Islands Project, in Reconciliation Projects, below*)

"The Museum regularly collaborates with Tribal staff on research projects. The Burke also facilitates cultural exchange among Northwest Tribes and Indigenous peoples of other nations." The Burke Museum also offers classroom and cultural programs and Tribal Consultation.
https://www.burkemuseum.org/
https://www.burkemuseum.org/about/our-work/tribal-consultation
http://www.burkemuseum.org/search/?q=Native%20American

Center of Southwest Studies at Fort Lewis College

"The Center of Southwest Studies at Fort Lewis College connects individuals and communities with opportunities to explore, study, and experience the Southwest's dynamic heritage." The Center provides support to Fort Lewis College through its fellowship program, student internships, in addition to offering library and research resources to the public.
http://swcenter.fortlewis.edu

D'Arcy McNickle Center at the Newberry Library, https://www.newberry.org/darcy-mcnickle-center-american-indian-and-indigenous-studies

Dartmouth Occom Circle, https://www.dartmouth.edu/occom/

Digital Atlas of Native American Intellectual Traditions (DANAIT)

> "The Digital Atlas of Native American Intellectual Traditions (DANAIT) is an IMLS-funded project to create a space for conversation and collaboration, with the goal of developing a framework for sharing, exploring, and visualizing Native-authored library and archival collections. The project will bring together Native Studies scholars; Native librarians; tribal historians; representatives from libraries with large Native-authored collections; metadata, digital humanities, and user interface specialists; and technologists to expand and improve culturally appropriate access to Native digital collections and to create collaborative digital humanities scholarship that accurately represents Native American intellectual networks."
>
> https://danait.wordpress.amherst.edu/

Thomas Gilcrease Institute of American History and Art (Gilcrease Museum)

> "The Thomas Gilcrease Institute of American History and Art, known as Gilcrease Museum, houses a comprehensive collection of the art, culture and history of North America." The Gilcrease Museum is adjacent to the Helmerich Center for American Research, which houses the Gilcrease Library and Archive.
>
> https://collections.gilcrease.org

Heard Museum

> "Dedicated to the advancement of American Indian art, the Heard successfully presents the stories of American Indian people from a first-person perspective, as well as exhibitions that showcase the beauty and vitality of traditional and contemporary art." In addition, the Heard Museum staffs a Tribal Relations Department.
>
> https://heard.org

Indigenous Digital Archive

> "The Museum of Indian Arts and Culture, in collaboration with the Indian Pueblo Cultural Center and the State Library Tribal Libraries Program, is creating a free online resource of interest to students, families, researchers, and communities. The Indigenous Digital Archive (IDA) is funded by a National Leadership Grant from the Institute of Museum and Library Services (IMLS), the New Mexico Historical Records Advisory Board, and the Knight Foundation." The IDA offers events and exhibitions, programs and education initiatives, as well as research resources, including documents relating to individual, family, and community history and records.
>
> http://indianartsandculture.org/indigenous-digital-archive

Indigenous Education Institute

"The Indigenous Education Institute (IEI) was created for the preservation and contemporary application of traditional Indigenous knowledge. The mission and goals were developed in order to provide awareness of the importance of cultural and linguistic diversity in the world today. Cultural and linguistic diversity provide strength and richness to individuals and nations. Indigenous ways of knowing contain knowledge that can provide greater sustainability and stewardship of the earth and cosmos, leading to a harmonious, balanced future." IEI conducts research in Indigenous science including: Navajo and Cherokee astronomy; Indigenous physics and ecology and sustainability, as well as GPS technology.

http://neatoeco.com/iei/wp/

Institute of American Indian Arts

The Institute of American Indian Arts (IAIA) offers degree-granting academic programs. In addition, IAIA has an on-campus museum that provides public education and events, as well as a Social Art Residency program for Native American artists.

https://iaia.edu

Minnesota Historical Society, Native American Initiatives

The Minnesota Historical Society (MNHS) has a variety of resources, including preservation and legacy grants, educational programming, and research resources for the public, including maps, library and archives catalogs, and genealogy and family history guides.

http://www.mnhs.org/visit/native-american

Montana State University, Indian People of the Northern Great Plains Digital Collection, https://arc.lib.montana.edu/indian-great-plains/

Museum of Indian Arts and Culture

"The Museum of Indian Arts and Culture, one of four museums in the Museum of New Mexico system, is a premier repository of Native art and material culture and tells the stories of the people of the Southwest from pre-history through contemporary art. The museum serves a diverse, multicultural audience through changing exhibitions, public lectures, field trips, artist residencies, and other educational programs."

http://www.miaclab.org

Native American Studies Collection, U.C. Berkeley

The Native American Studies Collection from the Ethnic Studies Library at U.C. Berkeley aims to help provide a service in accessing useful and reliable sources related to Native American issues. These include classroom resources, research materials, and databases.

http://eslibrary.berkeley.edu/NASsites

Oklahoma Historical Society

"The mission of the Oklahoma Historical Society is to collect, preserve, and share the history and culture of the state of Oklahoma and its people." The Oklahoma Historical Society (OHS) also offers a Heritage Preservation Grant Program and Historical Marker Program.

https://www.okhistory.org/historycenter/index

School for Advanced Research

"The School for Advanced Research is North America's preeminent independent institution advancing creative thought and innovative work in social sciences and humanities and fostering the preservation and revitalization of Native American cultural heritage."

https://sarweb.org

Sequoyah National Research Center

"The collections of the University of Arkansas at Little Rock's Sequoyah National Research Center constitute the largest assemblage of Native American expression in the world. Our mission, to acquire and preserve the writings and ideas of Native North Americans, is accomplished through collecting the written word and art of Native Americans and creating a research atmosphere that invites indigenous peoples to make the Center an archival home for their creative work. The Sequoyah National Research Center strives to ensure that the discussion of Native America accounts for the perspectives of the peoples themselves."

https://ualr.edu/
https://ualr.edu/sequoyah/

Society of American Archivists—Native American Archives

The goal of the Native American Archives is: "To serve as a forum to educate archivists on the complexities and beauty of Native American archives of the Western Hemisphere and as a source of communication and inspiration for archivists working with Native American collections."
Contact: NativeArchSAA@gmail.com

https://www2.archivists.org/groups/native-american-archives-section

University of Arkansas' Native Writers Digital Text Project, https://ualrexhibits.org/tribalwriters/

University of Washington, American Indians of the Pacific Northwest Collection, https://content.lib.washington.edu/aipnw/

Yale Indian Papers Project, http://catalog.yale.edu/div/research-outreach/yale-indian-papers-project/

Many state historical societies have useful information (for example, the Minnesota History Society [Native American Initiatives], https://www.mnhs.org/media/news/

9690, and the Oklahoma Historical Society, https://www.okhistory.org/research/americanindians).

Other innovative projects include the American Indian Film Institute and Festival, https://filmfreeway.com/AIFF, the Indian Country Media Network, and Tribesourcing Southwest Film [for films addressing Native American life in the 1940s–1960s with "culturally competent metadata" https://tribesourcingfilm.com].

## Tribal Archives and Institutions

Many Tribal Nations have associated archives, libraries, museums (ALM), and learning institutions, some open to the public. For ATALM research on Tribal archives, libraries, and museums, see "Sustaining Culture: The Structure, Activities, and Needs of Tribal Archives, Libraries, and Museums" (https://www.atalm.org/sites/default/files/sustaining_indigenous_culture.pdf). A list of the tribal archives, libraries, and museums which participated in the ATALM survey, along with information on selected institutions with resources at the time of this text's publication, follows.

Acoma Learning Center, Pueblo of Acoma, New Mexico
Agua Caliente Cultural Museum, Palm Springs, California
Ahtna Heritage Foundation, Glennallen, Alaska
Ak-Chin Him-Dak Eco Museum & Archives, Maricopa, Arizona
Aleut Community, St. Paul Island, Alaska
Aleut, Chignik Lake Village Council, Chignik Lake, Alaska
Alutiiq Museum & Archaeological Repository, Kodiak, Alaska
Aroostook Band of Micmacs, Presque Isle, Maine
Athabaskan, Native Village of Eagle, Eagle, Alaska
Bad River Public Tribal Library, Odanah, Wisconsin
Barona Cultural Center & Museum, Lakeside, California
Bay Mills Indian Community, Brimley, Michigan
Bear River Library/ Museum/ Archive, Loleta, California
Big Sandy Rancheria Band of Western Mono Indians, Auberry, California
Bishop Paiute, Bishop, California
Blue Lake Rancheria, Blue Lake, California
Cabazon Cultural Museum, Indio, California
Cachil Dehe Wintun, Colusa Indian Community Council, Colusa, California
Caddo Heritage Museum, Binger, Oklahoma
Capitan Grande Band of Diegueno, California
Mission Indians, Lakeside, California
Catawba Cultural Center, Rock Hill, South Carolina
Cherokee National Historical Society, Park Hill, Oklahoma
Cheyenne and Arapaho Tribes, Concho, Oklahoma
Chilkat Indian Village, Klukwan Community Library, Haines, Alaska
Chitimacha Tribe of Louisiana, Charenton, LA
Choctaw Nation Museum, Tuskahoma, Oklahoma
Chukchi Consortium Library, Kotzebue, Alaska
Cocopah Indian Tribal Library, Somerton, Arizona
Cocopah Indian Tribal Museum, Somerton, Arizona
Comanche National Museum and Cultural Center, Lawton, Okalhoma

Confederated Tribes of Coos, Lower Umpqua, and Siuslaw Indians, Coos Bay, Oregon
Confederated Tribes of Grand Ronde, Grand Ronde
Confederated Tribes of Siletz Indians, Siletz, Oregon
Confederated Tribes of Siletz, Cow Creek Band of Umpqua, Roseburg, Oregon
Confederated Tribes of the Goshute Reservation, Ibapah, Utah
Confederated Tribes of the Umatilla Indian Reservation Cultural Resources, Pendleton, Oregon
Coquille Indian Tribe, North Bend, Oregon Coushatta Tribe of Louisiana Heritage Department, Elton, Louisiana
Coyote Valley Band of Pomo Indians, Redwood Valley, California
Deg Hitan Athabaskan Alaska Native, Innoko River School and Tribal Library, Shageluk, Alaska
Delaware Nation, Anadarko, Oklahoma
Delaware Tribe of Indians, Bartlesville, Oklahoma
Dillingham Public Library, Dillingham, Alaska
Dine', Navajo Nation Museum, Window Rock, AZ
Dry Creek Rancheria Band of Pomo Indians, Healdsburg, California
Eastern Shawnee Tribe, George J. Captain Library, Wyandotte, Oklahoma
Eastern Shoshone, Fort Washakie School/ Community Library, Ft. Washakie, Wyoming
Elk Valley Rancheria, California, Crescent City, California
Fallon Paiute Shoshone Tribe, Fallon, Nevada
Fort Belknap College, Harlem, Montana
Fort Bidwell Paiute Reservation, Fort Bidwell Indian Community Council, Fort Bidwell, California
Fort Peck Assiniboine & Sioux, Fort Peck Tribal Library, Poplar, Montana
Gila River Indian Community, Huhugam Heritage Center, Chandler, Arizona
Hydaburg Cooperative Association, Hydaburg, Alaska
Ho-Chunk Nation, Black River Falls, Wisconsin
Hoopa Valley Tribal Council, Hoopa, California
Hopi Tribe, The Department of Education, Kykotsmovi, Arizona
Houlton Band of Maliseet Indians, Littleton, Maine
Huna Heritage Foundation, Juneau, Alaska
Igiugig Tribal Village Council, Igiugig, Alaska
Iipay Nation of Santa Ysabel (Kumeyaay), San Diego, California
Iowa Tribe of Kansas and Nebraska, White Cloud, Kansas
Iowa Tribe of Oklahoma, Iowa Tribe Library, Perkins, Oklahoma
Iroquois Indian Museum, Howes Cave, New York
Jamestown S'Klallam Tribal Library, Sequim, Washington
Kalispel Tribal Library, Usk, Washington
Kalispel Tribe of Indians, Camas Path, Usk, Washington
Karuk Tribe, Happy Camp, California
Karuk Tribe, People's Center, Happy Camp, CA Kaw Nation, Kaw City, Oklahoma
Kenaitze Indian Tribe, Ts'itsatna Archives, Kenai, Alaska
Ketchikan Indian Community, Ketchikan, Alaska
Klamath Tribes, Chiloquin, Oregon

Klawock Cooperative Association, Klawock, Alaska
Koyukon Athabascan, Top of the Kuskokwim School, Nikolai, Alaska
Lac Courte Oreilles Ojibwe College Community Library, Hayward, Wisconsin
Lac du Flambeau Band of Lake Superior Chippewa, George W. Brown Jr. Museum & Cultural Center, Lac du Flambeau, Wisconsin
Lac Vieux Desert, Watersmeet, Michigan
Laguna Pueblo, Laguna Public Library, Laguna, New Mexico
Lake Superior Band of Chippewa Ojibwa, Keweenaw Bay Indian Community, Baraga, Michigan
Lakota, Ute Indian Museum, Montrose, Colorado
Lovelock Paiute Tribe, Lovelock, Nevada
Lower Elwha Klallam Tribe, Port Angeles, Washington
Lummi Indian Business Council, Bellingham, Washington
Lummi Nation Library, Northwest Indian College, Bellingham, Washington
Lummi Nation, Northwest Indian College, Bellingham, Washington
Makah Cultural and Research Center, Neah Bay, Washington
Mandan, Hidatsa and Arikara, Three Affiliated Tribes Museum, Inc., New Town, North Dakota
Mashantucket Pequot Museum & Research Center, Mashantucket, Connecticut
Mechoopda Indian Tribe, Chico, California
Mendas Cha'ag Tribe of Healy Lake, Healy Lake Traditional Council, Fairbanks, Alaska
Menominee, College of Menominee Nation, Keshena, Wisconsin
Mescalero Community Library, Mescalero, New Mexico
Mescalero Cultural Center, Mescalero, New Mexico
Mille Lacs Indian Museum, Onamia, New Mexico
Mississippi Band of Choctaw, Choctaw, Mississippi
Modoc Tribe of Oklahoma, Miami, Oklahoma
Mohave, Colorado River Indian Tribes
Library/Archives, Parker, Arizona
Mohegan Indian Tribe, Uncasville, Connecticut
Mooretown Rancheria of Maidu Indians of California, Mooretown Rancheria, Oroville, California
Muckleshoot Tribe Preservation Program, Auburn, Washington
Muscogee (Creek) Nation, Muscogee (Creek) Nation, Okmulgee, Oklahoma
Myaamia Heritage Museum & Archive, Miami, Oklahoma
Narragansett, Tomaquag Indian Memorial Museum, Exeter, Rhode Island
Native Village of Buckland, Buckland, Alaska
Native Village of Eyak, Ilanka Cultural Center, Cordova, Alaska
Native Village of Mary's Igloo, Mary's Igloo Traditional Council, Teller, Alaska
Native Village of Noatak, Noatak, Alaska
Native Village of Scammon Bay, Scammon Bay Public Library, Scammon Bay, Alaska
Native Village of White Mountain, White Mountain, Alaska
Navajo Nation, Office of Navajo Nation Library, Window Rock, Arizona
Nenana Public Library, Nenana, Alaska
Nisqually Tribal Library, Olympia, Washington
Nooksack Tribal Library, Deming, Washington

Northern Arapaho Tribe, Northern Arapaho Archives/Records Management, Ethete, Wyoming
Northwestern Band of the Shoshone Nation, Cultural/Natural Resource Program, Brigham City, Utah
Nottawaseppi Huron Band of the Potawatomi, NHBP, Fulton, Michigan
Oglala Sioux Tribe, Oglala Lakota College, Kyle, South Dakota
Ohkay Owingeh, P'oe Tsawa Community Library, Ohkay Owingeh, New Mexico
Ojibwe—Leech Lake Reservation, Bug-O- Nay-Ge-Shig School, Bena, Minnesota
Ojibwe, Sault Ste. Marie Tribe of Chippewa Indians, Sault Ste. Marie, Michigan
Organized Village of Kasaan, Ketchikan, Alaska
Otoe-Missouria Tribe of Indians, Ernest Vetter Jr. Library & Museum, Red Rock, Oklahoma
Ottawa Tribe of Oklahoma, Miami, Oklahoma
Native Village of Ouzinkie, Ouzinkie, Alaska
PA'I, Honolulu, Hawaii
Paiute, Reno-Sparks Indian Colony, Reno, Nevada
Pala Band of Mission Indians, Cupa Cultural Center, Pala, California
Pamunkey Indian Tribe, King William, Virginia
Pascua Yaqui Tribe, Pascua Yaqui Tribe, Tucson, Arizona
Pauloff Harbor Tribe, Sand Point, Alaska
Pechanga Band of Luiseño Indians, Pechanga Cultural Resource Facility, Temecula, California
Peoria Tribe of Indians of Oklahoma, Miami, Oklahoma
Petersburg Public Library, Petersburg, Alaska
Picayune Rancheria of the Chukchansi Indians, Coarsegold, California
Pit River Tribe, Burney, California
Poarch Band of Creek Indians, Atmore, Alabama
Pokagon Band of Potawatomi Indians, Department of Education, Dowagiac, Michigan
Ponca Tribe of Nebraska, Niobrara, NB Ponca Tribe, White Eagle Library, Ponca City, Oklahoma
Prairie Band Potawatomi Nation, Mayetta
Pueblo de San Ildefonso, Santa Fe, New Mexico
Pueblo of Isleta, Cultural Affairs Office, Isleta, New Mexico
Jemez Pueblo Community Library, Jemez Pueblo, New Mexico
Pueblo of Pojoaque Public Library, Santa Fe, New Mexico
Pueblo of Sandia, Bernalillo, New Mexico
Quapaw Tribal Library, Quapaw, Oklahoma
Quechan Indian Tribe, Yuma, Arizona
Quinault Cultural Affairs, Taholah, Washington
Rincon Band, Valley Center, California
Ruby Tribal Council, Ruby, Alaska
Sac and Fox Nation of Missouri, Reserve, Kansas
Saginaw Chippewa Indian Tribal Library, Mt. Pleasant, Michigan
Salt River Pima Maricopa Indian Community Cultural Resources Department, Scottsdale, Arizona
Samish Indian Nation, Anacortes
San Carlos Public Library, San Carlos, Arizona

Santo Domingo Pueblo Library, Santo Domingo, New Mexico
Seldovia Village Tribe, Seldovia, Alaska
Seminole Nation of Oklahoma, Wewoka, Oklahoma
Seminole Tribe of Florida, Tribal Historic Preservation Office, Clewiston, Florida
Seneca Nation Library, Salamanca, New York
Seneca Nation of Indians Archives, Salamanca, New York
Seneca-Cayuga Tribe of Oklahoma, Grove, Oklahoma
Shawnee Tribe, Miami, Oklahoma
Sherwood Valley Rancheria, Willits, California
Shingle Springs Rancheria, Placerville, California
Shoalwater Bay Tribal Library, Tokeland, Washington
Shoshone-Paiute Tribes of the Duck Valley Indian Reservation, Owyhee, Nevada
Sisseton Wahpeton Oyate, Tribal Historical Preservation Office—Tribal Archives, Sisseton, South Dakota
Southern Ute Cultural Center & Museum (Archive), Ignacio, Colorado
Southern Ute Cultural Center, Ignacio, Colorado
Spirit Lake Tribe, Valerie Merrick Memorial Library, Fort Totten, North Dakota
Squaxin Island Museum Library and Research Center, Shelton, Washington
St. Regis Mohawk, Akwesasne Cultural Center, Hogansburg, New York
Standing Rock Sioux Tribe, Sitting Bull College Library, Fort Yates, North Dakota
Stillaguamish Tribe of Indians, Arlington, Washington
Stockbridge-Munsee College of Menominee Nation
Suquamish Museum, Poulsbo, Washington
Swinomish Tribal Community, LaConner, Washington
Tamastslikt Cultural Institute, Pendleton, Oregon
Telida Village, Fairbanks, Alaska
The Museum at Warm Springs, Warm Springs, Oregon
Thlopthlocco Tribal Town, Okemah, Oklahoma
Tikigaq School, Point Hope, Alaska
Timbisha Shoshone, Death Valley, CA
Tlingit, Haida, and Tsimshian Indian tribes, Sealaska Heritage Institute, Juneau, Alaska
Togiak Public Library and Cultural Center, Togiak, Alaska
Tohono O'odham Nation Cultural Center & Museum, Topawa/Sells, Arizona
Tohono O'odham, Venito Garcia Library, Sells, Arizona
Tolowa Dee-ni', Smith River Rancheria, Smith River
Trinidad Rancheria Library, Trinidad, CA
United Keetoowah Band of Cherokee Indians in Oklahoma, Tahlequah, Okalhoma
Ute Mountain Ute Tribe, White Mesa Library, White Mesa, Utah
Viejas Band of Kumeyaay Indians, Alpine, California
Village of Old Harbor, Old Harbor Tribal Council, Old Harbor, Alaska
Walker River Paiute Tribe, Schurz, Nevada
Wanapum Heritage Center, Beverly, Washington
Sierra Mono Museum, North Fork, California
White Earth Nation of Minnesota Chippewa, White Earth, Minnesotqa
White Mountain Apache Tribe, Nohwike' Bagowa Museum, Fort Apache, Arizona
Wrangell Cooperative Association, Wrangell, Alaska
Wyandotte Nation, Wyandotte, Oklahoma

Yakama Nation Library, Toppenish, Washington
Yankton Sioux Tribe, Ihanktonwan Community College, Marty, South Dakota
Yavapai, Fort McDowell Tribal Library, Fountain Hills, Arizona
Yavapai-Apache Cultural Resource Center, Camp Verde, Arizona
Yavapai-Prescott Indian Tribe, Prescott, Arizona 25
Yerington Paiute Tribal Library, Yerington, Nevada
Yocha Dehe Wintun Academy, Brooks, California
Yokuts, Towanits Education Center, Porterville, CA
Yurok Tribe, Klamath, California
Zuni Pueblo, Zuni Public Library, Zuni, New Mexico
Aaniiih Nakoda College

"The mission of Aaniiih Nakoda College is to provide quality post-secondary education for residents of the Fort Belknap Indian Reservation and surrounding communities. The college promotes individual and community development by maintaining and revitalizing the indigenous lifeways of the Aaniinen and Nakoda Tribes and by preparing students to succeed in an American technological society." Aaniiih Nakoda College (ANC) further aims to represent Native knowledge and culture throughout the curriculum, and to make these skills accessible through grants and sponsored programs.
Phone: (406) 353-2607

http://www.ancollege.edu/

Ah-Tah-Thi-Ki Museum (Seminole), https://www.ahtahthiki.com

Ak-Chin Indian Community, Maricopa, AZ

Akwesasne Cultural Center

"The Akwesasne Library and Cultural Center is a public library and museum that serves the people of Akwesasne, the surrounding communities and the visiting public by providing access to educational and cultural resources. Located in the heart of Akwesasne, the Akwesasne Cultural Center provides a positive space for educational purposes and is one of the cultural hubs of the community." Furthermore, the Library provides cultural classes for the purpose of education.
Phone: (518) 358-2240

http://www.akwesasneculturalcenter.org

Alutiiq Museum and Archaeological Repository

The Alutiiq Museum works to preserve and share the heritage and living culture of the Alutiiq people through educational programming and research, including "…archaeological studies, language documentation, and collections investigations."
Phone: (844) 425-8844

https://alutiiqmuseum.org/index.php

Arizona State Museum Library and Archives

"Arizona State Museum (ASM) is the oldest and largest anthropological research museum in the U.S. Southwest, with expansive collections that are exceptional resources for the teaching, study, and understanding of the region's 13,000-year human history. ASM serves the State of Arizona as its official archaeological repository and as the permitting authority for archaeological activity on state land. In addition to 38,000 cubic feet of archaeological research materials, ASM curates millions of archaeological, ethnographic, and modern objects created by the Indigenous peoples of the region."
Phone: (520) 621-4695

http://www.statemuseum.arizona.edu

Catawba Cultural Preservation Project

As early as 1989, a group of Catawba met in member's homes to organize a cultural preservation committee and began efforts to collect and preserve scrapbooks, letters, diaries and photographs. The committee formally incorporated in 1989, as the Catawba Cultural Preservation Project (CCPP) with the simple yet critical mission to "preserve, protect, promote and maintain the rich culture and heritage of the Catawba Indian Nation." CCPP offers an At-Risk after school and summer cultural immersion camp for Tribal youth, an archive repository for Tribal historic records, an archaeology department under the Tribal Historic Preservation office, cultural programming and a craft store supporting the work of Native artists.
Web: https://catawbaindiancrafts.com/
Phone: (803) 328-2427

Cherokee Heritage Center, http://www.cherokeeheritage.org

Chickasaw Cultural Center, https://www.chickasawculturalcenter.com

Chief Dull Knife College

"Dr. John Woodenlegs Memorial Library serves as both an academic and community library. The library loans materials to members of the Northern Cheyenne Community, Chief Dull Knife students, faculty and staff as well as the surrounding area. An interlibrary loan service is provided as well."
Web: http://woodenlegslibrary.us/
Phone: (406) 477-8293

Chiefswood National Historic Site

The Chiefswood National Historic Site is a historic landmark and museum featuring collections and works of Mohawk writer E. Pauline Johnson and her family.
Web: http://chiefswoodnhs.ca
Phone: (519) 758-5444 ext. 6025

Citizen Potawatomi Cultural Heritage Center

"Citizen Potawatomi Nation's Cultural Heritage Center exists to educate tribal members, the greater Native American community and other visitors about the historical and contemporary aspects of the tribe."

https://www.potawatomi.org/culture/cultural-heritage-center/

Choctaw Nation Historic Preservation Department

"For the Historic Preservation Department, preservation is a multiple-level effort that involves physically protecting Choctaw historic sites, repatriating and reburying Choctaw ancestors and sacred objects that have been removed from the ground, conducting research on Choctaw history and traditional life, and providing a variety of classes and presentations on Choctaw history and culture to Tribal members and the general public."

http://www.choctawnation.com/tribal-services/cultural-services/historic-preservation

Diné College Libraries

The Diné College Libraries have branches in Crownpoint, Shiprock, and Tsaile. The libraries offer access to a Native American Studies collection and include Native interest resources across all sections.

http://library.dinecollege.edu/index.html

Forest County Potawatomi Cultural Center, Library & Museum

"The Forest County Potawatomi Cultural Center, Library and Museum was primarily created to educate the public with a permanent exhibit outlining significant historical events and to pass the culture and traditions of the Bodewadmi to the next generations. The core of the museum is our collection of historical and contemporary photographs, audio/video, books, treaties, manuscripts, language material and other memorabilia."

Web: https://www.fcpotawatomi.com/culture-and-history/
Phone: (715) 478-7478

George W. Brown Jr. Ojibwe Museum and Cultural Center, https://ldfmuseum.com

Grand Traverse Band Heritage Library

"Provides a unique collection of books focusing on Native American community, self-help, child care, children's books, nature and arts, graphic novels, biographies, language and many more."

http://www.gtbindians.org/pages9258206.asp
Phone: (231) 534-7229

Great Lakes Research Alliance for the Study of Aboriginal Arts and Culture (GRASAC)

"The Great Lakes Research Alliance for the Study of Aboriginal Arts & Cultures (GRASAC) is a vibrant multi-disciplinary research network whose 500+

members have been jointly researching Anishinaabe, Haudenosaunee, and Huron-Wendat cultures of the Great Lakes region of Turtle Island since 2005. GRASAC researchers from Indigenous communities, universities, museums and archives have worked together to locate, study, and create deeper understandings of Great Lakes arts, languages, identities, territoriality and governance."

https://carleton.ca/grasac/

Green Earth Branch Library

The Green Earth Branch Library is a branch of the Oneida Community Library. The Green Earth Branch Library is a full service public library, and offers resources for finding Native Americans in Children's literature.
Web: https://oneida-nsn.gov/resources/library/green-earth-branch-library/
Phone: (920) 833-7226

Haines Borough Public Library

The Haines Borough Public Library provides e-learning tools, including online programs and research sections, such as a Native Health Database.
Web: www.haineslibrary.org
Phone: (907) 766-6420

Haskell Indian Nations University, Academic Support Center

"Haskell is a unique and diverse inter-tribal university committed to the advancement of sovereignty, self-determination, and the inherent rights of tribes. The mission of Haskell Indian Nations University is to build the leadership capacity of our students by serving as the leading institution of academic excellence, cultural and intellectual prominence, and holistic education that addresses the needs of Indigenous communities."
Web: http://www.haskell.edu/library/
Phone: (785) 740-8470

Indian Pueblo Cultural Center

The Indian Pueblo Cultural Center is a museum and cultural center. The center hosts Native American dances as well as cultural, educational, and community events.
Web: https://www.indianpueblo.org/
Phone: (505) 843-7270

Innoko River School and Tribal Library

The Innoko River School and Tribal Library is a combined public school and library.
Phone: (907) 473-8206

Institute of American Indian Arts/Museum of Contemporary Native Arts

The Institute of American Indian Arts (formally known as the Institute of American Indian and Alaska Native Culture and Arts Development) is a Tribal College which is also home to the IAIA Museum of Contemporary Native Arts (MOCNA). The IAIA hosts an artist-in-residence program for First Nations and Native American artists.
Web: https://iaia.edu/
Phone: (505) 424-2325

James E. Shanley Tribal Library

The James E. Shanley Tribal Library is a part of Fort Peck Community College. "The College offers an academic program that enables students to earn credits in college courses designed to transfer to other institutions of post-secondary and higher education. The College serves the constituency of the reservation populations by maintaining an occupational training program based on the needs of the people living on and near the reservation and on potential employment opportunities available in the region."
Web: http://www.fpcc.edu
Phone: (406) 768-6300

Jamestown S'Klallam Tribe, House of Seven Generations

"This 'virtual museum' site allows Indian and non-Indian communities to learn about and appreciate the cultural and historical lifeways that came before the present time by viewing imagery from those times. The history of our people is being preserved for generations to come as we convert written, pictorial and audio recordings of our history into digital format."
www.tribalmuseum.jamestowntribe.org

Joan Subieta Memorial Library/Delaware Nation Library

The Joan Subieta Memorial Library hosts a variety of public activities and events, and has special Native American/Lenape reading sections.
https://www.delawarenation-nsn.gov/

Kitsap Regional Library

"The Little Boston Branch has more than 1,200 books, DVDs and CDs in its collection about Native American history, art and tradition. Numerous original works of Northwest Coast-style artwork, many of which were created by the S'Klallam tribe, are on display, as well." The Kitsap regional library has a selection of research guides and resources for genealogy, Kitsap history, and more.
Web: http://www.krl.org/little-boston
Phone: (360) 297-2670

Lac Courte Oreilles Ojibwa Community College Library

"Lac Courte Oreilles Ojibwe College (LCOOC) is a non-profit Ojibwe tribal

community college." The college provides a continuing education and training program.
Web: https://www.lco.edu/library
Phone: (715) 634-4790

Leech Lake Tribal College

Leech Lake Tribal College provides associate degree programs and resources to further serve as a cultural center for the Tribal community.
Web: https://www.lltc.edu
Phone: (218) 335-4200

Lummi Library at Northwest Indian College

"The mission of the library is to support the Northwest Indian College and the Lummi Community with research, informational, and recreational resources that enhance life-long learning."
Web: http://www.nwic.edu/lummi-library/
Phone: (360) 392-4218

Medicine Spring Library, Blackfeet Community College

Medicine Spring Library serves as the academic library for Blackfeet Community College, the Blackfeet Archives, and as a community library for the Blackfeet Reservation. The library has a searchable online catalog as well as a research guide for the public.
Web: https://bfcc.edu/medicine-spring-library/
Phone: (406) 338-5441 ext. 2701

Mescalero Community Library

"The mission of the Mescalero Community Library is to identify, acquire, organize, publicize, and disseminate those resources which encourage and support access to cultural, governmental, Tribal, recreational resources of all materials; the community library will strive for lifelong learning of citizens, community, business, personal and recreational growth for a better quality of life."
Web: https://mescalerolibrary.org/
Phone: (575) 464-5010

Miami-Illinois Digital Archive

"The Miami-Illinois Digital Archive (MIDA) was created to assemble all known primary language sources into one location in order to facilitate further research and analysis of the Miami-Illinois language. MIDA users are able to search available fields within, and across, manuscripts."

http://www.ilaatawaakani.org/

Mille Lacs Indian Museum and Trading Post

"The Mille Lacs Indian Museum and Trading Post interprets the contribution of the Ojibwe people to the culture, history, and contemporary society of Minnesota and the nation."
Web: http://sites.mnhs.org/historic-sites/mille-lacs-indian-museum
Phone: (320) 532-3632

Museum of Casa Grande, http://www.tmocg.org

Nueta Hidatsa Sahnish College

The Nueta Hidatsa Sahnish College has several community outreach programs, including resources for food sovereignty, and the library has a Native American holdings collection.

https://nhsc.edu/

Oglala Lakota College

"The mission which emanates from the charter of the Oglala Sioux Tribe is to educate students for professional and vocational employment opportunities in Lakota country. The College will graduate well-rounded students grounded in Wolakolkiciyapi-learning Lakota ways of life in the community-by teaching Lakota culture and language as part of preparing students to participate in a multicultural world." Oglala Lakota College offers a variety of resources as part of their campus, including a distance learning program.
Web: http://olc.edu
Phone: (605) 455-6000

Ojibwa Community Library/Keweenaw Bay Ojibwa Community College

The Ojibwa Community Library is open to Tribal Members and members of the public, and offers access to an inter-library loan program and the Great Lakes Digital Library
Web: http://www.kbic-nsn.gov/index.php/departments/library
Phone: (906) 353-8163

Onohsagwede Cultural Center (Seneca/Iroquois), https://www.senecamuseum.org

Osage Nation Museum

"It is the mission of the Osage Nation Museum (ONM) to foster the education of the public about the history, culture, and artistic expressions of the Osage people by preserving and developing collections as well as through exhibitions and educational programs that nurture creativity and encourage active learning." ONM also offers resources related to job opportunities, natural and cultural resources, financial security and food distribution information, and health empowerment resources for Tribal Members.
Web: http://www.osagenation-nsn.gov/museum
Phone: (918) 287-5441

Papahana Kuaola (Hawaii), https://www.hauolimauloa.org/partner/papahana-kuaola

Pawnee Nation College

"The Pawnee Nation College seeks to meet the higher educational and cultural needs of the Pawnee Nation, other surrounding Indian nations, and all who have the desire to engage in a life of learning. The intent of the College is to foster an awareness of rich cultural diversity in the region and its inherent complexities and vibrancies, while at the same time encouraging students to fulfill their potential through challenging academic or training programs that prepare them to matriculate to four-year institutions or to enter the workforce. As a tribal community college, the Pawnee Nation College is dedicated to promoting social responsibility, service to community, and research that contribute to the cultural, social, and economic well-being of the Pawnee Nation, other Indian nations, and surrounding communities."
Phone: (918) 762-3343

https://www.facebook.com/PawneeNationCollege/

Pawnee Nation Higher Education Program

"The Higher Education program or scholarship program is to provide supplemental financial assistance to enrolled members of the Pawnee Tribe that are pursuing a bachelor's degree at an accredited institution of higher learning. Applicants must satisfy all Federal Bureau of Indian Affairs requirements as specified in the application packet in order to determine eligibility."

https://www.pawneenation.org/page/home/divisions/division-of-education/higher-education-program

Pechanga Cultural Resources Department, https://www.pechanga-nsn.gov/index.php/culture/cultural-center

Pojoaque Pueblo Community Library

"Located on Tribal land in New Mexico, the Pueblo of Pojoaque Public Library is a public facility serving the Pueblo of Pojoaque as well as surrounding communities."
Phone: (505) 4557511

http://www.puebloofpojoaquepubliclibrary.org/

Red Lake Nation College, Medweganoonind Library

"The Medweganoonind Library is the library for both Red Lake Nation College and the Red Lake community. The library houses books for both academic and recreational purposes, a small reference collection, a growing collection of media items such as audio books and DVDs, and an Ojibwe Language and Culture collection. There are also three quiet study rooms available for group or individual study and two computer labs connected to the library with Microsoft Office and Internet Access."
Phone: (218) 679-1014

https://www.rlnc.education/

Saginaw Chippewa Tribal Library

"The Saginaw Chippewa Indian Tribe provides library services at three locations: the Saginaw Chippewa Tribal Library, the Saginaw Chippewa Academy, and the Saginaw Chippewa Tribal College. We also manage satellite collections, available to use on the honor system, at Andahwod Continuing Care Community, Nimkee Memorial Wellness Center, Elijah Elk Cultural Center for Living Culture and Traditions Seventh Generation Program, and the Soaring Eagle Hideaway RV Park. We also administer the Little Free Library at Cardinal Pharmacy."
Web: https://youseemore.com/saginaw
Phone: (989) 775-4508

Salish Kootenai College

"The mission of Salish Kootenai College is to provide quality post-secondary educational opportunities for Native Americans, locally and from throughout the United States. The College will promote community and individual development and perpetuate the cultures of the Confederated Tribes of the Flathead Nation."
Web: http://www.skc.edu
Phone: (406) 275-4800

Sealaska Heritage Institute

"Sealaska Heritage Institute is a private nonprofit founded in 1980 to perpetuate and enhance Tlingit, Haida and Tsimshian cultures of Southeast Alaska. Its goal is to promote cultural diversity and cross-cultural understanding through public services and events. Sealaska Heritage also conducts social, scientific and public policy research that promotes Alaska Native arts, cultures, history and education statewide."
Web: http://www.sealaskaheritage.org/
Phone: (907) 463-4844

Sisseton Wahpeton College Library

"To provide higher education, research, vocational and technical education and continuing education to the members of the Sisseton Wahpeton Oyate of the Lake Traverse Reservation and others within the historical lands of the Sisseton Wahpeton Oyate. As a 1994 land grant institution, SWC will preserve and extend Dakota culture, language, and history while contributing to economic development through the provision of human capital and other resources."
Web: https://www.swcollege.edu/index.php, and https://www.facebook.com/Sisseton-Wahpeton-College-484284091620786/
Phone: (605) 698-3966 ext. 1320

Southern Ute Cultural Center and Museum, https://www.southernutemuseum.org

Southwestern Indian Polytechnic Institute, https://www.sipi.edu

Stone Child College

"Stone Child College (SCC) is a tribally chartered college established to deliver post-secondary educational opportunities through degrees, certificates and continuing education. SCC stresses the importance of preserving the Chippewa Cree language, culture and history. SCC will promote transfer students, professionally prepared and career-ready individuals."
Phone: (406) 395-4875

https://www.stonechild.edu/

Tantaquidgeon Museum, https://www.mohegan.nsn.us/explore/heritage/important-sites/tantaquidgeon-museum

Tohono O'odham Community College

"Tohono O'odham Community College's vision is to become the Tohono O'odham Nation's center for higher education, and to enhance the Nation's participation in the local, state, national, and global communities.' The College aims to help develop academic and adaptive learning, include Elders as 'primary resources, instructors, advisors and counselors as a means of reinforcing Tohono O'odham Himdag,' to ensure that the course offerings and opportunities reflect the needs of its communities, and to 'ensure the integration of appropriate Tohono O'odham Himdag in the physical environment, curriculum, and processes of the college."
Phone: (520) 383-0066

https://www.tocc.edu/

Trinidad Rancheria Library Saa-a-goch (Speak Yurok) Cultural Literacy Project

"The mission of the Cher-Ae Heights Indian Community of the Trinidad Rancheria is to preserve and promote our cultural and traditional beliefs; improve quality of life and self-sufficiency; uphold tribal sovereignty; create positive partnerships; and protect the environment in order to provide a healthy community, honor our elders, and guide our youth." The Trinidad Rancheria covers a Tribal Program invested in the environment, education, roads and land use, finance, and more. The contact information of these department coordinators are listed on the website.

https://trinidad-rancheria.org

Venita K. Taveapont Memorial Library

The Venita K. Taveapont Memorial Library is a part of the Uintah River High School, which offers activities for the public and special collections relating to local history.
Contact: cheryll@utetribe.com
Phone: (435) 725-4248

https://www.facebook.com/uitvktlibrary/

Yakama Nation Library

"The purpose of the Library is to provide Plateau cultural and mainstream American, educational-contemporary popular materials for its patrons. The mission of the educational branch of Human Services which includes the Yakama Nation Tribal School, Yakama Nation Library and its Computer Lab is to encourage lifelong learning, and promotes cultural awareness utilizing current technology."
Web: http://www.yakamamuseum.com/library.php
Phone: (509) 865-5121 ext. 4747

White Earth Tribal and Community College Library

"White Earth Tribal and Community College is an institution of higher learning dedicated to academic excellence grounded in Anishinaabe culture, values, and traditions."
Web: http://www.wetcc.edu/
Phone: (218) 935-0417

White Earth Land Recovery Project

"The mission of the White Earth Land Recovery Project is to facilitate the recovery of the original land base of the White Earth Indian Reservation while preserving and restoring traditional practices of sound land stewardship, language fluency, community development, and strengthening our spiritual and cultural heritage."
http://welrp.org

Wiyot Heritage Center

"The mission of the Wiyot Heritage Center is to promote the understanding, revitalization, and celebration of Wiyot culture. Recognizing culture as a dynamic process and the Wiyot as a living people, the Center's goals are to treasure the past, enrich the present, and meet the challenges of the future."
Web: https://www.wiyot.us/
Heritage Center: https://www.wiyot.us/155/Heritage-Center
Directory: https://www.wiyot.us/Directory.aspx?did=5
Phone: (707) 733-5055

Ziibiwing Center of Anishinabe Culture and Lifeways

"The Ziibiwing Center is a museum and cultural center built to share the history of the Saginaw Chippewa Indian Tribe of Michigan with the rest of the world."
www.sagchip.org/ziibiwing

## Tribal Institutions and Organizations

Institutions and organizations serving and supporting tribal Nations are numerous. A sample, but not exhaustive, listing of regional and national initiatives and organizations includes the Alliance for California Traditional Arts; American Indian Library Association; American Indian Movement, 1968–1978; American Indian Policy Institute; American Indian Science and Engineering Society; Central Coast Indigenous Resource Alliance; Institute for Tribal Environmental Professionals; National Congress of American Indians; National Indigenous Women's Resource Center; Native American and Indigenous Studies Association; North American Traditional Indigenous Food Systems (NāTIFS); New Mexico Tribal Libraries Program; Ojibwe People's Dictionary; Society for the Advancement of Chicanos and Native Americans in Science; Sustainable Heritage Network; Tribal College Librarians Institute; Tribal Impact Program through Little Free Libraries for Native Communities; Tribal Libraries, Archives, and Museums Project; Trinidad Rancheria Library Saa-a-goch (Speak Yurok) Cultural Literacy Project; and University of New Mexico Health Sciences Library and Informatics Center. Resources for Indigenous scholars include the: American Philosophical Society's Center for Native American and Indigenous Research; Indian Arts Research Center at the School for Advanced Research at the School for Advanced Research in Santa Fe; and Labriola National American Indian Data Center at Arizona State University.

Information on and assistance with repatriation in the U.S. is available through the Association on American Indian Affairs Repatriation Working Group, Native American Graves Protection and Repatriation Resources, and the Indigenous Repatriation Handbook. Protection of property is provided under NAGPRA, ARPA, 18 U.S.C. sections 1163, 2314-2315 and 68, and state and tribal laws. Legal assistance is available through the Advisory Council on Historic Preservations; Association on American Indian Affairs; National Lawyers Guild; National Indian Law Library; Native American Rights Fund, and mechanisms of the National Indian Child Welfare Act; as well as Local Contexts, an initiative to support Indigenous communities in managing their intellectual property and cultural heritage in the digital world, and Free, Prior, and Informed Consent, a resource providing knowledge for Indigenous people about industry and government negotiations. Individual colleges of law also provide useful legal guides: see, for example, the Native American Indian Law Research Guide at Marquette University, https://libraryguides.law.marquette.edu/c.php?g=318577&p=2127456.

## Grants and Fellowships

Many institutions and museums offer full and partial scholarships for tribal members, and/or free information on issues related to education, preservation, and repatriation. Ask or search for your school or institution online. Contact the American Indian College Fund, American Indian Graduate Center; American Indian Library Association; American Library Association Office for Diversity, Literacy, and Outreach Services; American Indian Higher Education Consortium; American Philosophical Society (Center for Native American and Indigenous Research); Association on American Indian Affairs, Generation Indigenous; Indian Arts Research Center at the School for Advanced Research at the School for Advanced Research, and/or the institution you are interested in, which may have unpublished scholarships, grants, or opportunities for tribal members. Fellowships for scholars and institutions are available at state and federal levels. See, for example, the American Philosophical Society Library; Institute of Museum and Library Services (IMLS); National Endowment for the Arts; National Endowment for the Humanities (search all grant opportunities, including the Digital Advancement Grant and Digital Native American Studies Project); Native American Museums Studies Institute at U.C. Berkeley; Native American Scholars Initiative at the Center for Native American and Indigenous Research; National Park Services Tribal Heritage Grants Program; the National Preservation Institute; and the Peabody Essex Museum Native American Fellowship Program.

## Children's Resources

There are many educational resources available for children, although they are not centralized or comprehensive, and many contain generic information. Find out which tribal band or Nation once occupied the region where your school is located, and search for information on that Nation. For suggestions on teaching, see the Burke Museum, https://www.burkemuseum.org/education/learning-resources/tips-teaching-about-native-peoples. For examples of resources, see the American Folklife Center at the Library of Congress classroom materials; American Indians in Childrens' Literature blogspot; Birchbark Books and Native Arts (bookstore carrying Native American books); California Indian History Curriculum; California Indian Storytellers' Association; California Education and the Environment Initiative; First Nations, Metis and Inuit Education Association of Ontario; Indian Education for All (Montana); Indigenous Teacher Education Project at the University of Arizona; Mapping Indigenous L.A. teaching materials; Native Land teaching materials and phone app; lessons and resources from the National Museum of the American Indian (see Native Knowledge 360°); Native American Heritage Month materials; Native Voices Concepts of Health and Medicine Lesson Plans through the National Library of Medicine; Oneida Community Education Center; Oyate; Pirurvik Center for Inuit culture; Project 562: Changing the Way We See Native America; Reclaiming Native Truth; Red Planet Books and Comics; Since Time Immemorial: Tribal Sovereignty (Washington State); Spirit Lines: Bringing Culture Home (Manitoba Museum); Talk Story: Sharing Stories, Sharing Culture; The History Project at U.C. Irvine; We Shall Remain: Utah Indian Curriculum Project; Wisconsin Department of Public Instruction (https://dpi.wi.gov/amind/resources/maps); and the Zinn Education Project: Teaching People's History. Maps of tribal Nations in the United States for classrooms are available from Aaron Carapella at Tribal Nations Maps; maps for First Nations place names are available from Coming Home to Indigenous Place Names in Canada. Many museums offer virtual tours, including Jamestown S'Klallam Tribe's House of Seven Generations.

For information on discussing Indigenous communities with respect, see Krol, Debra Utacia. "Covering Indigenous Communities with Respect and Sensitivity." *TheOpenNotebook* (June 18, 2019). https://www.theopennotebook.com/2019/06/18/covering-indigenous-communities-with-respect-and-sensitivity/?fbclid=IwAR3LlvbXU_giuYgN6mQoai5D3ScGk_N6SFpQTQcbht6kX93M6xoswOAPL6I.

Although this list is intended as a sample listing only, if you would like to see your digital or educational project, tribal ALM, or organization listed in future editions of this text, please contact the editor of "Digital Mapping and Indigenous America" (janet.hess@sonoma.edu).

Contact: janet.hess@sonoma.edu

# Index

Note: *Italicized* page numbers refer to figures, **bold** page numbers refer to tables